THE EIS BOOK

Managing and Preparing Environmental Impact Statements

THE EIS BOOK

Managing and Preparing Environmental Impact Statements

Charles H. Eccleston

CRC Press
Taylor & Francis Group
Boca Raton London New York

CRC Press is an imprint of the
Taylor & Francis Group, an **informa** business

CRC Press
Taylor & Francis Group
6000 Broken Sound Parkway NW, Suite 300
Boca Raton, FL 33487-2742

Printed on acid-free paper
Version Date: 20130923

International Standard Book Number-13: 978-1-4665-8363-4 (Hardback)

Library of Congress Cataloging-in-Publication Data

Eccleston, Charles H.
The EIS book : managing and preparing environmental impact statements / Charles H. Eccleston.
pages cm
Summary: "The National Environmental Policy Act (NEPA) requires that ALL federal agencies prepare Environmental Impact Statements (EIS) for any proposal that may significantly affect the quality of the human environment. They generally take more than a year to prepare, have an extensive public participation component, and require substantial collection of original data. An Assessment can generally be completed in a few months. A company, institution, dedicated environmental professional, or federal agency generally is involved with preparation of both assessments and impact statements. This book helps agencies prepare EISs by providing step-by-step direction for navigating the EIS process"-- Provided by publisher.
Includes bibliographical references and index.
ISBN 978-1-4665-8363-4 (hardback)
1. Environmental impact analysis--United States. 2. Environmental impact statements--United States. 3. United States. National Environmental Policy Act of 1969. I. Title.

TD194.65.E354 2013
333.71'40973--dc23 2013036866

Visit the Taylor & Francis Web site at
http://www.taylorandfrancis.com

and the CRC Press Web site at
http://www.crcpress.com

Contents

Preface

Figure 0.1 NEPA involves timely environmental issues. (Courtesy images.google.com.)

The US National Environmental Policy Act (NEPA) provides a systematic and comprehensive planning process for considering the consequences and alternatives of federal actions before a final decision is made to pursue a course of action (Figure 0.1). NEPA's most notable planning provision undoubtedly involves the preparation of an environmental impact statement (EIS). An EIS must be prepared for all federal actions that may significantly affect environment quality. Unfortunately, one can easily find examples where an EIS has been prepared in excruciating detail, sometimes *ad nauseam*, investigating the most trivial and remote issues in detail. Such EISs may examine every conceivable impact, significant or not, yet blatantly ignore the very purpose of preparing the EIS; the EIS is the federal government's planning and decision-making tool. Poor EIS practice leads to poorly planned projects, and ultimately poor environmental protection. Some poorly planned projects not only jeopardize environmental quality but pose severe risks to society as well. Perhaps no example better illustrates this problem than the case study described in Chapter 1 in which the Nuclear Regulatory Commission has prepared deficient EISs to justify its mission of relicensing the nation's fleet of antiquated nuclear reactors.

This book is not about preparing bigger EISs—but better ones. It focuses on the EIS process from a planning perspective. Rather than describing an EIS as a document preparation procedure, this book describes it from the context of a comprehensive framework for *planning* future actions. It presents a step-by-step guide to the management and preparation of EISs.

Objectives of this book

Emphasis is placed on providing direction for preparing defensible analyses that facilitate well-planed projects and improved decision making. Beginning with fundamental topics and advancing into successively more advanced subjects, this book can be used by beginners and experts alike. The reader is presented with a single compendium synthesizing and describing all relevant requirements and guidance for preparing a legally sufficient EIS.

Specific objectives

In this book

- All EIS document requirements (documentation requirements) are detailed, including the Council on Environmental Quality's NEPA regulations and related guidelines; Environmental Protection Agency (EPA) guidance and requirements; presidential executive orders; and case law. Emphasis is placed on addressing timely and controversial issues such as how to perform a legally sufficient cumulative impact assessment and how to evaluate greenhouse emissions and climate change.
- The EIS process (process requirements) for preparing the statement is detailed. A step-by-step approach for navigating the entire EIS process is described. All pertinent process requirements from issuing the notice of intent, through public scoping, to issuing the final record of decision (ROD) are detailed.

- Analytical requirements (analysis requirements) for preparing the EIS analysis are detailed. Guidance for performing various types of analyses is also described.
- Tools, techniques, and best professional practices for preparing the EIS and performing the analysis are detailed. Lessons learned from case law are integrated with the relevant requirements.
- To reinforce key EIS regulatory requirements, a case study is presented in Chapter 1. Lessons learned from this case study are integrated with appropriate regulatory requirements throughout this text.

Annotated outline

Chapters 1 and 2: The book begins with a case study of a faulty EIS process. The purpose is to show the types of problems that can be encountered and how to avoid repeating similar errors. Chapter 2 provides a brief introduction to the NEPA process.

Chapters 3 and 4: The first two chapters set the stage for presenting a step-by-step approach for navigating the complexities of the EIS process. To this end, Chapters 3 and 4 present the reader with all pertinent EIS procedural requirements (process requirements) from issuing the notice of intent, through public scoping, to preparing the EIS, and cumulating with the issuance of the record of decision (ROD).

Chapter 5: Chapter 5 presents the analytical requirements (analysis requirements), including guidance and direction for preparing an accurate, objective, rigorous, and legally sufficient analysis of impacts; tools, techniques, and best professional practices for performing a systematic and rigorous analysis are also introduced.

Chapter 6: Chapter 6 details all key EIS documentation requirements (document requirements). This chapter synthesizes and integrates a large, complex, and diverse body of guidance, direction, and requirements for preparing a legally sufficient EIS document.

On completing this book, the reader should have a firm grasp of the step-by-step process for preparing an EIS, including all key regulatory requirements that a legally sufficient EIS document must satisfy. No other book synthesizes all such requirements and guidance into a single source for easy and rapid access. This book is therefore unique in that it provides readers with all essential requirements as well as practical guidance for preparing an EIS.

Audience

Although this book is aimed toward NEPA professionals in government, consulting, and the private sector, the organization lends itself equally to individuals who desire only an introduction to certain selected aspects of the EIS planning process. Skilled practitioners may use the book as a resource for quickly reviewing complex issues. Individuals, professionals, and groups who will find this book of interest include

- NEPA practitioners
- Educators and students
- Project managers
- Scientists
- Planners
- Analysts
- Regulators
- Decision makers
- Environmental lawyers
- Public advocacy and watchdog organizations

Because it starts with elementary topics and progressively advances into more intricate subject matter, it is also an ideal book for undergraduate/graduate students in environmental, planning, and engineering curricula. Each chapter begins with a set of learning objectives and ends with a list of questions designed to test comprehension. Three capstone projects are also presented at the end of the book.

If you have technical questions or issues, or need assistance, the author can be contacted at NEPAservices@msn.com.

Author

Charles H. Eccleston is a NEPA consultant, environmental trainer, and author. His responsibilities include assisting contractors and agencies in preparing EISs that are legally sufficient and meet all key regulatory requirements; his expertise includes providing assistance on NEPA lawsuits, including identification of regulatory and legal flaws. He is currently developing a series of NEPA and environmental training videos.

With 30 years of experience, he has managed and prepared a diverse array of NEPA, environmental, energy, and planning assessments. He has served on two US White House–sponsored taskforces for resolving environmental policy problems. Eccleston is recognized in *Marquis' Who's Who in Science and Engineering, Who's Who in America*, and *Who's Who in the World* as a leading international expert for his NEPA and environmental impact assessment (EIA) achievements. He is the author of more than 75 professional papers and eight books on the NEPA process, EIA, and environmental and energy policy.

He was elected three times to the board of directors of the National Association of Environmental Professionals (NAEP) and received its national award for Outstanding Environmental Leadership. Currently, he serves as an elected representative to the International Organization for Standardization's 242 working group, responsible for developing an ISO 50001 Energy Management System (EnMS) standard for the worldwide use and management of energy. Eccleston developed and published the original concept (adopted by a number of US agencies and around the world) for integrating NEPA or a similar process with an ISO 140001 Environmental Management System (EMS).

Eccleston is fluent on a wide range of environmental and energy policy issues such as assessment of ecological effects, sustainability, climate change, water and food scarcity, radioactive/hazardous waste, peak oil, population issues, and energy generation. His energy-related experience includes investigating nuclear, gas-fired, and coal-fired plants, and renewable energy systems. His recent books include

- *Inside Energy: Developing and Managing an ISO 50001 Energy Management System* (CRC Press 2012)
- *Preparing NEPA Environmental Assessments: A Users Guide to Best Professional Practices* (CRC Press 2012)
- *Environmental Impact Assessment: A Guide to Best Professional Practices* (CRC Press 2011)
- *Global Environmental Policy: Principles, Concepts and Practice* (CRC Press 2011)
- *NEPA and Environmental Planning: Tools, Techniques, and Approaches for Practitioners* (CRC Press 2008)

Eccleston is currently developing a series of professional training videos on subjects ranging from NEPA and EISs, too environmental and regulatory requirements, and energy and EIA. This series includes an EIS video that encapsulates this book. The author's consulting services include assisting agencies and consulting companies on NEPA projects, resolving NEPA problems, reviewing EISs to ensure they meet regulatory and legal requirements, providing expert assistance in NEPA lawsuits, and conducting NEPA and environmental training. He can be contacted at NEPAservices@hotmail.com, or visit the NEPA website at http://www.NEPAservices.com. For advanced EIS and NEPA training, visit the author at http://campus.education.com/NEPAcampus. The NEPA website is being modified and http://campus.education.com/NEPAcampus is under construction. Both will be completed before the book is published.

List of acronyms

ACHP Advisory Council on Historic Preservation
ADREC administrative record
AEC Atomic Energy Commission
AIM action–impact model
AR associate reviewer
ARTS administrative record tracking system
BA biological assessment
BE biological evaluation
CEQ Council on Environmental Quality, also referred to as the "Council"
CERCLA Comprehensive Environmental Response, Compensation, and Liability Act
CFR United States Code of Federal Regulations
CIA cumulative impact assessment
CO carbon monoxide
Commission US Nuclear Regulatory Commission
COMTRACK comment tracking
CZMA Coastal Zone Management Act
DBS decision-based scoping
DEIS draft environmental impact statement
DIT decision-identification tree
DLR Division of License Renewal, Nuclear Regulatory Commission
EA environmental assessment
EC environmental concerns
EIS environmental impact statement
EJ environmental justice
EMS environmental management systems
EO executive order; also refers to environmental objections
EPA US Environmental Protection Agency
EU environmental unsatisfactory
FEIS final environmental impact statement
FOIA US Freedom of Information Act
FONSI finding of no significant impact
FR *Federal Register*

FWS US Fish and Wildlife Service
GEIS generic environmental impact statement
GIS geographic information system
HIA health impact analysis
IAJM interim action justification memorandum
IDT interdisciplinary team
ISO International Organization for Standardization
L-EIS legislative EIS
LO lack of objections
MAP management action plan; also refers to mitigation action plan
MOA Memorandum of Agreement
N/A not applicable
NAAQS National Ambient Air Quality Standards
NEPA National Environmental Policy Act
NHPA US National Historic Preservation Act
NMFS US National Marine Fisheries Service
NOA notice of availability
NOAA US National Oceanic and Atmospheric Administration
NOI notice of intent
NO*x* nitrogen oxides
NPDES National Pollutant Discharge Elimination System
NRC US Nuclear Regulatory Commission
NRHP US National Register of Historic Places
OFA US Office of Federal Activities
PA programmatic agreement
PERT program evaluation review technique
PEIS programmatic environmental impact statement
PM project manager
PM2.5 particulate matter 2.5 micrometers in diameter
PM10 particulate matter 10 micrometers in diameter
PR principal reviewer
PSD prevention of significant deterioration
RCRA US Resource Conservation and Recovery Act
Regulations The NEPA Implementing Regulations (40 CFR Parts 1500–1508)
ROD record of decision
RPB1 relicensing project branch 1
SAMA severe accident mitigation alternatives
S-EIS supplemental EIS
SEPA state environmental policy acts
SHPO State Historic Preservation Officer
SIP state implementation plan
SOW statement of work
SO*x* sulfur oxides

SPN statement of purpose and need
THPO Tribal Historic Preservation Officer
US United States
U.S.C. United States Code
USFWS United States Fish and Wildlife Service
VOC volatile organic compounds
WBS work breakdown structure

Introduction

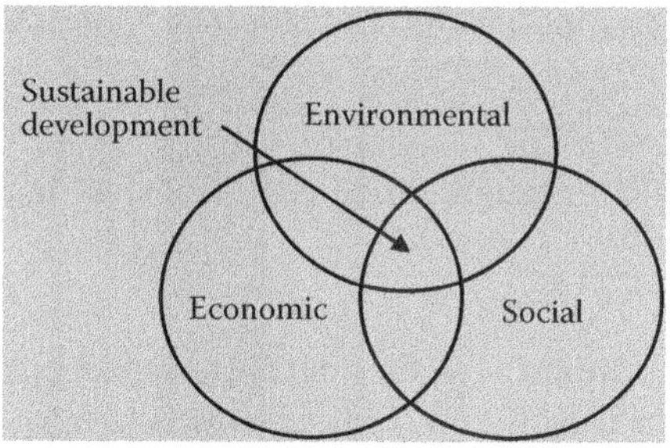

Figure 0.2 Economic and social goals need to be balanced with environmental concerns.

NEPA's history and precedent

To fully appreciate the forces that led to enactment of a national environmental policy, one must understand the context in which the National Environmental Policy Act (NEPA) was created. The American public and Congress were becoming increasingly troubled by the deterioration in environmental quality (Figure 0.2). Perhaps no event captured the public's imagination more than the nightly national news broadcasting scenes of the Cuyahoga River in Cleveland, Ohio (Figure 0.3). It was so polluted that it actually caught fire!

Figure 0.3 Cuyahoga River caught fire in 1969. (Courtesy images.google.com.)

Enactment of NEPA

Before NEPA's enactment, there was actually a precedent for preparing a study of possible environmental impacts from proposed projects. In the early 1960s, Congress required the Atomic Energy Commission, the predecessor to today's Nuclear Regulatory Commission, to prepare an "environmental report" on a disturbing proposal to blast a harbor along the Alaskan coastline using nuclear explosives (e.g., nuclear bombs). This project has since been criticized as potentially one of the most environmentally questionable proposals ever concocted. The project was ultimately canceled, in large measure because of the results of this environmental report. This report has been viewed as the world's first *de facto* environmental impact statement (EIS).[1] This trailblazing report would provide a model for NEPA's EIS requirement in the later 1960s.

The world's first environmental policy

By the late 1960s, Congress was increasingly hearing testimony from the scientific community regarding the alarming rate of environmental degradation and the potential for future calamity. Congress had many avenues available for addressing the nation's looming environmental problems. It chose to begin the long road to environmental recovery by adopting a national environmental policy regarding the nation's vision and commitment to a clean and healthy environment. As described below, NEPA would become the world's first national environmental policy. Many nations would follow suit with their own national policies.

THE NEPA STATUTE

The National Environmental Policy Act of 1969, as amended (Pub. L. 91–190, 42 U.S.C. 4321–4347, January 1, 1970, as amended by Pub. L. 94–52, July 3, 1975, Pub. L. 94–83, August 9, 1975, and Pub. L. 97–258, § 4(b), Sept. 13, 1982)

An Act to establish a national policy for the environment, to provide for the establishment of a Council on Environmental Quality, and for other purposes.

Be it enacted by the Senate and House of Representatives of the United States of America in Congress assembled, That this Act may be cited as the "National Environmental Policy Act of 1969."

PURPOSE

Sec. 2 [42 USC § 4321].

The purposes of this Act are: To declare a national policy which will encourage productive and enjoyable harmony between man and his environment; to promote efforts which will prevent or eliminate damage to the environment and biosphere and stimulate the health and welfare of man; to enrich the understanding of the ecological systems and natural resources important to the Nation; and to establish a Council on Environmental Quality.

TITLE I

CONGRESSIONAL DECLARATION OF NATIONAL ENVIRONMENTAL POLICY

Sec. 101 [42 USC § 4331].

(a) The Congress, recognizing the profound impact of man's activity on the interrelations of all components of the natural environment, particularly the profound influences of population growth, high-density urbanization, industrial expansion, resource exploitation, and new and expanding technological advances and recognizing further the critical importance of restoring and maintaining environmental quality to the overall welfare and development of man, declares that it is the continuing policy of the Federal Government, in cooperation with State and local governments, and other concerned public and private organizations, to use all practicable means and measures, including financial and technical assistance, in a manner calculated to foster and promote the general welfare, to create and maintain conditions under which man and nature can exist in productive harmony, and fulfill the social, economic, and other requirements of present and future generations of Americans.

NEPA's EIS provision

NEPA was a revolutionary statute and the subject of considerable debate. Because it is the single largest entity in the United States and as a result of the vast scope and nature of its actions, the federal government accounted for a disproportionally larger share of the nation's environmental degradation. Congressional leaders believed it was necessary for the US government to take a leadership role in protecting the environment. Congress began considering the need for a policy that would spell out a national commitment to environmental protection. They believed that passage of such a bill would demonstrate the seriousness with which Congress viewed environmental protection and set a precedent for stricter laws and standards that would follow in its wake.

NEPA'S EIS REQUIREMENT (SEC. 102, 42 USC § 4332)

All agencies of the Federal Government shall—

1. utilize a systematic, interdisciplinary approach which will insure the integrated use of the natural and social sciences and the environmental design arts in planning and in decisionmaking which may have an impact on man's environment;
2. identify and develop methods and procedures, in consultation with the Council on Environmental Quality established by title II of this Act, which will insure that presently unquantified environmental amenities and values may be given appropriate consideration in decisionmaking along with economic and technical considerations;
3. include in every recommendation or report on proposals for legislation and other major Federal actions significantly affecting the quality of the human environment, a detailed statement by the responsible official on—
    i. the environmental impact of the proposed action,
    ii. any adverse environmental effects which cannot be avoided should the proposal be implemented,
    iii. alternatives to the proposed action,
    iv. the relationship between local short-term uses of man's environment and the maintenance and enhancement of long-term productivity, and
    v. any irreversible and irretrievable commitments of resources which would be involved in the proposed action should it be implemented.

Senator Henry "Scoop" Jackson sponsored NEPA in the Senate. He was adamant that it must be more than simply a policy statement. Some provision needed to be incorporated into the statute to ensure that it would not be simply a paper tiger. An action-forcing mechanism, of some type, was needed to ensure that federal agencies truly integrated NEPA's policy into their day-to-day operations. With respect to an action-forcing mechanism, the following key provision was added:

> All agencies of the Federal Government shall... include in every recommendation or report on proposals for legislation and other major Federal actions significantly affecting the quality of the human environment, a detailed statement by the responsible official. (Sec.102(2)(C) of NEPA)

This action-forcing "detailed statement" would later become known as the environmental impact statement—the subject of this book. Among other requirements, the detailed statement was required to investigate "alternatives to the proposed action."

Passage and implementation of NEPA

The NEPA statute received the unanimous vote of the Senate Interior Committee and enjoyed widespread support among members of Congress. To reinforce the significance of this act, President Richard Nixon chose to sign NEPA into law on New Year's Day of 1970, proclaiming this as "my first act of the decade." Thus, NEPA has the distinction of being the first law enacted during the new decade of the 1970s. From this point on, agencies would be expected to balance the goal of protecting the environment with other competing factors and policies, such as economic growth. In short, agencies would be required to infuse NEPA into their traditional decision-making processes. Following on the cusp of NEPA, Congress created the Environmental Protection Agency (EPA) in 1970. The world's first "Earth Day" was celebrated on April 22, 1970.

First and foremost, NEPA is a statement of our national will to protect the environment. Less a regulatory statute than a policy statement, NEPA establishes a fundamental principle by which the federal government is to conduct its operations. Perhaps, NEPA's single greatest contribution has been that it expects federal agencies to consider environmental issues in reaching decisions, just as these agencies consider other factors that fall within their domain.

At first glance, NEPA appears to be a rather weak law. It established no substantive environmental standards and defines no enforcement mechanisms beyond a federal agency's discretion. In practice, however, NEPA's weaknesses are tempered with complementary strengths. The brilliance of its creators was their vision of the Act's power in the face of such weaknesses. Lack of substantive standards can actually provide planners with great flexibility in planning actions and is more than compensated for by an array of such standards in other environmental laws. Lack of an enforcement mechanism has been partially compensated by the courts as parties have challenged agency actions based on NEPA.

NEPA's implementing regulations and global precedent

By the mid-1970s, complaints about NEPA were increasingly making their way to the Oval Office in the White House. Such criticism prompted President Jimmy Carter to issue an executive order in 1977, directing the Council on Environmental Quality (CEQ) to issue formal regulations for implementing NEPA, including direction for streamlining preparation of EISs. Eight years after NEPA was enacted, the CEQ promulgated its formal NEPA implementing regulations (40 CFR 1500–1508) in November 1978.

Beyond American shores, NEPA has established a global precedent that has been emulated by scores of other nations. Today, NEPA has permeated virtually every corner of the globe. It has the distinction of being one of the most copied statutes in the world. By one count, NEPA has been emulated, in one form another, by more than 25 states in the US and more than 100 countries worldwide.

By now, most agencies have made substantial strides in learning to comply with NEPA and its EIS requirement. Despite Congress's clear congressional intent and more than 40 years of operating experience, however, a few agencies have still not learned the lessons. For instance, the US Forest Service has led the nation in terms of being sued for failure to adequately comply with NEPA's requirements. One lawyer noted that it was not simply the shear number of lawsuits but also the agency's failure to learn from past mistakes that is so troubling. Chapter 1 provides a case study of NEPA practice that is particularly troubling. By reviewing this case study, it is hoped that other agencies can avoid repeating such portentous errors.

Three laws of the environmental movement

Before introducing Chapter 1, the author wishes to propose three corollaries to Newton's famous Three Laws of Motion (Figures 0.4 and 0.5). These corollaries are cited for more than humor's sake. The reader may note that each has a subtle, yet tangible implication in terms of safeguarding environmental quality and the manner in which agencies choose to conduct their EIS planning responsibilities. The author loosely refers to these corollaries as Eccleston's *Three Laws of Environmental Movement*:

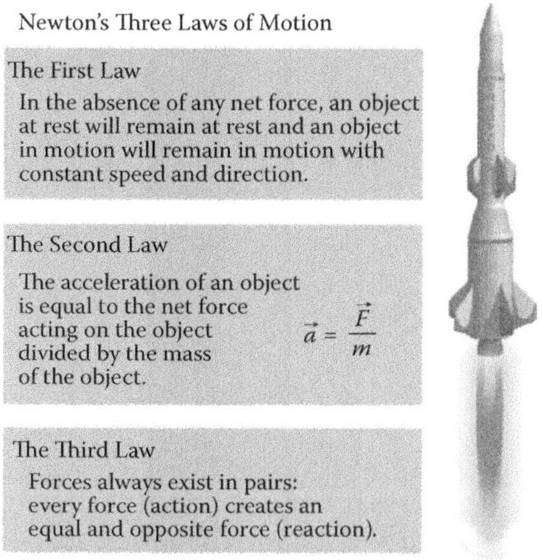
Newton's Three Laws of Motion
The First Law
In the absence of any net force, an object at rest will remain at rest and an object in motion will remain in motion with constant speed and direction.
The Second Law
The acceleration of an object is equal to the net force acting on the object divided by the mass of the object.
$\vec{a} = \frac{\vec{F}}{m}$
The Third Law
Forces always exist in pairs: every force (action) creates an equal and opposite force (reaction).

Figure 0.4 Newton's Three Laws of Motion.

Figure 0.5 Isaac Newton discovered the Three Laws of Motion.

1. *First law of environmental movement:* A top-level commitment to environmental quality tends to continue in the direction of quality, lack of commitment to environmental excellence, promotes environmental degradation.
2. *Second law of environmental movement:* The force that an EIS brings to bear in protecting the environment is equal to the mass of forethought expended in the planning process, multiplied by the decision maker's commitment to environmental protection.
3. *Third law of environmental movement:* For every project proponent attempting to side step the EIS process, there is an equal and opposite adversary waiting to contest the project.

Note

1. O'Neil D. Project Chariot: How Alaska escaped nuclear excavation. *The Bulletin of the Atomic Scientist* (December 1989): 35.

Frogs and the EIS planning process

Before introducing the subject of federal planning, let us stop and consider what the parable of Felix the frog and federal planning have in common.

> Once upon a time, there lived a man named Clarence who had a pet frog named Felix. Clarence lived a modestly comfortable existence on what he earned working at the local mall; but he always dreamed of being rich.
>
> "Felix!" he exclaimed one day, "We're going to be rich! I'm going to teach you how to fly!" Felix, of course, was terrified at the prospect: "I can't fly, you idiot! I'm a frog, not a canary!"
>
> Clarence, disappointed at the initial reaction, told Felix: "That negative attitude of yours could be a real problem. I'm sending you to class." So Felix went to a three day class and learned about problem solving, time management, and effective communication... but nothing about flying.
>
> On the first day of "flying lessons," Clarence could barely control his excitement (and Felix could barely control his bladder). Clarence explained that their apartment had 15 floors, and each day Felix would jump out of a window starting with the first floor eventually getting to the top floor. After each jump, Felix would analyze how well he flew, focus on the most effective flying techniques, and implement the improved process for the next flight. By the time they reached the top floor, Felix would surely be able to fly.
>
> Felix pleaded for his life, but it fell on deaf ears. "He just doesn't understand how important this

is..." thought Clarence, "but I won't let nay-sayers get in my way." So, with that, Clarence opened the window and threw Felix out (who landed with a thud).

Next day (poised for his second flying lesson) Felix again begged not to be thrown out of the window. With that, Clarence opened his pocket guide to Managing More Effectively and showed Felix the part about how one must always expect resistance when implementing new programs. And with that, he threw Felix out the window. (THUD!)

On the third day (at the third floor) Felix tried a different ploy: stalling, he asked for a delay in the "project" until better weather would make flying conditions more favorable. But Clarence was ready for him: he produced a time line and pointed to the third milestone and asked, "You don't want to slip the schedule do you?" From his training, Felix knew that not jumping today would mean that he would have to jump TWICE tomorrow... so he just said: "OK. Let's go." And out the window he went.

Now understand that Felix really was trying his best. On the fifth day he flapped his feet madly in a vain attempt to fly. On the sixth day he tied a small red cape around his neck and tried to think "Superman" thoughts. Try as he might, though, Felix couldn't fly.

By the seventh day, Felix (accepting his fate) no longer begged for mercy... he simply looked at Clarence and said, "You know you're killing me, don't you?" Clarence pointed out that Felix's performance so far had been less than exemplary, failing to meet any of the milestone goals he had set for him.

With that, Felix said quietly, "Shut up and open the window," and he leaped out, taking careful aim on the large jagged rock by the corner of the building. With this jump, Felix went to that great lily pad in the sky.

Clarence was extremely upset, as his project had failed to meet a single goal that he set out to accomplish. Felix had not only failed to fly, he didn't even learn how to steer his flight as he fell like a sack of cement... nor did he improve his productivity when Clarence had told him to "Fall smarter, not harder."

> The only thing left for Clarence to do was to analyze the process and try to determine where it had gone wrong. After much thought, Clarence smiled and said: "Next time... I'm getting a smarter frog!"

As in the case of Clarence, more is needed than simply *demanding* that federal agencies meet the nation's increasingly complex environmental challenges. Correctly used, the EIS process provides a powerful tool for planning federal actions. Consistent with the Three Laws of Environment Movement described in the Introduction, the intent of this book is to provide a comprehensive step-by-step guide for preparing such EISs.

> The only thing left for Clarence to do was to analyze the process and try to determine where it had gone wrong. After much thought, Clarence smiled and said, "Next time, I'm getting a map first."

As in the case of Clarence, more is needed than simply "going through the motions" to ensure that federal agencies meet the nation's increasingly complex environmental challenges. Correctly used, the EIS process provides a powerful tool for planning federal actions. Consistent with the Three Laws of Environment described in the introduction, the intent of this book is to provide a comprehensive, step-by-step guide to preparing such EISs.

chapter one

Scientific facades—how not to prepare an EIS

A case study on how a flawed EIS process can imperil society

> Don't tell fish stories where the people know you; but particularly don't tell them when they know the fish.
>
> **Mark Twain**

As detailed in this book, the US National Environmental Policy Act (NEPA) was enacted to ensure that federal decision makers devote appropriate consideration to environmental factors during the decision-making process. One of NEPA's most important provisions is the requirement to prepare an environmental impact statement (EIS) for all major federal actions that may significantly affect the quality of the human environment. As briefly described in the Introduction to this book, an EIS evaluates potentially significant effects of a proposed action, including alternatives and mitigation measures for avoiding those impacts. As we will see, this requirement is intended to force federal agencies to objectively evaluate and disclose the implications of their proposals to the decision maker and public.

While the enactment of this congressional mandate was a landmark achievement, it is equally fair to ask how well the EIS process is actually working. In the author's judgment, most agencies are making concerted efforts to comply with NEPA's EIS requirements. As we will see, some agencies may even have exceeded NEPA's original expectations while in other quarters there is considerable room for improvement.

Unfortunately, much of NEPA practice is plagued by deficient and defective EISs that contribute little or no value to the federal decision-making process. This chapter provides a lucid case study of unsound, flawed, and even deceptive EIS management practices. As illustrated below, many stumbling blocks encountered in NEPA practice can be traced directly back to poor management, direction, and oversight.

Figure 1.1 An aging nuclear power plant.

Although NEPA has been emulated by more than 100 nations around the world, a few US agencies still view it with disdain and as a hindrance to implementing their projects. No process more vividly illustrates this than does the US Nuclear Regulatory Commission (aka NRC or "Commission") and it's controversial and widely contested nuclear reactor licensing process (Figure 1.1). This chapter explains why critics charge that the Commission's management pays "lip service" to NEPA's intent, essentially treating it as "window dressing."[1] The author has rarely witnessed EISs that are plagued with such egregious regulatory and legal deficiencies, weaknesses, and flaws. The Commission's EIS process provides an ideal case study illustrating how to avoid similar errors and pitfalls. As we will see, not only can a flawed EIS expose an agency to litigation, but it can also undermine the agency's credibility, standing, and long-term trust with the public. Worse yet, it can even endanger the livelihood of its citizens.

Lessons learned from the errors documented in this case study will be revisited throughout this book. As we will see, such deficiencies and flaws can often be traced back to shortcoming in management and oversight.

This chapter cites regulatory requirements that are routinely violated. References to the Council on Environmental Quality's (CEQ's) NEPA implementing regulations (Regulations) are provided to assist the reader in understanding these regulatory errors and flaws.* For the

* Specific provisions referenced from the NEPA implementing regulations are abbreviated so as to cite the specific "part" of the NEPA implementing regulations in which it is found. For example, a reference to a provision in "40 Code of Federal Regulations (CFR) 1501.1" is simply cited as "§1501.1."

reader's convenience, a copy of the NEPA regulations (40 Code of Federal Regulations [CFR] parts 1500–1508) is provided in Appendix B.

Chapter 2 of this book provides an overview of NEPA and its three levels of compliance, including the EIS process. This is followed by Chapters 3 and 4, which describe the step-by-step process requirements for preparing the EIS. Chapter 5 describes the general process for performing the EIS analysis, while Chapter 6 describes the actual requirements (documentation requirements) that the statement must comply with.

1.1 Learning objectives

- How to avoid legal deficiencies, weaknesses, and flaws in the EIS process
- How the EIS process facilitates federal decision making
- Project Chariot, the first de facto EIS
- Understanding the risks of taking actions without adequately evaluating the impacts and alternatives
- Technological mismanagement and how it can imperil society
- How to prepare EISs that meet NEPA's legal and regulatory requirements
- How to use the EIS process to instill public trust

1.2 A human and environmental disaster of epic proportions

It is etched indelibly into our minds: the image of a ferocious earthquake slamming Japan followed by an earth-shattering tsunami. An even more terrifying picture began to emerge as the Japanese Fukushima Daiichi nuclear power station went into meltdown mode. This initiated a long chain of events that would threaten hundreds of thousands, if not millions, of Japanese men, women, and children.

1.2.1 Thinking about the unthinkable

As the tsunami crashed into Fukushima, it severed the "life-sustaining" connection to the electrical grid, leaving the station's six nuclear power plants with no means of cooling their superhot reactor cores. The station's six nuclear reactors began to overheat. A series of chemical explosions spewed vast amounts of radiation into the air, water, and the surrounding area. Much of the area is now contaminated with radiation. Radiation has seeped into Japan's food sources, including rice, fish, and beef. Radioactive isotopes of cesium 137 and 139 have even been found in baby milk formula.[2]

This calamity is particularly disconcerting given that concerns had been raised about the effects of a potentially catastrophic earthquake and tsunami. In response, Japanese engineers arrogantly dismissed concerns about a gigantic tsunami crushing the nuclear complex with an effortless analysis—a simple unsubstantiated memo. Engineers should have prepared a detailed and comprehensive analysis similar to that of an EIS to evaluate the impacts of a catastrophic tsunami as well as mitigation measures for minimizing such impacts. The Japanese paid dearly in terms of impacts and treasure for taking this trouble-free shortcut (see Figure 1.2).

Events similar to those witnessed in Japan could also happen in the United States, which hosts 104 outdated commercial nuclear power reactors. As detailed shortly, experienced scientists, engineers, and stakeholders charge that the NRC's focus on relicensing this aging fleet of nuclear reactors on an accelerated schedule has led to woefully inadequate and flawed analyses of potential impacts and safety concerns. Just as we saw with the Japanese experience, the NRC has likewise dismissed and even refused to consider impacts for serious accidents such as a large tsunami at many nuclear power plants.[3] NRC management steadfastly maintains that US nuclear power plants can be safely operated without a serious threat of a "severe accident" similar to that which struck Japan. On the basis of this bizarre claim, NRC management prepares EISs that publicly conclude that the risk of a severe nuclear power plant accident is "small." As we will see, this assumption cannot be scientifically defended particularly in light of the Fukushima disaster. Before investigating the flaws that plague NRC's EIS process, it is instructive to consider its history and the management culture that has led to such brazen assertions. We will then examine how to avoid some of the most egregious flaws in their EISs.

Figure 1.2 Aftermath of Fukushima Daiichi nuclear power station disaster. (Courtesy images.google.com.)

1.3 *Calvert Cliffs—NEPA's first major lawsuit*

The NRC proudly declares itself to be an "independent agency," meaning that it operates outside the normal sphere of the federal government's executive branch. As an independent agency, the Commission does not report to a cabinet secretary. As noted below, such independence has led to a long history of flouting NEPA's congressional intent. Over the years, the Commission has faced a continuous string of legal battles over its EIS process. In fact, actions taken by NRC's predecessor agency, the Atomic Energy Commission (AEC), triggered the now infamous *Calvert Cliffs'* lawsuit.[4] Calvert Cliffs' was the first major lawsuit filed on behalf of NEPA.* In this landmark case, the Calvert Cliffs Coordinating Committee filed a lawsuit against AEC for treating the requirements of NEPA as mere formalities.

Among other things, AEC's NEPA implementing regulations provided that, when considering an application for a nuclear power plant construction permit or operating license, its hearing board did *not* need to consider environmental factors unless an outside party or staff member affirmatively raised them. Given such a brazen interpretation of NEPA's intent, one was left to wonder what the purpose was for even preparing an EIS. The AEC suffered a resounding and embarrassing defeat. The court was harsh in its criticism of the agency. In rebuking the AEC, the court noted that the agency was effectively arguing that it is sufficient enough for environmental evaluations to "… *merely accompany* an application through the review process, but receive no consideration whatever from the hearing board." In other words, the AEC would prepare and then essentially ignore the results of the analysis. The court described AEC's action as a "crabbed interpretation of NEPA" that "makes a mockery of the Act." Quoting from NEPA, the court said this about AEC's responsibility to comply with NEPA[4]:

> [NEPA compliance duties] … are qualified by the phrase "to the fullest extent possible." We must stress as forcefully as possible that this language does not provide an escape hatch for footdragging agencies; it does not make NEPA's procedural requirements somehow "discretionary." Congress did not intend the Act to be such a paper tiger. Indeed, the requirement of environmental consideration "to the fullest extent possible" sets a high

* More information on the Calvert Cliffs' case can be found in two books by the author: *NEPA and Environmental Planning: Tools, Techniques, and Approaches for Practitioners* (CRC Press 2008) and *Global Environmental Policy: Concepts, Principles, and Practice* (CRC Press 2011).

> standard for the agencies, a standard which must be rigorously enforced by the reviewing courts.

The court went on to admonish AEC's actions, noting for example that the

> Commission's approach to statutory interpretation is strange indeed—so strange that it seems to reveal a rather thoroughgoing reluctance to meet the NEPA procedural obligations in the agency review process, the stage at which deliberation is most open to public examination, and subject to the participation of public interveners.

In language that seemed to foretell the NRC's attitude in spite of the Fukushima calamity, the court went on to note:

> It seems an unfortunate affliction of large organizations to resist new procedures and to envision massive roadblocks to their adoption.

The *Calvert Cliffs* lawsuit was precedent setting as it firmly established that NEPA is the law of the land and that its requirements are binding on *all* federal agencies. By essentially flouting basic EIS requirements, critics now charge that NRC is displaying the same defiant attitude that plagued its sister agency, the AEC, some 40 years ago. Because it is an "independent agency," NRC's managers have even made crass statements such as "NRC is not obligated to follow NEPA. We only do so because we wish to comply."[5] Such eye-opening statements are made despite the fact that in enacting NEPA, Congress expressly stated that the EIS requirement applied to "*all* agencies of the federal government"[6] and that NEPA's implementing regulations go on to state that its requirements are "… applicable to and binding on *all* Federal agencies. …"[7] In the following sections, we will explore the potential ramifications of disregarding NEPA's congressional mandate.

1.3.1 *A perilous legacy*

The NRC and its predecessor, the AEC, have a long history of involvement with dubious and potentially dangerous projects. One particularly disturbing example dates back to 1958. Here the AEC proposed a plan, dubbed Project Chariot, to create an artificial harbor along the coastline of Alaska (Figure 1.3). But it was not the construction of the harbor that

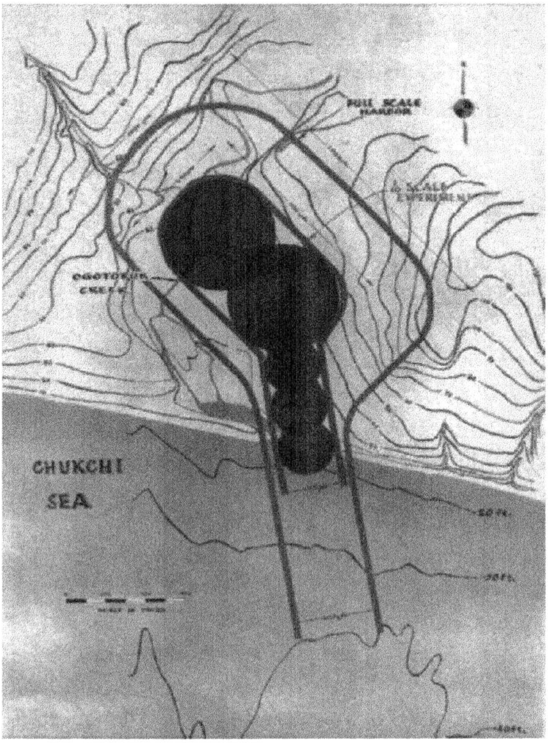

Figure 1.3 Project Chariot plan for using nuclear explosions to excavate an artificial harbor at Cape Thompson, Alaska. The thickly drawn outer border represents the original full-scale plan, involving five nuclear detonations totaling 2.5 megatons of energy. The inner border represents a scaled-down version of the plan.

led to public outcry. The proposal involved excavation of the harbor by detonating multiple nuclear devices (e.g., nuclear bombs).

In response to public uproar and fears, Congress took the then-unprecedented step of ordering AEC to prepare an "environmental report" to study the potential for environmental harm. As one commentator noted[8]:

> Chariot was possibly the first government project challenged on ecological grounds, and occasioned the first integrated bioenvironmental study—the progenitor of the modern [NEPA] environmental impact statement.

Congress effectively ordered the AEC to prepare the world's first de facto EIS.[9] On the basis of this report and the public outcry that followed, Project Chariot was abandoned in 1962. As documented in the next section, NRC unfortunately appears to have inherited some of the AEC's legacy practices. One of the reoccurring themes in this book (demonstrated by the following case study) is that NEPA compliance problems can frequently be traced to mismanagement and sometimes even arrogance toward the public.

1.4 NRC's flawed EIS process

As will be described in Chapter 2, an EIS must be prepared for any federal action that may significantly affect the quality of the human environment. Among other things, an EIS is required to investigate[10]

- Environmental impacts of the proposed action
- Any adverse environmental effects that cannot be avoided if the proposed action is implemented
- Alternatives to the proposed action

NEPA is unique, unlike any other US environmental statute. It is virtually the only planning process that opens the federal decision-making door to the public and allows them to learn about and provide input into the decisions that may affect their lives. NEPA's purpose is *not* to prepare an EIS to satisfy a regulatory requirement so that the agency can merrily "dance onward" with its predetermined decision. Instead, its purpose is to perform a rigorous and objective assessment of impacts and alternatives, and provide that information to the public and decision maker for use in making a reasoned decision to proceed with a course of action.

1.4.1 NRC's troubled license renewal program

The United States has a fleet of 104 aging and antiquated commercial nuclear power reactors. These nuclear reactors were originally licensed for a 40-year operating period. The nuclear operating licenses are now expiring or nearing expiration. The individual commercial operators want to extend their operating licenses. License renewal applications are streaming into the NRC. The NRC's license renewal (LR) program has been extending these operating licenses for an additional 20-year period, effectively pushing the operating window to a full 60 years. This is being done despite the fact that these reactors are based on antiquated designs using 1960s and 1970s technology. These operating licenses have been renewed under the NRC's Division of License Renewal (DLR) directed by Mr. Brian Holian. The division's project branch (RPB1), managed by Mr. Bo Pham, prepared the EISs and safety assessment for relicensing many of these plants. A project-specific EIS (technically a supplemental EIS) is prepared for each individual license renewal application that NRC grants to a commercial operator.

Before venturing further, it should be noted that the author is not anti-nuclear. To the contrary, much of his career has been devoted to preparing EISs for nuclear projects. However, as the reader will soon see, there is room for serious concern when it comes to renewing operating licenses for a national fleet of aging nuclear reactors based on outdated designs. More to the point, relicensing an aging reactor poses considerably greater threat than that posed by constructing a modern plant built to the latest technological design standards.

1.5 When mismanagement threatens society

Holian's relicensing division experienced documented management and morale problems. According to NRC staff, the division had the highest turnover rate in the entire agency. The division's own project managers (PMs) were critical of management practices and complained of EIS safety-related issues.[5,11] These problems became so visible that a decision was finally made to conduct internal focus group meetings to determine the root cause of these morale and management problems.[11] An independent company was hired to facilitate these focus group meetings. The focus group meeting for the PMs was held on September 14, 2010. To facilitate candid responses, no managers were permitted to attend this meeting. The independent facilitator presented the group of PMs with a series of management and safety-related questions. Comments voiced by the PMs were captured on flip charts and then categorized into summary statements. Comments were particularly critical of the division's management. The comments were so negative and the results so embarrassing that management restricted the final report and did not even circulate it to the division's own staff. A synopsis of the focus group statements voiced by the PMs is presented in Table 1.1.[12]

These are indeed grim if not disturbing allegations, particularly given the fact that the very mission of Holian's division is to safely relicense the nation's fleet of nuclear power plants. Even more upsetting is the source of these comments. They were not lodged by political organizations or anti-nuclear critics, but by the very staff and PMs responsible for preparing these safety evaluations and EISs. Particularly disconcerting are comments that Pham and other managers are "bypassing the regulatory process and compromising the safety mission to impress upper management" and DLR is "sacrificing quality for schedule" and that "poor management decisions" are being made. A degree of arrogance may also be involved when NRC's own staff report that "Managers don't listen—they act like know-it-alls."

Table 1.1 Summary of Key Criticisms Voiced by the License Renewal Division's Own Staff at the Project Manager's Focus Group Meeting

- Managers are "bypassing the regulatory process and compromising the safety mission to impress upper management."
- "Poor management decisions" are being made.
- Some managers are "very condescending."
- "Managers don't listen—they act like know-it-alls."
- "Managers are arrogant."
- There are "strained relations between project managers and management."
- Managers are "schedule driven."
- Managers have "dominant personalities"—they place pressure on project managers to shortcut the process.
- Managers are "sacrificing quality for schedule."

Source: 2010 DLR Safety Culture Focus Group Summary. NRC internal report regarding results of license renewal focus group meeting held on September 14, 2010. Includes supplemental statements supplied by project managers that attended focus group meetings.

These are troubling comments given that it is difficult to envision any program with potentially more serious ramifications than a full-scale nuclear power plant accident. Such comments are particularly unsettling given the fact that these same managers are supposed to play a critical link in safety and quality control. As public officials, Pham and Holian act as final reviewers for each license renewal EIS. In doing so, they sign and "certify" the accuracy, rigor, and thoroughness of each license renewal approval. Holian was assigned decision-making authority by the Commission to approve the EISs; yet his own staff indicated that "poor management decisions" are being made. His official management responsibilities include reviewing each EIS for adequacy and accuracy; yet his own staff indicated that management is "sacrificing quality for schedule." Given the veracity of these comments, how can Congress and the public have confidence that management provided the critical stop-gap checks to ensure that each and every license renewal has been properly investigated and fully vetted? This raises a more troubling question: how can the public be confident that there are no nuclear reactors operating with renewed licenses based on flawed assessments?

1.6 Nuclear power and black swans

It was once believed that all swans were white. But a surprising sighting overturned this widely held belief. A rare and previously unknown black species was discovered. The *black swan theory*, developed by Nassim Nicholas Taleb, is a metaphor referring to surprising events that are highly unlikely but which have catastrophic repercussions.[13] The list of potential black swan events is "damningly diverse"[14]:

> Nuclear reactors and their spent-fuel pools are targets for terrorists piloting hijacked planes. Reactors may be situated downstream from dams that, should they ever burst, could unleash biblical floods. Some reactors are located close to earthquake faults or shorelines exposed to tsunamis or hurricane storm surges. Any one of these threats could produce the ultimate danger scenario like the ones that emerged at Three Mile Island and Fukushima—a catastrophic coolant failure, the overheating and melting of the radioactive fuel rods.

A black swan event—particularly one that has never occurred—can be very difficult to anticipate and evaluate, and as we will see below, easy to discount with statistics. The Japanese learned a bitterly painful lesson—simply because something is deemed to be remote does not mean that it will not happen tomorrow. As we will see, US nuclear power plants are and always will be vulnerable to black swan events. Yet, as discussed below, the NRC has brazenly neglected to identify, evaluate, and plan for black swan events in its EISs. This may also be the case for EISs prepared by some other agencies for high-risk projects.

1.6.1 The NRC's disingenuous NEPA process

Fukushima may be a warning shot "across the bow." Chapters 4–6 detail some, but by no means all, of the flaws and errors plaguing EISs prepared for NRC's license renewal program. Like the case of Fukushima, some of these flaws pose potentially serious threats to environmental quality and public safety.

Recall the lackadaisical attitude expressed by Japanese engineers when they rebuffed concerns about a tsunami with a simple memo. A report prepared by Fukushima's management assessed the potential for a severe accident this way:

> The possibility of a severe accident occurring is so small that from an engineering standpoint, it is practically unthinkable.[15]

An assessment of this accident later concluded that the plant's operators and nuclear regulators "fail[ed] to envision the kind of worst-case scenario that [eventually] befell Japan…."[15] We will now examine how the NRC has managed similar issues.

1.6.2 Failed programmatic decision making

The CEQ's NEPA implementing regulations (Regulations) state that a programmatic EIS (PEIS) is to be prepared for

> … broad Federal actions such as the adoption of new agency programs or regulations … Agencies shall prepare statements on broad actions so that they are relevant to policy and are timed to coincide with meaningful points in agency planning and decisionmaking. (§1508.18[b][3]; §1502.4 [b])

The NRC initiative to relicense the nation's fleet of nuclear reactors certainly meets the definition of "new program" for a "broad action." For example, the US Department of the Interior recently issued a programmatic EIS for the development of a solar energy program on public land.[16] Critics are quick to point out that the potential impacts of proposals, such as this solar energy program, pale in comparison to the Commission's program to relicense the nation's fleet of nuclear power plants. While the Department of the Interior fully met its legal responsibility, the NRC violated its statutory duty from the first day it initiated its nuclear relicensing program. Consistent with its goal to relicense the nation's fleet of nuclear power plants on the fastest track possible, it flouted the requirement to prepare a programmatic environmental impact statement (PEIS) for this program.

The purpose of the PEIS is to evaluate the impacts of broad or programmatic alternatives so that an agency and public can understand the risks, impacts, trade-offs, and ramifications of pursuing a particular programmatic course of action. It also allows the decision maker to question the wisdom of pursuing a programmatic course of action. As an example, the PEIS for pursuing NRC's relicensing program should have investigated programmatic and alternative courses of action such as those outlined in Table 1.2.

Such a programmatic examination would have clearly and openly illuminated the trade-offs between safety, cost, and environmental impacts. It would have allowed the public, stakeholders, decision maker, and Congress to readily understand the optimum path forward, as well as the ramifications and risks associated with relicensing an entire generation of antiquated nuclear reactors. But this was not case. Instead, the Commission and its management bypassed the mandate required under the NEPA regulations and by case law, and prepared what it refers to as a

Table 1.2 Programmatic Alternative Courses of Action That NRC Should Have Rigorously Investigated

- Not renewing the operating licenses (no-action) for the fleet of nuclear reactors
- Renewing the operating licenses for the fleet of reactors (proposed action)
- Renewing some types or classes of aging reactors, but not other more hazardous designs
- Pursuing alternative energy sources, including renewable energy
- Replacing the existing antiquated plants with a generation of modern and safer ones

Note: This table illustrates the types of alternative courses of action that an NRC programmatic EIS should have rigorously investigated before reaching a decision to renew the operating licenses for an entire fleet of aging nuclear reactors. The results of the programmatic alternatives analysis should have established a programmatic path forward. NRC was negligent in its regulatory duty to prepare a programmatic EIS.

generic environmental EIS (GEIS).[17,18] The GEIS is *not* a PEIS. The purpose of the GEIS was never to rigorously investigate programmatic impacts or alternatives to the relicensing program. Instead, as its name implies, it examines a set of environmental issues that are considered to be either *common* or *generic* to plant-specific EISs that will later be prepared for each individual license renewal application. Its purpose was to speed up the relicensing process by investigating generic environmental issues so that they would not have to be reinvestigated in later plant-specific EISs. The goal was to speed up plant relicensing by reducing the number of generic issues that would have to be investigated in plant-specific EISs that would follow.

1.6.2.1 *Neglecting to seriously consider programmatic alternatives*

As just witnessed, the assessment of alternatives is the single most important function of any PEIS. According to the Regulations, the chapter on alternatives is the "heart"[19] of an EIS. With this in mind, consider the Commission's program to relicense the nation's fleet of aging nuclear reactors. The alternatives chapter in the GEIS utterly fails to seriously examine broad, programmatic courses of action. This defect is plainly evidenced by the fact that the alternatives chapter is relegated to the second to the last chapter (just before the chapter describing the conclusions) of the main body of the GEIS. Contrast this arrangement with CEQ's format, which places the alternatives chapter front and center as the second chapter in a typical EIS.[20] Placing the alternatives chapter at the end of the GEIS clearly demonstrates the lack of importance that the Commission's management placed on examining programmatic courses of action.

Consistent with the objective of establishing a programmatic course of action, the alternatives chapter should have been one of the longest and most exhaustive chapters in the GEIS. Yet, it is one of the shortest chapters (i.e., it hardly constitutes the "heart" of the GEIS). The GEIS devotes little more than lip service to determining a programmatic path forward.

1.6.2.2 *Failure to evaluate the alternative of not renewing licenses*

The lip service devoted to investigating programmatic alternatives is further witnessed by the fact that the no-action alternative (one of the central sections of any PEIS) is a ridiculous length—two brief paragraphs (only one of which is devoted to actually describing the alternative of taking no action). In contrast, the no-action alternative in even a *simple* NEPA environmental assessment generally runs many paragraphs in length.

This scanty two-paragraph description states that the "no-action alternative is denial of a renewed license." In other words, it is referring to the denial of a license for an individual applicant. Had the GEIS investigated a real programmatic alternative, it would have considered no action from the perspective of taking no action to relicense the nation's *entire* fleet of nuclear reactors. Thus, the GEIS never even questioned the wisdom of pursuing the license renewal program. In violation of NEPA's regulatory direction, NRC has effectively assumed that the relicensing program would be initiated and that the nation's fleet of aging nuclear plants would be relicensed. This flaw is contrary to one of the fundamental purposes of enacting NEPA—to investigate and question the validity of pursuing a proposed course of action.

In circumventing its legal responsibility to prepare a PEIS for its national program, the NRC erred from the "get go." Thus, it violated both regulatory direction and NEPA case law (§1508.18[b][3]; 40 CFR §1502.4[b]). Recall from Table 1.1 that NRC's own staff complained that management was "schedule driven." The problems witnessed in this section are the result of an agency mindset that places more emphasis on speeding along its licensing initiative than it does in taking the time to prepare a PEIS to develop a comprehensive programmatic direction. The lesson to be learned is that a properly scoped PEIS needs to be prepared for new and broad-based programs to determine a path forward.

1.6.3 *A meaningless public comment process*

The Regulations define many requirements in terms of an agency's public review and comment responsibilities. One such provision requires the agency to

> … make every effort to disclose and discuss at appropriate points in the draft statement all major points of view on the environmental impacts of the alternatives including the proposed action. (§1502.9[a])

The NEPA regulations and case law clearly demonstrate that one of the central purposes of NEPA is the duty to inform the public and allow them to provide input into the decision-making process. For each of its relicensing projects, NRC hosts a public scoping meeting and a second meeting to receive comments on the draft EIS. However, a careful review of public comments shows that much, if not most, of this input is given "lip service" and then simply dismissed, often based on the agency's argument that the EIS comment is outside the plant's "licensing basis."[5]

1.6.3.1 *Dismissing issues outside the plant's licensing basis*

So what is meant by the term "licensing basis"? Each nuclear power reactor is licensed on the basis of a given set of design and operating requirements. This set of requirements is called the plant's "licensing basis."[21] These requirements can vary with the type of plant. More specifically, a plant's licensing basis consists of[22]

> ... the set of NRC requirements applicable to a specific plant and a licensee's written commitments for ensuring compliance with and operation within applicable NRC requirements and the plant specific design basis (including all modifications and additions to such commitments over the life of the license) that are docketed and in effect.

Thus, the licensing basis contains such things as applicable NRC regulations, license conditions, technical specifications, and plant-specific design requirements. In essence, the licensing basis includes the requirements that the plant was originally designed to meet plus any modified commitments that NRC later added to the plant's original design basis.

If a member of the public submits a comment that the NRC deems to lie outside of the plant's design and operating requirements, the NRC routinely dismisses it as being "outside the plant's licensing basis." Now consider this dubious practice from the perspective of the Japanese Fukushima Daiichi nuclear power station disaster. Japanese management was warned about potentially catastrophic events such as a mega-quake and tsunami. Yet, they arbitrarily dismissed these concerns. Japan ultimately paid the price for such inept practices.

Similarly, many EIS comments and concerns raised by the American public have also involved new and potentially risky issues beyond the plant's licensing basis. The NRC simply dismisses most of these concerns; critics counter that there is no legitimate basis for dismissing relevant and potentially significant EIS comments simply because they lie outside a plant's licensing basis.

1.6.3.2 Beyond design basis accidents

Plants are designed to withstand certain reasonably foreseeable events. But that does not mean that they are designed to withstand all events. Some events referred to as "beyond design basis accidents" exceed what the plant is designed to withstand. As Fukushima revealed, external and natural threats—earthquakes, tsunamis, fires, flooding, tornadoes, and terrorist attacks—pose some of the greatest risk factors. Forecasting events such as the location or size of the next earthquake or tsunami is a scientific art. Designers of the Fukushima nuclear plant did not anticipate that an earthquake-generated tsunami would disable the backup power systems designed to protect the reactor. In reality, nuclear reactors are[23]

> … inherently complex, tightly coupled systems that, in rare, emergency situations, cascading interactions will unfold very rapidly in such a way that human operators will be unable to predict and master them.

Yet, NRC relicensing EISs have typically considered beyond-design basis accidents so unlikely that they are simply disregarded, often with little or no mitigation safeguards.[24] This defeats the congressional purpose of NEPA, which is to protect the environment and safeguard the public from unexpected black swan events.

1.6.3.2.1 Dismissing the issue of a catastrophic solar flare. Let us examine just how remiss NRC management has been in dismissing such comments, even those submitted by technically qualified experts. An electrical engineer with the US Department of Energy raised a concern that a gigantic solar flare could destroy a nuclear power plant's cooling system, resulting in a nuclear accident such as a full-scale meltdown. This engineer had direct knowledge of the catastrophic impacts that could result from a major solar storm.[25] A major solar storm constitutes a real threat and is not merely a remote or theoretical possibility.[26] The NRC's response was to respectfully dismiss the consequence as mitigatable without any solid or documented evidence that such an event could be properly mitigated. So a potentially life-threatening comment that affected not only that specific plant but potentially all 104 commercial nuclear reactors in the United States was essentially dismissed with a scant reply. The reason for this dismissal was simple. Holian and Pham deemed it more important to complete the project on schedule than to take the time and accept the possibility of a schedule delay to adequately investigate the potential for catastrophic impacts and how they could be properly mitigated. No effort was even made to alert the plant's management so that they could study the problem and institute mitigation measures. Nor was any effort made to notify other US plants of a potential threat. Nor was any study instituted to investigate this engineer's concern.[25]

Apparently, NRC management's focus is on completing licensing approvals on a near assembly line basis—not on seeking tangible input that may help it identify and prevent a future nuclear accident such as Fukushima. This is an egregious violation of the NEPA regulations, which direct agencies to seek out and seriously consider input and concerns from the public (§1501.7). It is particularly disconcerting given the fact that Japanese engineers casually dismissed concerns about the potential of a gigantic tsunami crashing into the Fukushima Daiichi reactors with a similar mechanism—a simple unsubstantiated memo.

1.6.3.3 *Dismissing stakeholder and public concerns*

The NRC has a long and defiant history of battling the public and stakeholders in court. Watchdogs charge that the NRC has hoodwinked courts into accepting their arguments and dismissing stakeholders' concerns. For instance, in issuing an operating license for the Diablo Canyon nuclear power plant (see Figure 1.4), NRC management insisted that the odds of an earthquake triggering a severe nuclear accident were negligible. The Commission was sued over this and other related safety concerns. The court deferred to the NRC, ruling that

> The Commission has determined that the chance of such a bizarre concatenation of events occurring is extremely small.[27]

Figure 1.4 Diablo Canyon nuclear power station.

Despite fierce opposition from a significant portion of the scientific community, NRC management argued its geologically questionable stance for more than two decades. Yet, in 2011, the Commission finally found itself in the awkwardly embarrassing dilemma of having to announce that it would be conducting new seismic risk assessments at 17 nuclear power plants. Apparently, the agency erred in its sweeping dismissal of public concerns more than two decades earlier.

Now consider this behavior with respect to a recent issue of concern. The NRC has failed to learn from lessons like the Diablo Canyon nuclear plant. NRC management argues that unless required by the courts, it will not evaluate potential impacts of terrorist attacks on nuclear reactors. This is true despite the fact that federal agencies are required to conduct an open and transparent EIS process that discloses all potentially significant impacts. Catastrophic scenarios involving terrorist attacks are indeed conceivable.[28] The NRC has continued this defiant stance even in light of the terrorist tragedy of 9/11 and the Fukushima calamity. Watchdogs charge that the NRC does not want to expose the true risk and implications to the public. To do so would alarm the public, jeopardizing its nuclear relicensing mission.

This is no theoretical possibility, as nuclear reactors make ideal military and terrorist targets. For instance, in 1972, three hijackers overtook a domestic passenger flight along the east coast of the United States and threatened to crash the plane into a US nuclear weapons plant in Oak Ridge, Tennessee. The plane got as close as 8000 ft before the hijackers' demands were met.[29] The public is often surprised to learn that over the past three decades, nuclear reactors have been repeatedly attacked during military air strikes, occupations, and invasions. Between 1977 and 2003, there have been no less than eight attacks, some of them destroying the entire reactor.[30] Despite such statistics, NRC defiantly refuses to evaluate and plan for such events in its EISs unless the project is located in a jurisdiction in which the courts have required such assessments. Added to this is the fact that NRC's "design-basis threat" criteria for plants are classified and so the size of an attacking force that the plants are able to defend against is unknown. With respect to the risks, the public is left in the dark.

Such behavior opens larger issues. How many other safety- and environmentally related issues has NRC management ignored? As indicated in Section 1.2, NRC's staff complained that the management is "very condescending" and "managers are arrogant." Such attitudes may partly explain why the Commission dismisses stakeholder concerns, and appears to take a cavalier attitude toward its public participation responsibilities. The lesson to be learned here is that an EIS is a "public" process, and federal officials are required to seriously consider and, where appropriate, investigate issues of concern in the EIS analysis.

1.6.4 Can the consequences of a "serious nuclear accident" really be "small"?

The issue of nuclear safety and the potential for a catastrophic accident are clearly the most dominant issues of concern to most independent experts and the public. Rather than use a descriptive term like "catastrophic" (as is the case for a number of other agencies) in referring to an accident such as a full-scale nuclear meltdown, NRC prefers the less disconcerting term "severe accident." Regardless of the terminology used, a severe nuclear accident such as that witnessed at Fukushima or the Chernobyl nuclear power plant is widely regarded to be one of the, if not the most, calamitous technological events that can be imagined (see Figure 1.5).[31]

The issue of a severe nuclear reactor accident is described in Chapter 5 of each relicensing EIS. Regrettably, the analysis of "severe accidents" utterly fails the test of providing a detailed scientifically based analysis of potentially severe nuclear accidents. This is evidenced by the fact that a typical NRC relicensing EIS falls in the range of about 150,000 words or so. Out of about 150,000 words, NRC management typically devotes less than 500 words to the actual assessment of "severe accidents," which constitutes a paltry 0.4% of the EIS; in other words, less than about one-half of 1% of the

Figure 1.5 Chernobyl nuclear power plant accident. The battle to contain the spread of radiation and avert a greater catastrophe ultimately involved more than 500,000 workers and cost an estimated 18 billion rubles, crippling the Soviet economy. (From interviews with Mikhail Gorbachev, Hans Blix and Vassili Nesterenko. *The Battle of Chernobyl*. Discovery Channel. Relevant video locations: 31:00, 1:10:00.)

entire EIS is devoted to the assessment of an accident that could result in a national catastrophe.

A relicensing EIS typically runs hundreds of pages in length, examining virtually every conceivable environmental impact, from biota to air emissions and water usage; these reviews are routinely performed in near myopic detail. Yet, when it comes to examining the real issue that most people are intimately concerned with—the issue that lies at the heart of the entire relicensing process—Holian's management team provides nothing but a scant 500-word statement about the consequences of an accident such as a full-scale nuclear meltdown. For example, Pham dismissed these flaws when they were brought to his attention.[32]

It is little wonder that public stakeholders and watchdog organizations charge that NRC's assessment of nuclear accidents is the equivalent of a scientific façade. They charge that the NRC has prepared a lopsided analysis (hundreds of pages) to divert attention away from the predominant concern—the issue of overriding concern—and refocused it on other rather mundane and less controversial "bug-and-bunny" issues. It is not difficult to understand why so many stakeholders and grassroots organizations charge that the Commission is "hiding" the impacts of an accident from the public who would have to live with the consequences; if the public understood the true risk they are faced with, it could kill the issuance of some, and perhaps many, renewed nuclear operating licenses.

The Commission and its management attempt to justify its analysis by arguing that it prepares a detailed assessment known as Severe Accident Mitigation Alternatives (SAMA), which might theoretically mitigate some of the impacts of a severe accident. The SAMA assessment is typically included in Chapter 5 of a relicensing EIS. But here is what the NRC conveniently neglects to tell the public:

1. Only infrequently does NRC conclude that any of these SAMAs are cost beneficial and therefore justified.
2. The NRC rarely imposes a duty on the commercial applicant to adopt a mitigation measure evaluated as part of its SAMA assessment.
3. Most SAMAs would have only a small improvement in reducing the risk or impacts of a severe nuclear accident.

From the standpoint of truly protecting the environment and public safety, the SAMA analysis is largely hollow. So why perform it? Scientific and engineering critics, including public stakeholders, explain that by doing so, the Commission can at least claim to the public that they have looked at means for reducing the impacts of a severe accident, but concluded such SAMAs were not worth the cost and effort. As one engineering critic quipped, "It is a slick public relations ploy."

1.6.4.1 When the impacts of a severe accident are "small"

As we have seen, the impact of a major nuclear accident could be felt by millions and could sweep across many states; radiation released from a single accident could menace millions of citizens. Then there are the paralyzing effects of widespread panic, psychological trauma, socioeconomic costs, emergency evacuation disruptions, and relocations that would follow in the aftermath. This is no mere theoretical possibility; at least two, if not three, of the five most expensive accidents in the world involved the Chernobyl, Three Mile Island, and Japanese Fukushima Daiichi nuclear power station accidents.[33]

Despite the potentially catastrophic impacts of a severe accident, Chapter 5 of each relicensing EIS typically repeats the following "canned" 50-word statement:

> The probability weighted consequences of atmospheric releases, fallout onto open bodies of water, releases to ground water, and societal and economic impacts from **severe accidents** are **small** for all plants. However, alternatives to mitigate severe accidents must be considered for all plants that have not considered such alternatives.

This dismal 50-word "assessment" of the effects of a severe accident is all that NRC management has to say on the matter. There is no analysis or discussion of the radiation contamination, radiation-caused deaths, health effects such as cancer and birth defects, species and habitats that would be contaminated, or widespread socioeconomic costs and impacts (see Figures 1.6 and 1.7).

Moreover, there is a presidential executive order mandating that the effects on low-income and minority populations be evaluated to determine if they would be disproportionately affected by an agency action

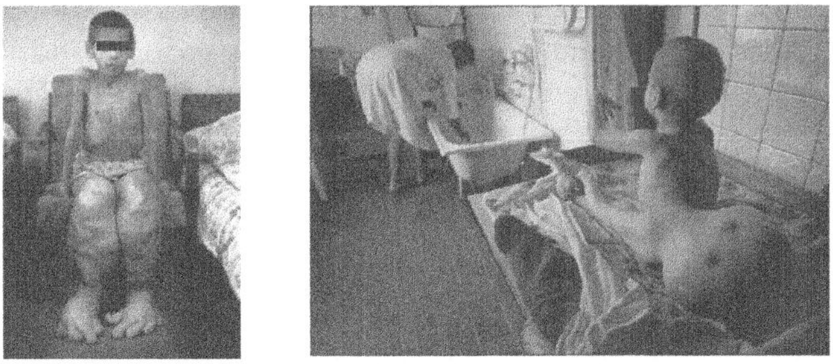

Figure 1.6 Victims of Chernobyl nuclear power reactor accident.

Figure 1.7 Aftermath of Chernobyl nuclear power reactor accident.

such as renewing a nuclear operating license. Again, no NRC EIS accident analysis has ever complied with this requirement.

Now consider this two-sentence, 50-word assessment from the following perspective. A typical relicensing EIS contains perhaps 150,000 words or so. Out of this total, the decision maker, stakeholders, and public are provided with a cursory 50-word "canned" statement assessing the impacts of a potentially catastrophic accident that could jeopardize the lives and property of millions. Framed another way, the Commission devotes a mere 0.03% of each relicensing EIS to describing the impacts of a severe accident (described above), including their bewildering and implausible conclusion that the risk is "small." To frame this yet another way, the same relicensing EIS devotes on the order of 25,000 words to the chapter describing the affected environment and 20,000 words to the chapter describing nonnuclear environmental impacts, yet only a paltry 50 words on the consequences of a potentially catastrophic nuclear scenario such as a full-scale nuclear meltdown.

Recall that a report prepared by Fukushima's management assessed the potential for a severe accident this way: "The possibility of a severe accident occurring is so small that from an engineering standpoint, it is *practically unthinkable*."[15] There are troubling and uncanny parallels in the way NRC management dismisses the risk of a severe accident and the manner in which the management at the Japanese Fukushima reactor denied the potential for a catastrophic event.

1.6.4.2 Obscuring the risk of a catastrophic nuclear accident

As just witnessed, NRC management steadfastly maintains that the risk posed by a "serious accident" is "small" in each and every one of its relicensing EISs. Critics, including a sizeable portion of the scientific community, charge that the Commission maintains this incredulous position despite the fact that the livelihood of tens of thousand, if not millions, of citizens would be irreversibly affected by an accident such as a full-scale nuclear meltdown. You might ask how officials of any agency could be so rash as to claim that the risk of a catastrophic nuclear accident, such as a full-scale nuclear meltdown, could possibly be "small." The Commission employs an analytical technique that from the public's standpoint is something akin to mathematical "smoke and mirrors."

Here is how it is done. A concept known as *risk analysis* is employed. Risk is most typically defined as the consequences (impacts) of an accident multiplied by the frequency or the probabiltuty of the accident. The NRC's risk analysis essentially takes the frequency (which it maintains is very small) of an accident, multiplies it by the consequences (assume it is large), and then concludes that the human, environmental, and socioeconomic impacts of a severe (catastrophic) accident are small because the probability is so small. Let us examine this bewildering conclusion in more detail.

1.6.4.3 Concealing the probability of a catastrophic accident

Is the potential effect that a severe accident would have in terms of contaminated air and water, human radiation poisoning and resulting deaths, long-term health effects such as cancer, genetic mutations including birth defects, affected species, contaminated food chains, evacuation of tens or hundreds of thousands of downwinders, property damage in the hundreds of billions of dollars, and possible contamination of hundreds or thousands of square miles really "small"? Were the effects of Chernobyl, Three Mile Island, and more recently the Japanese Fukushima Daiichi power station disasters "small"? Not surprisingly, public stakeholders and watchdogs have suggested that perhaps the Commission should consult with officials in Japan to determine if they believe that the risk of the Fukushima Daiichi accident was "small."

In essence, NRC management has employed a mathematical "trick" to conceal the risk and impacts, and rationalize its claim to the public that the risk of a catastrophic nuclear accident is "small." All management had to do was assume that the frequency of such an accident is *exceptionally small*. Such a conclusion might be plausible as long as the assignment of a small frequency is scientifically defensible. We will now examine the validity and defensibility of such a small frequency value.

If the probability of an accident is as small as management claims, why have there been four other near-catastrophic nuclear reactor accidents (near misses) in the United States, in addition to Three Mile Island? These four other "near misses" are actually acknowledged and taught to NRC staff as part of its board qualification and training program[34]:

- Browns Ferry nuclear reactor incident
- Vogtle nuclear reactor incident

- Davis–Besse nuclear reactor incident
- Salem nuclear reactor incident

While these "near misses" are documented in the technical literature, the NRC avoids advertising or drawing public attention to them. In reality, there have been numerous other near misses in the United States as well. While a full-scale nuclear reactor accident was averted in each of the four cases cited above, each accident came perilously close to a calamity. These near misses are not necessarily restricted to incidents decades ago; the Davis–Besse nuclear power station event occurred as recently as 2002. The point of this discussion is that the three major nuclear accidents, Chernobyl, Three Mile Island, and Fukushima Daiichi, were not necessarily isolated events.

This is only half the story. Sovacool[35] reports that worldwide there have been 99 accidents at nuclear power plants in which 57 nuclear-related accidents occurred after the Chernobyl disaster. Out of this total, 57% (56 of 99) of these accidents occurred in the United States. The US General Accountability Office reported than between 2001 and 2006 alone, there were more than 150 incidents in which nuclear plants had not operated within "acceptable safety guidelines."[36]

NRC management has discounted and consistently argued that the likelihood of severe multiple incidents at nuclear reactors is small.[37] Using its best mathematical techniques, the agency concluded that the simultaneous failure of both emergency shutdown systems designed to prevent a core meltdown is so unlikely that it would happen only once every 17,000 years. Yet, a mere 20 years ago, it happened twice within a period of 4 days at a pair of nuclear reactors in southern New Jersey.[38] If NRC cannot even accurately assess the probability of a potential accident over a 20-year period, why should the public believe their EISs, which routinely conclude that the impacts of a severe accident are small?

To date, there have been five serious accidents worldwide, including Three Mile Island, Chernobyl, and three at Fukushima Daiichi. This leads to, on average, one serious accident occurring every 8 years worldwide.[39] A Massachusetts Institute of Technology (MIT) study concluded that *at least* four serious nuclear accidents can be expected in the period between 2005 and 2055.[40,41] Other critics conclude that even MIT's study has underestimated the seriousness of such an event, particularly in light of the fact that these aging reactors are based on antiquated technology.

1.6.4.4 *When the risk of a catastrophic nuclear accident is really "large"*

Clearly, when the United States has suffered a partial meltdown of Three Mile Island, the Japanese battled the aftermath of three nuclear meltdowns at Fukushima, and the former Soviet Union struggled through the Chernobyl catastrophe, it is publicly indefensible to claim that the probability of a severe accident is as small as NRC professes. Add to this the fact that the United States experienced at least four (if not dozens) other "near misses," and it becomes clear that NRC's position is scientifically incredulous. We are led to the inescapable conclusion that the frequency of a catastrophic accident is much greater than the agency's management publicly admits.

Recall that the NRC bases its claim on the premise that a very small frequency or probability multiplied by a large impact results in a "small" risk. However, it was just demonstrated that the frequency is much larger than NRC openly admits. Thus, the risk of a severe accident cannot be "small" and is probably on the order of "large."

As we will see in Chapters 4 and 5, such obstinate practices violate any number of NEPA regulatory provisions, including requirements to perform an "open," "objective," "scientific," and "accurate" analysis. The lesson to be learned here is that officials and practitioners need to ensure that they prepare EISs that comply with these requirements.

1.6.4.5 Concealing impacts of a severe accident

The NEPA regulations require that the EIS thoroughly assess the impacts (consequences) of its actions, including potential accident scenarios. In fact, a lengthy section of the Regulations is devoted to detailing requirements for performing an analysis of environmental consequences.[42] Nowhere in the 35 pages of requirements do the Regulations even acknowledge, let alone allow, an agency to substitute an analysis of risk for an assessment of the impacts. For instance, the Regulations require the EIS to

> … present the environmental impacts of the proposal and the alternatives in comparative form, thus sharply defining the issues and providing a clear basis for choice among options by the decision-maker and the public. (§1502.14)

As required under this provision, neither the risk nor the consequences of a potential nuclear accident involving the proposed action (relicensing) are compared with other alternatives. This makes it impossible to sharply define issues and provide a clear basis for choice among options by the decision maker and the public.

But there are other more menacing problems with the Commission's analysis. Even if one accepts the anemic argument that the risk (in terms of probability) is "small," it is clear that the consequences of an actual accident could be "large," even disastrous. While the EIS can certainly include an analysis in terms of "risk," the regulatory provisions (§1502.14) clearly indicate that the potential *consequences* must be described and disclosed. Yet, the NRC's EISs lack an analysis or description of the impacts (consequences) of a severe accident. For instance, there is no description of the potential consequences, which could include the nature, scope, and extent of radioactively contaminated air, land, and water bodies; human radiation poisoning, including deaths; long-term health effects such as cancer; genetic mutations, including birth defects; affected species and habitats; contaminated food chains; evacuation of tens or hundreds of thousands of downwinders; property damage and economic losses in the tens or hundreds of billions of dollars; widespread psychological trauma; and possible long-term contamination of hundreds or thousands of square miles.

Why do the NRC's EISs fail to comply with this requirement? As noted in Table 1.1, NRC's own staff charges that its management is "sacrificing quality for schedule" and "bypassing the regulatory process and compromising the safety mission to impress upper management." Many environmental organizations maintain that management does not want to publicly disclose the catastrophic implications of a major accident; disclosing the true consequences of an accident could ignite public concerns and opposition, delay licensing projects, or even jeopardize the Commission's entire relicensing program. As a result, the Commission has not met its legal responsibility to adequately investigate the consequences of a catastrophic accident. The lesson is that officials and practitioners need to evaluate and disclose the true impacts regardless of how severe they may appear to the public and stakeholders.

1.6.5 Concealing cumulative risk from the public

As we have seen, an EIS must thoroughly investigate direct and indirect impacts of a proposal. As will be described in Chapter 5, Section 5.8, an EIS must also rigorously evaluate *cumulative impacts* (i.e., the combined impacts of other past, present, and reasonably foreseeable actions).[43] As witnessed earlier, NRC typically uses a flawed 50-word canned paragraph to describe the risk posed by a severe nuclear accident. In addition to providing a woefully inadequate "assessment," this paragraph is also flawed in terms of investigating the cumulative impacts of an accident; this stems from the fact that NRC only considers the probability and consequences of an accident from a single nuclear power plant station. But this is not the case at all. There are actually 104 commercial nuclear reactors in 31 states. Moreover, many nuclear sites host multiple nuclear power plants. The actual cumulative risk to the American public from an entire fleet of operating reactors is much greater than that posed by a single reactor. Again, NRC's relicensing EISs have presented the public with a misleading assessment of the "true" or cumulative risk.

For example, suppose that the NRC computes the frequency of a severe accident at a particular nuclear reactor. What NRC management neglects to disclose is that this same segment of the public is also affected by the risk of a severe accident at other nuclear stations or by multiple plants at a single station. A potentially affected population is actually exposed to multiple risks from many nuclear plants. The Commission's EISs have utterly failed to acknowledge, compute, or disclose this total or cumulative risk as well as the cumulative impacts of a severe accident from more than 100 operating reactors. In disregarding its legal responsibility to evaluate cumulative risk or the cumulative consequences of a severe accident from multiple nuclear stations, the public is exposed to much greater risk than many realize. Officials and practitioners need to understand that an EIS must evaluate and *disclose all* impacts, including cumulative impacts, openly and accurately.

1.6.6 *Failure to assess significance*

As we have seen, NRC management consistently makes the implausible conclusion that the risk of severe accidents is "small." Recall that they reach this conclusion by computing the risk of an accident (multiplying the probability by the consequence). But this assessment is flawed. The Commission has completely ignored the extensive regulatory direction, which requires the assessment of the impact in terms of "significance."[44] An EIS is required to assess and disclose the significance of the impacts.[45] The Regulations do not permit agencies to simply dismiss the significance of the impacts of a potential accident simply because the "risk" is deemed to be "small."

The Regulations require that the assessment of significance be considered in terms of both the *context* and *intensity* of the impact.[46] Interestingly, the relicensing EISs approved by Holian's division assess the significance of every environmental impact except that of a severe accident. While these EISs state that the "risk" of an accident is "small," they fail to assess the significance of the consequences of an accident.* For instance, there is not a shred of evidence indicating that any consideration has even been given to considering significance in terms of the context in which the accident would occur. But the problems do not end here.

1.6.6.1 *Intensity factors to be used in assessing significance*

The EIS must likewise consider significance in terms of the severity or intensity of an impact. The Regulations specify 10 intensity factors that must be considered in reaching a conclusion regarding the significance of

* Moreover, as required by this same regulatory provision, NRC's EISs fail to address long-term consequences of issues such as latent cancers, genetic defects, and the permanent dislocation or contamination of affected areas.

an impact.[47] Triggering even a single intensity factor is typically deemed sufficient to reach a determination that the action poses a "significant impact." As noted in Table 1.3, the consequences of a severe nuclear accident trigger at least 5 of the 10 intensity factors. We will consider the significance of a severe nuclear accident in terms of each of these five intensity factors.

Intensity factor #1: As indicated by the first intensity factor in Table 1.3, a severe nuclear accident could undoubtedly affect public health or safety. The potential for a severe accident therefore poses a significant impact because its consequences could gravely affect public health and safety.

Intensity factor #2: A severe nuclear accident is unquestionably controversial in terms of both its political and scientific ramifications. Because of its controversial nature, the consequences of a severe accident are significant.

Intensity factor #3: Potential effects such as the radiation dose received by members of the public, number of fatalities, long-term health effects such as cancers, and genetic effects such as birth defects undoubtedly involve uncertain, unique, or unknown risks. A severe accident is therefore significant from the standpoint of uncertain, unique, or unknown consequences.

Intensity factor #4: As we have seen, the Commission's EISs only consider the potential risk posed by a single operating station. In fact, there are 104 nuclear power plants operating in the United States. Collectively, these relicensing actions pose a cumulatively significant impact on the environment and society at large.

Table 1.3 The Consequences of a Severe Accident Trigger at Least Five NEPA Intensity Factors for Determining Significance

1. The degree to which the proposed action affects public health or safety.[48]
2. The degree to which the effects on the quality of the human environment are likely to be highly controversial.[49]
3. The degree to which the possible effects on the human environment are highly uncertain or involve unique or unknown risks.
4. Whether the action is related to other actions with individually insignificant but cumulatively significant impacts. Significance exists if it is reasonable to anticipate a cumulatively significant impact on the environment. Significance cannot be avoided by terming an action temporary or by breaking it down into small component parts.[50]
5. Whether the action threatens a violation of federal, state, or local law or requirements imposed for the protection of the environment.[51]

Note: As indicated in this table, the consequences of a severe nuclear accident such as a nuclear meltdown trigger at least five intensity factors used in reaching a determination that an action poses a significant impact (§1502.27[b]).

Intensity factor #5: In addition to exceeding radiation limits, a severe nuclear accident would violate any number of federal and local laws and regulations such as restrictions on radiation releases. From this standpoint, a severe nuclear accident poses a significant impact on the regional and local society and environment.

No further evidence is needed to prove that the consequences of a severe accident are anything but "small" and, in fact, constitute a very significant impact on the environment and society that has yet to be properly investigated. This conclusion is consistent with the criticism noted in Table 1.1, in which NRC's own staff charges that their management makes "poor management decisions."

1.6.7 Reaching the final decision before the EIS process has been completed

As stated in the Regulations, an EIS

> [Must be completed and made] available to public officials and citizens before decisions are made and before actions are taken. (§1500.1[b])
>
> … shall be used by Federal officials in conjunction with other relevant material to plan actions and make decisions. (§1502.1)
>
> [shall] include the alternative of no action. (§1502.14[d])
>
> No decision on the proposed action shall be made [until the EIS is completed and reviewed by the decision-maker]. (§1506.10[b])

As these provisions show, the EIS provides the decision-making tool for reaching a decision to pursue a course of action. A final decision regarding the course of action to be taken may not be made until the final EIS has been prepared and reviewed by the decision maker. Consistent with this direction, an EIS must include the alternative of taking no action with respect to the proposal. In terms of renewing nuclear operating licenses, the no-action alternative means that the Commission would not issue a renewed operating license to the plant applicant. If the Commission were to choose the no-action alternative, the plant would be shut down at or before the end of the current license.

Project-specific relicensing EISs consider alternatives to license renewal; however, most of the public do not realize that none of these alternatives, not even the no-action alternative, have ever been *seriously* entertained by the decision maker. In direct violation of the law, even the NRC's own staff has openly and publicly admitted that the Commission does not seriously consider an alternative beyond the option of relicensing a nuclear power plant. Consider what one of Pham's own project managers had to say at a NEPA public meeting on a draft EIS to renew the Cooper nuclear reactor's operating license. When asked by a member of the press about the choice of taking no action and shutting down the reactor on or before its operating license expired, Ms. Bennett Brady, the project manager, candidly and publicly admitted[52]:

> ... that option wasn't even considered because of the important role which Cooper Nuclear Station plays in providing energy.

In other words, the project manager in charge of relicensing the Cooper nuclear reactor frankly admitted in a public meeting that while the no-action alterative (in addition to all the other alternatives) was evaluated, it would not even be considered by the final decision maker. Perhaps this admission best depicts the Commission's defiant and disingenuous attitude toward stakeholders and the public.

NEPA's regulations in tandem with case law clearly state that a final decision is *not* to be made until the final EIS process has been completed.[53] Moreover, the courts have made it crystal clear that agencies must seriously consider the "... adoption of *all* reasonable alternatives" and that the no-action alternative may not be ruled out until the EIS process has been completed. Despite these legal requirements, one of the Commission's own project managers publicly admitted that the decision to renew the operating license had already been made. One is left to ponder what the point of preparing the EIS even was. Again, this is a systemic misstep in the Commission's NEPA process that is ripe for legal challenge.

1.6.7.1 Failure to adequately evaluate the no-action alternative

The purpose of investigating the no-action alternative is to determine the effects on society and the environment if the proposal is not implemented. Under the Commission's no-action alternative, if a plant is shut down, the threat of a severe accident is essentially eliminated. Yet, the no-action alternative does not even consider or evaluate the beneficial reduction in potential impacts and risk that would result from denying (shutting down the plant) a renewed operating license. This constitutes a serious omission that could profoundly shape the decision maker's and public's perceptions of nuclear power. Moreover, the decision maker cannot even weigh the benefits of this risk reduction because it was never considered, computed, or disclosed. This is yet another deficiency that most of the public is largely unaware of.

Recall from Table 1.1 that NRC staff state that managers have "dominant personalities—they place pressure on project managers to shortcut the process." Such criticism may partly explain why the Commission can reach its decision before the EIS has even been prepared. The lesson here is that under no circumstances is a decision to be made that biases or predetermines the decision before the EIS process has been completed; moreover, the no-action alternative must openly disclose all impacts, beneficial and adverse, that could occur as a result of taking no action.

1.6.8 Failure to adequately evaluate reasonable alternatives

As witnessed earlier, the underlying purpose for preparing an EIS is to evaluate *alternatives* and *mitigation measures* that can be implemented to avoid or reduce potential impacts.[54] The assessment of alternatives must be sufficient to allow the decision maker and public to understand and discriminate between the proposed action and various alternatives. Yet, the NRC's relicensing EISs provide only a cursory review of alternatives and mitigation measures—just enough so that the Commission can claim they were not ignored.

Added to this, the Commission intentionally skews the presentation of the alternatives to make them appear unreasonable. For instance, the relicensing EISs generally provide nothing more than a scant one-paragraph description of the alternative for constructing a new reactor (i.e., "New Reactor Alternative"). This one-paragraph discussion neglects to even mention the important consideration that replacing the aging reactor with a modern one would substantially reduce the risk of a nuclear accident. How can a decision maker make a rational and reasoned decision based on a one-paragraph description?

The NRC justifies such practice based on the following rationale. It routinely dismisses the New Reactor Alternative because it could not be completed in time to meet the expiration date of the operating license. This reasoning is bogus. For instance, NRC management totally neglects to consider the very realistic option that the operating license could be *conditionally extended* for a period sufficient to allow the applicant to construct a new and safer reactor or an alternative form of power; the condition would be that the operator seek funding and begin preparations for constructing a new plant or power source. This option would allow the present reactor to continue operating for a defined period until an alternative power source is secured; if the operator could not comply with this conditional extension, they would be forced to shut down their nuclear reactor. Not only would this provide a "middle ground" alternative, but it would also enhance safety by replacing an antiquated reactor with a more modern and safer plant or alternative energy source; yet, this option is not even considered. So it goes for most of the alternatives that are "examined" within these EISs.

One of the single most important reasons that led Congress to enact NEPA was to force agencies to consider reasonable alternatives. Bogus rationales that dismiss reasonable courses of action, biasing the analysis in favor of the proposed action, are inexcusable and illegal. As noted in Table 1.1, Holian's own staff complains that "managers don't listen—they act like know-it-alls." Such criticism may partly explain why NRC refuses to openly perform a rigorous investigation of all reasonable alternatives. The lesson to be learned is that officials and practitioners need to ensure that a full range of reasonable alternatives have been rigorously and objectively evaluated; the alternatives analysis is not to be skewed in favor of the proposed action.

1.6.9 *Issues never considered or disclosed*

The NRC has steadfastly refused to evaluate some of the greatest risks associated with renewing operating licenses. For instance, consider the following two issues:

1. For years, the Commission has refused to consider the impacts of the huge inventories of highly radioactive nuclear waste piling up at nuclear power plants around the nation. The ultimate disposition method and site for this radioactive waste has been uncertain. Yet, NRC simply dismissed one of the most prominent and controversial nuclear issues, stating that it would be "addressed in the future." Such an indefensible position violates numerous NEPA requirements and case law. For years, the Commission was able to get away with such fly-by-night NEPA practices. That was until it was sued in 2012 by 24 organizations. The Commission lost.[55] The NRC was forced to suspend licensing until this waste issue has been adequately investigated. Finally, the agency is being forced by the courts to assess a key and critical issue that it should have been evaluating all along. The suspension has lowered its public credibility and shaken the confidence of many lawmakers. At least one congressman called for the end of new nuclear power in response to the order.[56]
2. The uranium fuel used to power a nuclear plant is eventually "burned up" to produce a waste product known as spent fuel. The highly radioactive spent-fuel rods are withdrawn, and then stored and cooled in spent-fuel pools located at the reactor site. Being highly radioactive, it is very dangerous, particularly from the standpoint of a fire. Most of the public does not understand that this waste may pose a greater threat than a severe accident involving the reactor itself.

Many of the cooling pools that house spent-fuel rods are filled nearly to capacity. A fire could spew voluminous amounts of deadly radioactive isotopes into the biosphere. Yet, as significant as these issues are, the NRC refused to investigate or disclose the impacts in their relicensing EISs. But this too is coming to an end. The Commission was forced to suspend licensing until the current practices of storing spent fuel onsite are examined and publicly disclosed.

Recall from Table 1.1 that NRC's staff complains that management is "arrogant" and "very condescending." Such criticisms may partly explain why the Commission has steadfastly refused to investigate and disclose vital decision-making information to the public. The lesson to be learned is that all pertinent issues and information must be addressed in an EIS. A court should not be compelled to step in and force an agency to do what it should have been doing all along.

1.7 Making the EIS process work

Many of the deficiencies encountered in EISs can be traced directly back to poor management and oversight. For instance, consider the criticisms voiced by the NRC's own project managers (Table 1.1), including statements such as management sacrifices "quality for schedule" and that the Commission bypasses "the regulatory process and compromises the safety mission." As a public official, Pham was responsible for reviewing and certifying the accuracy of these EISs. There is little excuse for the types of flaws witnessed in this chapter. It should therefore come as no surprise that the NRC's relicensing approvals are virtually rubber stamped on a near assembly-line basis. The NEPA process is merely a hurdle to jump so that NRC management can proceed with its relicensing initiative. But there may be problems on the horizon. As just outlined in Section 1.6.9, the NRC may be witnessing the first phase of future NEPA lawsuits. A management process unencumbered by the problems depicted in Table 1.1 could have saved the Commission the public embarrassment of losing a major lawsuit. Mismanagement practices have bred suspicion and doubt over the Commission's credibility. Worst yet, such practices may have harmed the very future of nuclear power in the United States. At least three lessons can be distilled from this case study:

1. Major EIS errors and deficiencies can frequently be traced directly to management problems.
2. Misrepresenting the results of an EIS analysis can cast doubt on the integrity and credibility of an agency's actions.
3. Egregious errors in the EIS process can result in undesirable repercussions, including lawsuits (see *Three Laws of the Environmental Movement* in the Introduction to this book).

1.7.1 A re-review of license renewal

In February 2012, the NRC's commissioners voted 4–1 to approve the application to construct two new nuclear power reactors at the existing Vogtle nuclear power plant site in Georgia. What was extraordinary was the fact that the official casting the lone dissenting vote was the chairman of the Commission. Citing concerns over the Japanese Fukushima nuclear disaster, Chairman Gregory Jaczko stated, "I cannot support issuing this license as if Fukushima never happened."[57] Perhaps this a red flag: if the Commission's licensing process for new nuclear reactors is anything similar to that for relicensing aging reactors, then the NRC's chairman and the American public have reason to be concerned.

Critics are quick to point out that if the Commission's mission involved less hazardous actions such as forest harvesting, rural development, wetlands development, highway construction, or siting of renewable energy projects, what we have just witnessed would be a sad state of affairs. But the NRC's EISs involve one of the most precarious technologies on the planet—renewing the operating licenses of antiquated nuclear reactors. The lives of millions hinge on how adequately and thoroughly these actions are considered and vetted.

Do the public and affected stakeholders fully appreciate the extent to which most public concerns have been routinely dismissed with terse or even "canned" responses? Do they understand that the risk of a catastrophic nuclear accident located upwind of them is actually "large" rather than "small"? Does the public understand that the true cumulative risk was never even examined or publicly vetted, and that they are at greater risk of a devastating nuclear accident than each individual relicensing EIS would have them believe?

How many of the approved licenses are flawed simply because NRC management sacrificed "quality for schedule." Has NRC management already relicensed a ticking time bomb—the equivalent to the devastating Japanese Fukushima Daiichi nuclear reactor meltdown? The answer is that nobody knows. It will require a full and comprehensive review of every single renewed license to even remotely begin to answer this question. Meanwhile, a catastrophic meltdown may be a week away or 10 years into the future.

In 2012, Japanese Prime Minister Yoshihiko Noda acknowledged that his government and the Fukushima management had been blinded by a "safety myth" that led to their belief in the country's "technological infallibility."[58] Jeffrey Loman, deputy regional director for the Bureau of Ocean Energy Management, Regulation and Enforcement, has similar insight. Loman stated that before the Deepwater Horizon oil spill disaster, the former Minerals Management Service had come to a belief that it had a "gold-plated" safety system—a belief that had led to dangerous levels of complacency; the consequences of a major accident involving the Deepwater Horizon project had likewise been evaluated and deemed to be small. The lesson is simple—accidents and calamities having very grave consequences *can* and *do* occur.[59]

The ultimate lesson is that every effort must be made to ensure that "poor management decisions" are not made. NEPA managers never want to be placed into a position where their own staff accuses them of "bypassing the regulatory process and compromising the safety mission to impress upper management." Moreover, they need to exercise diligence so they are never accused by their own staff of sacrificing "quality for schedule." Managers need to avoid any perception of fly-by-night practices or condescension that would cause their staff to charge that "Managers don't listen—they act like know-it-alls."

PROBLEMS AND EXERCISES

1. Explain how improper management practices can taint the credibility of the EIS process.
2. What was the importance of the case of Calvert Cliffs in terms of ensuring that the requirements of NEPA must be complied with?
3. What is the typical definition of risk?
4. Suppose you prepare an analysis that concludes that the impacts of a project would significantly affect the water quality of a nearby stream. However, your manager tells you to write that, "Waste stream releases into the nearby stream would be minimal and would pose no significant impact to water quality." What should you do?
5. Imagine that you are preparing an EIS on replacing a water treatment facility that is in disrepair. If the facility is not replaced, it could force nearby residents to drink contaminated water. You are preparing a description of the no-action alternative (i.e., not replacing the water treatment plant). Your manager tells you that there is an urgent need to replace the plant and therefore to not include a description of the no-action alternative. Do you think your manager is correct? Defend your answer.
6. Under NEPA, is it correct to prepare an analysis of alternatives but not to seriously consider the results of this assessment in the decision-making process? Explain your answer.

Notes

1. Personal communications, NRC licensing engineer (2011).
2. Radioactive cesium found in baby milk in Japan, MSNBC.com (December 6, 2011), retrieved from http://worldnews.msnbc.msn.com/_news/2011/12/06/9252051-radiactive-cesium-found-in-baby-milk-in-japan.
3. 42 United States Code section 4321 et seq.

4. *Calvert Cliffs' Coordinating Committee v. United States Atomic Energy Commission*. 449 F.2d 1109 (D.C. Cir. 1971).
5. Personal communications with NRC staff (2011).
6. US Congress, *The National Environmental Policy Act of 1969*, 42 U.S.C. 4331, Section 102.
7. 40 CFR §1500.3 and §1508.12.
8. O'Neill D. Project chariot: How Alaska escaped nuclear excavation. *Bulletin of the Atomic Scientists* 45(10): 28–37 (December 1989).
9. Eccleston C. & Doub J.P. *Preparing NEPA Environmental Assessments: A User's Guide to Best Professional Practices*. CRC Press, Boca Raton, FL, p. 16 (2012).
10. NEPA section 102(2), codified at 42 United States Code section 4332(C).
11. Internal DLR NRC project managers focus group meeting, held September 14, 2010.
12. 2010 DLR Safety Culture Focus Group Summary. NRC internal report regarding results of license renewal focus group meeting held on September 14, 2010. Includes supplemental statements supplied by project managers that attended focus group meetings.
13. Taleb, N.N. *The Black Swan: The Impact of the Highly Improbable* (Second ed.). Penguin, p. 374–378 (2010).
14. Piore A. Nuclear energy: Planning for the black swan. *Scientific American* 32 (June 2011).
15. Dvorak P. & Landers P. Japanese plant had barebones risk plan. WSJ.com (March 31, 2011), http://online.wsj.com/article/SB10001424052748703712504576232961004646464.html.
16. US Department of the Interior. Final Programmatic Environmental Impact Statement for Solar Energy Development on Public Lands. (2012), http://news.yahoo.com/seia-lsa-statement-department-interior-release-final-programmatic-214636655.html.
17. Generic Environmental Impact Statement for License Renewal of Nuclear Plants: Main Report (NUREG-1437, Volume 1).
18. The GEIS can be accessed at: http://www.nrc.gov/reading-rm/doc-collections/nuregs/staff/sr1437/v1/index.html.
19. 40 CFR §1502.14.
20. 40 CFR §1502.10.
21. Fact Sheet on Reactor License Renewal, http://www.nrc.gov/reading-rm/doc-collections/fact-sheets/fs-reactor-license-renewal.html.
22. 10 CFR 54.3 (a) [Title 10—Energy; Chapter I—Nuclear Regulatory Commission; Part 54—Requirements for Renewal of Operating Licenses for Nuclear Power Plants; General Provisions].
23. Gusterson H. The lessons of Fukushima. *Bulletin of the Atomic Scientists* (March 16, 2011).
24. Butler D. Reactors, residents and risk. *Nature* (April 21, 2011).
25. Public scoping comment. Name withheld (2009).
26. Eccleston C.H. & Stuyvenberg A. The perfect electrical storm? *Journal of Environmental Quality Management* (Spring 2011).
27. San Luis Obispo Mothers for Peace et al., v. Nuclear Regulatory Commission, 751 F.2d 1287, 1984.
28. Jacobson M.Z. & Delucchi M.A. Providing all global energy with wind, water, and solar power, part I: Technologies, energy resources, quantities and areas of infrastructure, and materials. *Energy Policy* 6 (2010).

29. Newtan S.U. *Nuclear War 1 and Other Major Nuclear Disasters of the 20th Century*. AuthorHouse, p. 146 (2007).
30. Sovacool B.K. *Contesting the Future of Nuclear Power: A Critical Global Assessment of Atomic Energy*, World Scientific, p. 192 (2011).
31. From interviews with Mikhail Gorbachev, Hans Blix and Vassili Nesterenko. *The Battle of Chernobyl*. Discovery Channel. Relevant video locations: 31:00, 1:10:00.
32. Personal communications with NRC staff (2012).
33. Sovacool B.K. A preliminary assessment of major energy accidents, 1907–2007. *Energy Policy* 36: 1802–1820 (2008).
34. The US Nuclear Regulatory Commissions established a board qualification and training program for its professional staff. The near misses cited in this reference are acknowledged and discussed in NRC's training program: Nuclear Regulatory Commissions board qualification requirements, NRR Office Instruction ADM 504 Rev. 1, General Qualification Requirements and Forms, GEN-SA-14, "Major Events and Regulatory Implications."
35. Sovacool B.K. A critical evaluation of nuclear power and renewable electricity in Asia. *Journal of Contemporary Asia* 40(3): 393–400 (August 2010). http://en.wikipedia.org/wiki/Nuclear_and_radiation_accidents, accessed November 27, 2012.
36. http://en.wikipedia.org/wiki/Nuclear_reactor_accidents_in_the_United_States, accessed November 27, 2012.
37. This denial pertained to non-NEPA safety concerns.
38. At US nuclear sites, preparing for the unlikely. *New York Times*, http://www.nytimes.com/2011/03/29/science/29threat.html?pagewanted=1&_r=1&emc=eta1.
39. Diaz M. & François D.M. Fukushima: Consequences of systemic problems in nuclear plant design. *Economic & Political Weekly* 46(13): 10–12 (2011).
40. Sovacool B.K. *Second Thoughts About Nuclear Power*. National University of Singapore, p. 8 (January 2011).
41. *The Future of Nuclear Power*, Massachusetts Institute of Technology, p. 48 (2003).
42. 40 Code of Regulations 1502.16.
43. 40 Code of Regulations 1508.7.
44. 40 Code of Regulations 1508.27.
45. 40 Code of Regulations 1502.27 (a) and (b).
46. 40 Code of Regulations 1502.27.
47. 40 Code of Regulations 1502.27 (B).
48. 40 Code of Regulations 1502.27 (b)(2).
49. 40 Code of Regulations 1502.27 (b)(4).
50. 40 Code of Regulations 1502.27 (b)(7).
51. 40 Code of Regulations 1502.27 (b)(10).
52. Comment documented in the public transcripts for this meeting. The comment was also published in *The Nemaha County Herald*, "Only Positive Remarks Presented Regarding Cooper Nuclear Station's License Renewal" April 15, 2010, http://www.anewspaper.net/index.php?option=com_content&view=article&id=354:only-positive-remarks-presented-regarding-cooper-nuclear-stations-license-renewal&catid=1:local&Itemid=2.
53. 40 Code of Regulations 1506.10(b).

54. 40 Code of Regulations 1502.14.
55. United States Court of Appeals, District of Columbia Circuit, June 8, 2012.
56. Reardon S. US nuke plant delay fails to solve storage conundrum. New Scientist (August 2012), http://www.newscientist.com/article/dn22165-us-nuke-plant-delay-fails-to-solve-storage-conundrum.html.
57. NRC OKs Georgia nuclear reactors—First in generation. Newsmax (February 9, 2012), http://www.newsmax.com/Newsfront/Nuclear-Power-reactors-NRC/2012/02/09/id/428958.
58. Tabuchi H. Japanese prime minister says government shares blame for nuclear disaster. *The New York Times* (March 3, 2012). Retrieved 2012-04-13.
59. Eccleston C. & March F. *Global Environmental Policy: Principles, Concepts and Practice*. CRC Press Inc. (Lewis Press).

chapter two

Overview of NEPA and the EIS process

> Always do right; this will gratify some and astonish the rest.
>
> **Mark Twain**

As described in the Introduction, Congress chose to rectify the nation's deteriorating environment by enacting the US National Environmental Policy Act (NEPA) process (Figure 2.1). Some were gratified, others astonished, and a few dismayed. This chapter introduces NEPA and provides a brief overview of the environmental impact statement (EIS) process. Chapters 3 through 6 detail the step-by-step process for preparing an EIS. Chapter 3 describes prescoping, including tools for managing and administrating the EIS process. Chapter 4 details the step-by-step requirements and process for preparing and filing the EIS. Chapter 5 describes requirements and introduces tools for analyzing impacts, while Chapter 6 details all regulatory requirements that the EIS document must meet.

A copy of the NEPA statute is provided in Appendix A, while a copy of the NEPA Implementing Regulations (40 Code of Federal Regulations [CFR] parts 1500–1508) is provided in Appendix B.* A comprehensive checklist for preparing an EIS is provided in Appendix C.

2.1 Learning objectives

- The NEPA statute
- NEPA's purpose
- NEPA's threshold question
- Significance
- Overview of the NEPA process
- Three levels of NEPA compliance
- Sliding scale, rule of reason, and nomenclature

* A citation referencing a specific provision in the NEPA Implementing Regulations is abbreviated in this book so as to cite the specific "part" of NEPA Implementing Regulations in which it is found. For example, a reference to a provision in "40 CFR 1501.2" is simply cited as "§1501.2."

Figure 2.1 The purpose of NEPA is to prevent environmental degradation and preserve environmental quality. (Courtesy images.google.com.)

2.2 The development of NEPA and the EIS requirement

The author's book, *Global Environmental Policy*,[1] is dedicated to Professor Lynton Keith Caldwell, who has been called the father of NEPA and the principal architect of its EIS provision. In 1963, Caldwell published his groundbreaking article, "Environment: A new focus for public policy?" in *Public Administration Review*. He was called a "lone voice in the wilderness" as he worked to promote a new environmental awareness and ethic. Caldwell was criticized by some of his peers who told him that there was "nothing you can do with environment."[2] Undeterred, Caldwell developed proposals for environment-related research and study, and spoke around the nation at seminars promoting his environmental policy ideas. He helped to pave the groundwork for the passage of NEPA in 1969.

2.2.1 The prelude to NEPA

As public concern over increasing pollution and environmental degradation intensified in the mid-1960s, several concerned congressmen and senators introduced bills aimed at protecting the environment. Senator Henry "Scoop" Jackson chaired the powerful Senate Interior and Insular Affairs Committee. By late 1965, Senator Jackson was becoming concerned about the deterioration in environment quality. He was alarmed to learn of proposals such as one by the Corps of Engineers to construct a series of dikes and canals to drain parts of the Everglades that would then be sold off.

By 1967, Jackson decided it was time for some type of national environmental legislation to protect the environment. But what would be its scope and who would take ownership for developing it? Jackson's senior staff lacked environmental expertise. Jackson learned of Dr. Caldwell's pioneering efforts and invited him to take the lead in crafting a bill. By March 1967, Representative John Dingell introduced a bill on environmental quality in the House of Representatives. Dingell's bill called for the establishing of a Council on Environmental Quality (CEQ). By December of that year, Jackson introduced his own version of an environmental bill in the Senate.

2.2.2 Lynton Caldwell—the architect of the EIS

Meanwhile, Caldwell was expanding a proposal for Jackson that defined the key objectives of a "National Program for Environmental Quality." By June 1968, he completed his 60-page draft.[3] It recommended passage of a comprehensive environmental policy. It spoke of the need for the United States to take a leading role in international cooperation on environmental concerns; and it recommended establishing a CEQ modeled after the Council of Economic Advisors established under the 1946 Employment Act. Caldwell also wrote position papers for Jackson in which he noted there was a need to establish an action-forcing provision to force implementation of the proposed policy. On April 15, 1969, Caldwell acted as principal witness in a Senate hearing on Jackson's bill. Caldwell testified

> that a statement of policy by the Congress should at least consider measures to require the Federal agencies … to contain within [their] proposals an evaluation of [the proposals' effect] upon the state of the environment….

When Caldwell finished, Jackson said, "[I believe what you are getting at is that] what is needed … is to legislatively create those situations that will bring about an action-forcing procedure the departments must comply with." Caldwell replied, "Exactly so." An action-forcing (i.e., EIS) provision was added into Jackson's bill. NEPA was passed by an overwhelming majority of the Senate and House in 1969, and was signed into law by President Richard Nixon on New Year's Day, 1970. Thus, NEPA was the first bill signed into law in the new decade of the 1970s. It blazed the trail for many environmental statutes that would implement its policy goals.

In a 1993 interview, Russell Train, who was first to chair the CEQ and later went on to become the second EPA administrator, looked back on the "environmental decade." He had this to say about Caldwell[2]:

> ... he [Caldwell] became the principal architect of the Environmental Impact Statement (EIS) and the National Environmental Policy Act (NEPA) ... I don't think this is a story and an association that has been particularly well-known.

2.3 The NEPA statute

Mark Twain shared his years of experience with his audience: "The truth is a precious commodity. That's why I use it so sparingly." He went on to add, "We have the best government that money can buy." NEPA was enacted to instill openness and objectivity into the federal decision-making process. In doing so, it set forth an environment *planning* process with the expressed purpose of forcing federal agencies to "look before they leap." In the hands of experienced environmental planners, an EIS provides federal officials with a powerful tool for planning future actions. However, as described in the case study (Chapter 1), an improperly implemented EIS can be an expensive, largely ineffective, and burdensome obstacle. One of the goals of this book is to break some of the myths surrounding the EIS process and present practitioners with a framework for effectively and efficiently planning federal actions.

NEPA's intent is not to control and regulate activities but to require agencies to consider the environmental consequences of their decisions during the early proposal stage, before a decision has been made to pursue the action. As a planning tool, the EIS allows agencies to account for environmental factors, yet it does not set performance standards or place burdensome restrictions on what federal agencies can do. By properly planning future actions, an agency can even avoid triggering other expensive and lengthy permitting and regulatory requirements.

2.3.1 Titles I and II of the NEPA statute

The NEPA statute is short, containing just three parts: a statement of purpose followed by two titles (Title I and Title II). The reader is referred to Appendix A, which contains a copy of the NEPA statute. Despite its brevity, NEPA has had a profound effect on the federal decision-making process. It is noted for three key elements:

1. Title I establishes an environment policy for the entire nation.
2. Title I also creates an "action-forcing" mechanism (i.e., EIS) for implementing the national environmental policy.
3. Title II creates a council for implementing NEPA and elevating environmental concerns directly to the presidential level.

While Title I contains five sections, it is most widely known for its first two sections:

Section 101 of Title I declares the nation's environmental policy. For this reason, it is sometimes referred to as the "spirit of the law."

Section 102 is sometimes referred to as the "letter of the law" because it provides the legal procedural or action-forcing mechanism (i.e., EIS requirement) for carrying out the policy established in Section 101.

2.3.2 *Title I of NEPA*

The Act begins with a statement of NEPA's purpose:

> declare a national policy which will encourage productive and enjoyable harmony between man and his environment; promote efforts which will prevent or eliminate damage to the environment and biosphere, and stimulate the health and welfare of man; enrich the understanding of the ecological systems and natural resources important to the Nation….

This statement of purpose is followed by Title I, which is the heart of the Act. Title I establishes both a national environment policy and creates an "action-forcing" mechanism (i.e., EIS) for implementing the national environmental policy. Its two most important provisions are Sections 101 and 102.

2.3.2.1 *Section 101*

As just noted, Section 101 declares the following national environmental policy[4]:

> The Congress, recognizing the profound impact of man's activity on the interrelations of all components of the natural environment, particularly the profound influences of population growth, high-density urbanization, industrial expansion, resource exploitation, and new expanding technological advances and recognizing further the critical importance of restoring and maintaining environmental quality to the overall welfare and development of man, declares that it is the continuing policy of the Federal Government, in cooperation with State and local governments, and other concerned public and private organizations, to

> use all practicable means and measures, including financial and technical assistance, in a manner calculated to foster and promote the general welfare, to create and maintain conditions under which man and nature can exist in productive harmony, and fulfill the social, economic, and other requirements of present and future generations of Americans.

Beyond announcing a national environmental policy, Section 101 also spells out six specific responsibilities that federal agencies are to exercise in carrying out the national policy (Table 2.1).[5]

2.3.2.2 Section 102

Section 102 requires that "to the fullest extent possible ... all agencies of the federal government shall" comply with the following two requirements[6]:

- Utilize a systematic, interdisciplinary approach that will ensure the integrated use of natural and social sciences and environmental design arts in planning and in decision making that may have an impact on man's environment.
- Include in every recommendation or report on proposals for legislation and other major federal actions significantly affecting the quality of the human environment, a *detailed statement* by the responsible official—on the environmental impact of the proposed action.

Table 2.1 Six Specific Responsibilities That Federal Agencies Are to Exercise in Carrying Out the National Policy

1. Fulfill the responsibilities of each generation as trustee of the environment for succeeding generations
2. Assure for all Americans safe, healthful, productive, esthetically and culturally pleasing surroundings
3. Attain the widest range of beneficial uses of the environment without degradation, risk to health or safety, or other undesirable or unintended consequences
4. Preserve important historic, cultural, and natural aspects of our national heritage and maintain, wherever possible, an environment that supports diversity and variety of individual choice
5. Achieve a balance between population and resource use that will permit high standards of living and a wide sharing of life's amenities
6. Enhance the quality of renewable resources and approach the maximum attainable recycling of depletable resources

Source: US National Environmental Policy Act, Public Law 91-190, 42 U.S.C. 4321, Title I, Section 101 (A) and (C).

Note: The federal government is to use all practicable means in achieving the goals listed in this table.

Table 2.2 Requirements Cited in the Statute for Preparing an EIS

... include in every recommendation or report on proposals for legislation and other major federal actions significantly affecting the quality of the human environment, a detailed statement by the responsible official ... on

i. The environmental impact of the proposed action
ii. Any adverse environmental effects which cannot be avoided should the proposal be implemented
iii. Alternatives to the proposed action
iv. The relationship between local short-term uses of man's environment and the maintenance and enhancement of long-term productivity
v. Any irreversible and irretrievable commitments of resources which would be involved in the proposed action should it be implemented

(Section 102[2][c] of NEPA)

This second provision is the one that NEPA is most famous for. The "detailed statement" is, of course, now known as the EIS. Table 2.2 lists the requirements cited in the statute for preparing an EIS.[7]

As noted in Table 2.2, the Act provides surprisingly little detail regarding the specific scope and content of the EIS. These details are provided in the NEPA implementing regulations. Despite claims by some agency officials that the Nuclear Regulatory Commission (see Chapter 1) is an independent agency and only complies with NEPA because it wants to, Congress clearly stipulated that the EIS requirement applies to "*all* agencies of the federal government."

2.3.2.3 "Evidence-based" decision-making process

The author is fond of stating that NEPA is an "evidence-based" decision-making process. But what does this mean? Before NEPA, many federal decisions were traditionally based on prevailing assumptions, opinions, or even anecdotal evidence. We see this even today. The issue of climate change is a prominent environmental issue of wide interest to policy makers and decision makers.

NEPA was enacted to provide a scientific basis for making rational decisions. As an "evidence-based" decision-making process, NEPA stands in stark contrast to the traditional decision-making process. While scientific definitions vary a bit, all definitions of the phrase "evidence-based" refer to the use of scientific investigations and established evidence as the basis for informing decision-makers and forging policies. One of the principles of evidence-based management is being "... committed to 'fact-based' decision making—which means being committed to getting the best evidence and using it to guide actions."[8] In passing NEPA, Congress effectively developed an evidence-based decision-making process long before the term "evidence-based" was in vogue.

2.3.3 Title II of NEPA

Title II establishes the CEQ (or "Council"). Patterned after the Council on Economic Advisors, the Council's purpose is to oversee the implementation of NEPA. The Council also advises the president on environmental matters. It reports directly to the president, who has ultimate authority for ensuring that federal agencies conduct their actions in a manner consistent with the national policy.

2.3.3.1 CEQ NEPA implementing regulations

The CEQ has issued regulations (Regulations) for implementing the procedural aspects of the NEPA Act. The Regulations consist of nine "parts," summarized on Table 2.3. The Regulations are focused on how to implement Title

Table 2.3 Summary of CEQ NEPA Regulations

Part 1500—Purpose, Policy, and Mandate: Discusses the purpose of the procedures, defines policy, and cites the laws and executive orders constituting the mandate; details means by which agencies shall reduce paperwork and delay.
Part 1501—NEPA and Agency Planning: Requires that NEPA be integrated with other planning processes at the earliest possible time; defines the duties of the lead and cooperating agencies when more than one is involved; defines when an EA or EIS is required; details the scoping process and specifies its requirements.
Part 1502—Environmental Impact Statement: States the purpose of an EIS and provides instructions for their preparation; specifies statutory requirement that the EIS must meet; defines actions requiring preparation of an EIS; defines the programmatic EIS and provides instructions on scoping and tiering to avoid duplication and delay.
Part 1503—Commenting: Requires proposing agencies to invite comments from other federal agencies, from state and local agencies, Indian tribes and the public, and specifies procedures for doing so.
Part 1504—Predecision Referrals to the Council: Defines the procedure for referring interagency disagreements to the CEQ.
Part 1505—NEPA and Agency Decision Making: Requires agencies to adopt internal procedures for implementing NEPA; defines requirements for the record of decision (ROD); provides guidance for implementing a decision after completing the ROD.
Part 1506—Other Requirements: Specifies restrictions on the types of activities that may be undertaken while an EIS is under way; discusses methods for eliminating duplication with state and local procedures; provides guidance for combining and adopting documents; requires public involvement; reviews the process governing the preparation of EISs for legislative proposals; defines NEPA exemptions for emergencies.
Part 1507—Agency Compliance: Requires all federal agencies to comply with these regulations but states that agencies have flexibility in implementing NEPA.
Part 1508—Terminology and Index: Defines 28 NEPA terms; several of these definitions are critical in defining how to carry out NEPA procedures.

I, Section 102(2) of the Act. These regulatory requirements will be referenced throughout this book. A copy of the Regulations is provided in Appendix B.

2.4 The threshold question

As just witnessed, Section 102 of NEPA requires federal agencies to prepare and include an EIS in every recommendation or report on

> ... proposals for legislation and other major federal actions significantly affecting the quality of the human environment....

Often referred to as the *threshold question* of significance, this mandate provides the linchpin for determining whether an EIS needs to be prepared. The outcome of this threshold determination can have profound ramifications in terms of project planning (e.g., cost, schedule, and potential litigation). As described in Section 2.4.1, the threshold determination decides whether a short environmental assessment will be prepared or a full-blown EIS. The threshold question can essentially be dissected into eight discrete criteria that must be satisfied before the EIS requirement as a whole is triggered:

- Proposals
- Legislation
- Federal
- Actions
- Significantly
- Affecting
- Quality
- Human environment

Each of these threshold criteria is briefly described in the following sections. An in-depth description of the threshold question is beyond the scope of this book. The reader is referred to the author's companion book, *NEPA and Environmental Planning*, for a more detailed review of this issue.[9]

2.4.1 Proposals

An EIS is prepared for federal "proposals." A proposal might exist in actual fact, even though the agency has not officially declared one to exist. A proposal is considered to exist either officially or unofficially when (§1508.23)

1. A federal agency has a goal.
2. The agency is actively preparing to make a decision on one or more alternative means of accomplishing the goal.
3. The effects can be meaningfully evaluated.

2.4.2 *Legislation*

A proposal can include submittals for congressional legislation. A *legislative proposal* involves

> … a bill or legislative proposal to Congress developed by or with the significant cooperation and support of a federal agency, but does not include requests for appropriations. (§1508.17)

2.4.3 *Major*

Most courts, and also the CEQ, have interpreted "major" to reinforce but not to have a meaning independent from the term "significantly" (described below). Under this interpretation, the size of a proposal has little bearing on whether the action requires preparation of an EIS. It is the degree of the *environmental impact* (i.e., the significance of the impact) that largely determines whether an EIS is required, not nonenvironmental metrics such as the size of the project, cost, or employment (although such factors might influence the level of environmental impact).

2.4.4 *Federal*

The term "federal" includes *all* agencies of the federal government; a federal agency does not include

> … the Congress, the Judiciary, or the President, including the performance of staff functions for the President in his Executive Office. (§1508.12)

As described in the next section, the courts have determined that actions undertaken by a nonfederal agency can be "federalized" for the purposes of NEPA. The companion book, *NEPA and Environmental Planning*, provides a general purpose tool for determining when a nonfederal action is subject to the requirements of NEPA.[10]

2.4.5 *Actions*

For the purposes of NEPA, the definition of an "action" is broad and pervasive. This term includes both programs and projects, including both new actions and changes to continuing activities.

2.4.6 *Significantly*

In reviewing the threshold question, the concept of "significance" is perhaps both the most important and elusive of the eight threshold criteria. Both the *intensity* and the *context* in which an impact would occur must be considered. For additional information on significance and its interpretation, the reader is directed to the companion book, *NEPA and Environmental Planning*.[11]

2.4.6.1 Context

The significance of an action is a function of the setting (i.e., context) in which an impact would occur. The term "context" recognizes potentially affected resources, as well as the location and setting in which an environmental impact would occur. The NEPA implementing regulations defines "context" to mean that the significance of an action must be analyzed in

> … several contexts such as society as a whole (human, national), the affected region, the affected interests, and the locality. Significance varies with the setting of the proposed action. For instance, in the case of a site-specific action, significance would usually depend upon the effects in the locale rather than in the world as a whole. Both short- and long-term effects are relevant. (§1508.27[a])

Table 2.4 presents examples of categories of environmental effects and the corresponding context that might be appropriate for the assessment of significance. Like most other considerations in NEPA, no formulaic approach can be used for considering context; effective evaluation of context relies on the professional judgment of NEPA practitioners.

2.4.6.2 Intensity

The term "intensity" is a measure of the degree or severity of an impact. The Regulations define 10 factors (i.e., significance factors) that are to be used in assessing intensity. These significance factors are indicated in Table 2.5.[12]

An impact cannot necessarily be deemed nonsignificant simply because the action is temporary. Moreover, agencies cannot *segment* or

Table 2.4 Examples of Effects and the Corresponding Context That Might Be Most Appropriate for Assessing Significance

Land use	Planning district or area covered by a land use plan
Visual impacts	Viewsheds that include the site
Soil impacts	Site and adjoining properties
Wetlands impacts	Site and remainder of subwatershed and watershed
Socioeconomic impacts	Political jurisdictions such as countries and municipalities
Noise	Areas where estimated noise levels generated by the action could be audible

Table 2.5 Ten Significance Factors to Be Used in Assessing the Intensity of an Environmental Impact

1. Impacts that may be both beneficial and adverse. A significant effect may exist even if the federal agency believes that on balance the effect will be beneficial.
2. The degree to which the proposed action affects public health or safety.
3. Unique characteristics of the geographic area, such as proximity to historic or cultural resources, park lands, prime farmlands, wetlands, wild and scenic rivers, or ecologically critical areas.
4. The degree to which the effects on the quality of the human environment are likely to be highly controversial.
5. The degree to which the possible effects on the human environment are highly uncertain or involve unique or unknown risks.
6. The degree to which the action may establish a precedent for future actions with significant effects or represents a decision in principle about a future consideration.
7. Whether the action is related to other actions with individually insignificant but cumulatively significant impacts. Significance exists if it is reasonable to anticipate a cumulatively significant impact on the environment. Significance cannot be avoided by terming an action temporary or by breaking it down into small component parts.
8. The degree to which the action may adversely affect districts, sites, highways, structures, or objects listed in or eligible for listing in the National Register of Historic Places or may cause loss or destruction of significant scientific, cultural, or historical resources.
9. The degree to which the action may adversely affect an endangered or threatened species or its habitat that has been determined to be critical under the Endangered Species Act of 1973.
10. Whether the action threatens a violation of federal, state, or local law or requirements imposed for the protection of the environment.

piecemeal an action by "breaking a project down into smaller component parts" that are individually nonsignificant.[13]

2.4.7 *Affecting*

The Regulations define "affecting" to mean "will or may have an effect on" the environment (§1508.3). With respect to NEPA, an action affects the environment if it produces a change in one or more environmental resources. A reasonably close connection must exist between a disturbance and its resulting effect on the environment. An environmental impact analysis considers the *potential* for an impact and is not limited to effects that will certainly occur.

2.4.8 Human environment

As some relationship exists between humans and virtually every aspect of the physical and natural environment, the courts have viewed the term "human environment" broadly. To be significant, an action must substantially affect the quality of the human environment. In reality, there is little distinction between the terms "environment" and "human environment." The "environment" can be divided into two broad categories: (1) natural and physical environs and (2) man-made or built environs (§1502.16[g], §1508.8[b], and §1508.14).

2.5 Overview of the NEPA process

A *federal* action is considered potentially subject to preparation of an EIS until it can be shown that it

1. Can be exempted from NEPA's EIS requirement; or
2. Does not *significantly* affect environmental quality

Most federal actions are subject to some level of NEPA review.[14] However, there are a number of restricted situations, including categorical exclusions, in which an action can be exempted from the requirement to prepare an EIS. The following section provides a general synopsis of the entire NEPA process, with emphasis on describing the three levels of NEPA compliance. A detailed accounting of every aspect and intricacy inherent to the NEPA process is beyond the scope of this text. The reader is referred to the companion text, *NEPA and Environmental Planning*, for a detailed review of the threshold question.[15]

2.5.1 Three levels of NEPA compliance

The NEPA process, including preparation of an EIS, must commence

> ... as close as possible to the time the agency is developing or is presented with a proposal. (§1508.23)

NEPA can be viewed as consisting of three principal levels of planning or environmental compliance. The three levels, defined from the least to the most demanding, are as follows:

- **Categorical Exclusion (CATX)**—Certain rather innocuous or "small" federal actions might qualify for a CATX, thus excluding them from further NEPA review and documentation requirements.
- **Environmental Assessment (EA)**—Where a federal action does not qualify for a CATX, an EA may be prepared to determine whether

a federal action qualifies for a Finding of No Significant Impact (FONSI), thus exempting it from the requirement to prepare an EIS.

- **Environmental Impact Statement (EIS)**—An EIS is generally prepared for proposed federal actions that do not qualify for either a CATX or a FONSI.

2.5.1.1 *Initiating the NEPA process*

A simplified overview of the entire NEPA process (including preparation of the EIS) is depicted in Figure 2.2. It is important to note that this is a

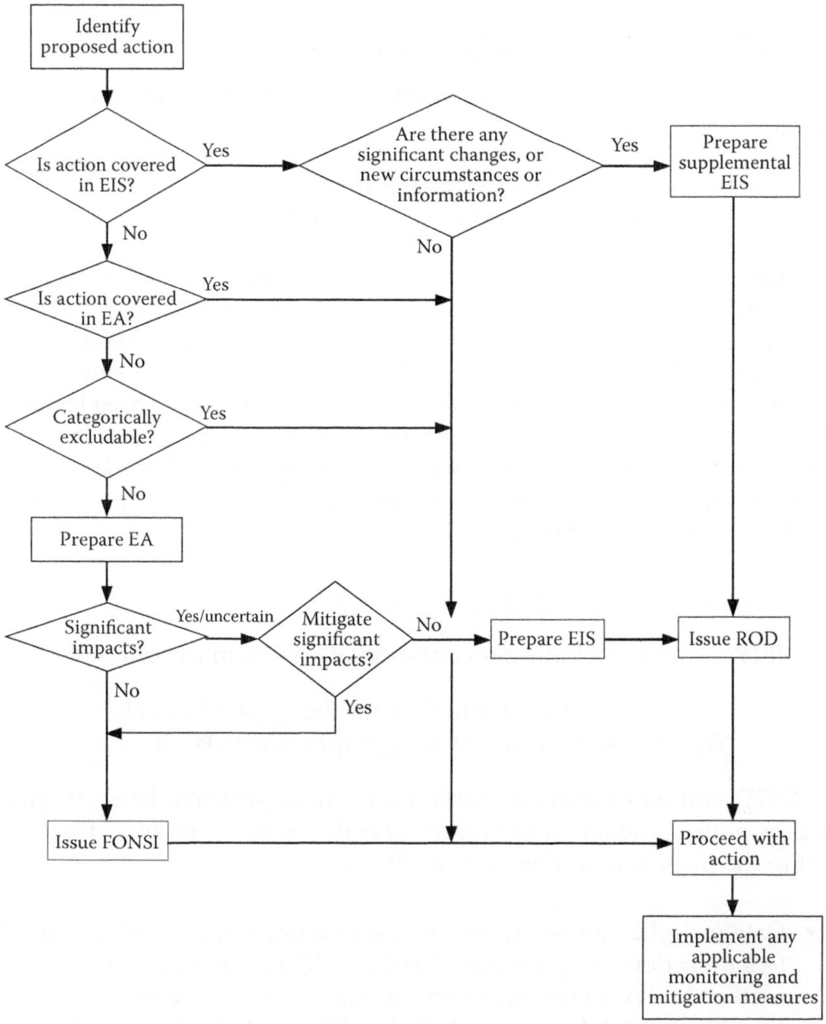

Figure 2.2 NEPA planning process, including three levels of NEPA compliance.

general purpose flow chart and does not show every nuance or circumstance in the NEPA process. As illustrated by Figure 2.2, NEPA is normally triggered when a need for taking action (i.e., proposal) has been identified (see the box in the upper left-hand corner of Figure 2.2).

The proposal may be covered in the scope of an existing EIS or EA. Existing NEPA documentation should be examined to determine whether the proposal has been subject to a previous NEPA review (the first and second decision diamonds on the left-hand side of Figure 2.2). If the proposal is sufficiently investigated under existing NEPA documentation, the agency can proceed with the proposal (with respect to NEPA's requirements).

2.5.1.2 *Categorically excluding actions*

Individual agencies are required to prepare a list of CATXs in their agency-specific NEPA implementation procedures. These CATXs are typical agency actions that have been studied and found to pose no significant environmental impact (cumulative or otherwise) and for which preparation of an EA/EIS is therefore not required. If a CATX is applicable to the proposed action, no further NEPA review is required and the agency is free to proceed with that action with respect to NEPA requirements (see the third decision diamond on the left-hand side of Figure 2.2). As noted in the companion text, *NEPA and Environmental Planning*, there are also a limited number of circumstances where an action is not subject to NEPA's requirements.[16]

2.5.1.3 *The environmental assessment*

If the action is not eligible for an exemption such as a CATX, the agency can choose to prepare an EA to determine whether the proposal could significantly affect the quality of the human environment (see the box labeled "Prepare EA" in Figure 2.2). An agency may choose to prepare an EIS without first preparing an EA.

2.5.1.3.1 *Finding of No Significant Impact (FONSI).* The results of the EA are reviewed by the decision maker to determine if the proposed action would result in a significant environmental impact (see the fourth decision diamond labeled "Significant impacts?" in Figure 2.2). The agency prepares a FONSI if the decision maker concludes that no significant impact would occur (see the box labeled "Issue FONSI," Figure 2.2). A FONSI can also be prepared if the proposal would result in a significant impact, but that effect can be mitigated to the point of nonsignificance (see the diamond labeled "Mitigate significant impacts?" in Figure 2.2).

If a FONSI is issued, the agency is free to proceed with the action (with respect to NEPA's requirements), in accordance with any applicable mitigation or monitoring measures. The reader is referred to the companion text, *Preparing NEPA Environmental Assessments*, for a detailed discussion of EAs.[17]

2.5.1.4 Environmental impact statement

If an action would result in a significant impact and does not qualify for some type of exemption, such as a CATX or FONSI, an EIS must be prepared to investigate environmental consequences and alternatives for pursuing a proposal (see the box labeled "Prepare EIS" in Figure 2.2). The EIS investigates alternative courses of action. The EIS is used by the decision maker(s) in reaching a final decision regarding the course of action to be taken (§1502.1). Table 2.2 lists requirements cited in the statute for preparing an EIS.

2.5.1.4.1 Record of decision. On completing the EIS, the agency prepares a record of decision (ROD). The agency is free to pursue the course of action that is adequately evaluated in the EIS. The ROD publicly documents the course of action that the agency has chosen to pursue. Once a ROD has been issued, the agency is free to proceed with the action (with respect to NEPA's requirements), in accordance with any applicable mitigation or monitoring measures.

2.5.1.4.2 Supplemental EIS. Sometimes an action, covered under an existing EIS, involves changes, or new information or circumstances (see the right diamond in the upper portion of Figure 2.2 labeled "Are there any significant changes, or new circumstances or information?"). In such cases, the agency may need to prepare a supplemental EIS (see the right box in the upper portion of Figure 2.2 labeled "Prepare supplemental EIS"). The supplemental EIS assesses the effect of these changes, including new circumstances or information.

Once the supplemental EIS process has been completed, the agency issues a ROD documenting its decision with respect to the course of action that will be taken. The agency is then free to proceed with its decision, in accordance with any applicable mitigation or monitoring measures.

2.6 Introduction to the EIS process

The principal steps followed in preparing a typical EIS are illustrated in Figure 2.3. As described in Chapter 3, a preliminary or prescoping effort is frequently performed before formally initiating the EIS. Chapter 4 provides a detailed discussion of the entire EIS process beginning with the formal scoping process, while Chapter 6 describes the detailed requirements that must be met in writing the statement.

2.6.1 Initiating the EIS process

The formal EIS process is publicly initiated with publication of a notice of intent (NOI) in the *Federal Register.* The NOI notifies the public of the

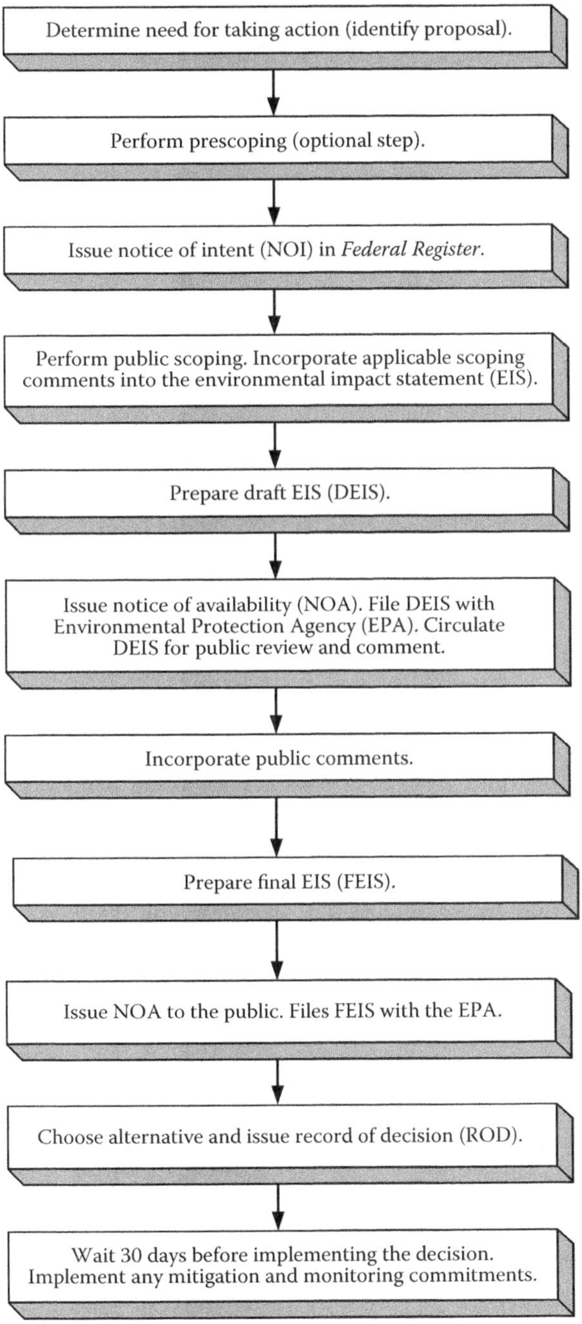

Figure 2.3 Principal steps in the EIS process.

agency's intent to prepare an EIS and invites them to participate in the public scoping process. This notice sets the stage for the formal public scoping process that follows. The public scoping process is used to determine the scope of analysis (i.e., range of alternatives, impacts, and actions) to be investigated in the EIS.

As part of the planning effort, the agency identifies and defines a set of alternatives for detailed analysis. The impacts associated with each of the "analyzed alternatives" are then analyzed. The standard EIS process typically consists of two distinct phases: a *draft EIS* followed by a *final EIS*. Thus, the term "EIS" describes a "statement" in either its draft or final state.

2.6.2 The draft and final EIS

The draft EIS should comply as close as possible with all of NEPA's applicable regulatory requirements. The completed draft EIS is circulated for public review and comment, and filed with the US Environmental Protection Agency (EPA). The EPA assigns a public rating to the draft EIS.

Comments received from the public are addressed in the final EIS. Once comments have been incorporated, the final EIS is circulated to the public and filed with the EPA.

2.6.2.1 Record of decision

The responsible decision maker reviews the impacts and alternatives described in the final EIS. The final EIS is to be used by the decision maker, in conjunction with other relevant decision-making material, in choosing the course of action to be taken (§1502.1). Unlike the EA, the decision maker has latitude to select any alternative regardless of its environmental consequences, so long as it is adequately analyzed in the EIS.

An ROD is issued, recording the agency's final decision. With respect to NEPA's requirements, the agency is now free to pursue the action described in the ROD. Any applicable monitoring and mitigation measures committed to in the EIS must be implemented.

2.6.3 Why an EIS protects human life and the environment

As described in the companion text, *NEPA and Environmental Planning*, the US Department of Energy made a cost-cutting decision (before it had even started the NEPA process) to discontinue an employee commuter bus system used to transport hundreds of workers daily from the surrounding community to a government installation.[18] Some of these workers were transported to work over round-trip distances of nearly 100 miles.

Discontinuing the mass transit system meant that workers would be forced to use private vehicles to travel the route. This action involved potentially significant impacts in terms of increased use of petroleum (a strategic natural resource), toxic air emissions, noise, wear and tear on the road infrastructure system, traffic congestion, and an increased risk of accidents due to greater vehicular traffic, resulting in potential injuries or even loss of life.

On the basis of the CEQ's list of 10 significant factors (Table 2.5), any number of the environmental and safety issues could be viewed as significant, requiring preparation of an EIS to investigate alternatives, including that of taking no action to discontinue this mass transit system. Yet, Mr. Paul Dunigan, the NEPA compliance officer, exempted this action from preparation of an EIS based on misapplication of a categorical exclusion. This case is a perfect example of a failure not only to implement NEPA correctly, but more important to protect the human environment against potentially adverse significant impacts, which is the very goal of NEPA. An EIS might well have shown this cost-cutting proposal to be very unwise when all impacts were properly accounted for. This illustrates what can happen when agency officials shortcut the NEPA planning process.

2.7 Sliding scale, rule of reason, and nomenclature

Professional judgment in conjunction with common sense must be exercised in determining the appropriate scope of an environmental impact analysis. Together, the following two principles provide federal officials, planners, and analysts with powerful tools for reducing cost, delays, and effort expended in the EIS planning process. These principles will be drawn on extensively throughout this text.

2.7.1 The sliding scale

A number of agencies have adopted a *sliding-scale approach* for use in preparing an EIS. A sliding-scale approach recognizes that the amount of effort expended on an analysis is a function of the particular circumstances and potential for significant impacts. Use of a sliding-scale approach is justified and based on the following regulatory direction:

> NEPA documents must concentrate on the issues that are truly significant to the action in question, rather than amassing needless detail. (40 CFR 1500.1[b])
>
> Impacts shall be discussed in proportion to their significance. There shall be only brief discussion of other than significant issues. (40 CFR 1502.2[b])

The US Department of Energy has written that[19]

> ... agency proposals can be characterized as falling somewhere on a continuum with respect to environmental impacts. The level of effort devoted to the assessment of an impact should be commensurate with the degree of its potential significance. This approach implements CEQ's instruction that in EISs agencies "focus on significant environmental issues and alternatives" (40 CFR 1502.1) and discuss impacts "in proportion to their significance" (40 CFR 1502.2[b]).

The reader should note that under CEQ's regulations and judicial rulings, a factor in determining significance involves the degree to which environmental effects are likely to be controversial with respect to technical issues. Consistent with this guidance, the author recommends that

> The amount of effort expended on the EIS analysis should vary with the significance of the potential impacts. A sliding-scale approach should be applied in which environmental impacts and issues are investigated, and other related regulatory requirements are addressed with a degree of effort commensurate with their importance to the decision-making process.

2.7.2 *Rule of reason*

An overly strict or unreasonable application of a regulatory requirement may lead to decisions, a course of action, or a level of effort that is wasteful, ridiculous, or absurd. The *rule of reason* provides a second mechanism, used by the courts, for injecting reason into the EIS process. *Common sense* must therefore be exercised in determining the scope and detail accorded to issues, alternatives, and impacts considered in the analysis.

2.7.3 *Nomenclature*

The term "proposal" as used in this text, means the set of reasonable alternatives, including no-action and the proposed action (if one is defined). Throughout this book, the terms "Act" or "statute" will frequently be used in referring to "NEPA." Similarly, the term "CEQ NEPA regulations" is abbreviated to simply "Regulations." For conciseness, references to a particular section of the CEQ regulations (40 Code of Federal Regulations [CFR] Parts 1500–1508) are abbreviated so that they simply cite the specific section number in the Regulations where the provision can be found. For instance, a reference such as "40 CFR 1500.1" is shortened to the expression "(§1500.1)."

PROBLEMS AND EXERCISES

1. What does "evidence-based" decision-making mean?
2. What are the three levels of NEPA compliance?
3. Describe two of NEPA's significance criteria (§1508.27[b]).
4. What does the term "sliding scale" mean?
5. Name two of the requirements cited in the NEPA statute that must be addressed in an EIS (Section 102[2][c] of NEPA).
6. As used in Section 102, does the word "major" have a meaning of and by itself?
7. With respect to determining "significance" what does "context" mean?
8. Why is the rule of reason of such importance in preparing an EIS?

Notes

1. Eccleston C.H. and March F. *Global Environmental Policy: Concepts, Principles, and Practice*. CRC Press, Boca Raton, FL (2011).
2. Read W. The journey to NEPA: How Lynton Keith Caldwell became a principal architect of the National Environmental Policy Act and its Environmental Impact Statement Provisions. *NAEP Newsletter* (November 2012).
3. *A National Policy for the Environment*. Published as a Special Report to the Committee on Interior and Insular Affairs in July 1969.
4. National Environmental Policy Act, Public Law 91-190, 42 U.S.C. 4321, Title I, Section 101(A).
5. US National Environmental Policy Act, Public Law 91-190, 42 U.S.C. 4321, Title I, Section 101 (a) and (c).
6. US National Environmental Policy Act, Public Law 91-190, 42 U.S.C. 4321, Title I, Section 101(b).
7. US National Environmental Policy Act, Public Law 91-190, 42 U.S.C. 4321, Title I, Section 102(C).
8. Pfeffer J. and Sutton R.I. Evidence-based management. *Harvard Business Review* 84(1): 62–74, 133 (January 2006).
9. Eccleston C.H. *NEPA and Environmental Planning: Tools, Techniques, and Approaches for Practitioners*. CRC Press, Boca Raton, FL (2008).
10. Eccleston C.H. *NEPA and Environmental Planning: Tools, Techniques, and Approaches for Practitioners*. CRC Press, Boca Raton, FL (2008).
11. Eccleston C.H. *NEPA and Environmental Planning: Tools, Techniques, and Approaches for Practitioners*. CRC Press, Boca Raton, FL (2008).

12. 40 Code of Federal Regulations §1508.27(b).
13. 40 Code of Federal Regulations §1508.27(b)(7).
14. Eccleston C.H. *NEPA and Environmental Planning: Tools, Techniques, and Approaches for Practitioners*. CRC Press, Boca Raton, FL (2008).
15. Eccleston C.H. *NEPA and Environmental Planning: Tools, Techniques, and Approaches for Practitioners*. CRC Press, Boca Raton, FL (2008).
16. Eccleston C.H. *NEPA and Environmental Planning: Tools, Techniques, and Approaches for Practitioners*. CRC Press, Boca Raton, FL (2008).
17. Eccleston C.H. *Preparing NEPA Environmental Assessments: A Users Guide to Best Professional Practices*. CRC Press, Boca Raton, FL (2012).
18. Eccleston C.H. *NEPA and Environmental Planning: Tools, Techniques, and Approaches*. Introduction, CRC Press, Boca Raton, FL (2008).
19. United States Department of Energy. *Recommendations for the Preparation of Environmental Assessments and Environmental Impact Statements*. Office of NEPA Oversight (May 1993).

chapter three

Preliminaries and prescoping

Initiating the EIS and tools for managing the process

Figure 2.1 provided an overview of the National Environmental Policy Act (NEPA) process, while Figure 2.2 illustrated the basic steps followed in preparing an environmental impact statement (EIS). This chapter describes the process of initiating the EIS, including prescoping tasks and activities.

An assortment of tools, techniques, and tasks that have been successfully used by NEPA practitioners for managing EISs are introduced. Not all of the tools, techniques, and tasks described in this chapter are applicable to every EIS, particularly those involving smaller and noncontroversial proposals. The user is encouraged to briefly review these tools and techniques to identify those that might be most useful on a particular EIS project; discretion and professional judgment must be exercised in determining which tasks are most applicable to any given EIS project. Those well versed in the EIS process may choose to forgo reading this chapter and move directly to Chapter 4.

Chapter 4 describes all key regulatory requirements and the step-by-step process for preparing a draft and final EIS, including the formal scoping process. Chapter 5 describes how to perform the environmental analysis, while Chapter 6 details all regulatory requirements that the EIS document must meet. A copy of the NEPA statute is provided in Appendix A, while a copy of the NEPA implementing regulations (Regulations) is provided in Appendix B. A citation referencing a specific provision in the NEPA regulations is abbreviated in this book so as to cite the specific "part" of the NEPA regulations in which it is found. For example, a reference to a provision in "40 CFR 1501.2" is simply cited as "§1501.2."

3.1 Learning objectives

- Initiating the EIS process
- Identifying the lead and any other cooperating agencies
- Forming an interdisciplinary team

- Defining the agency's statement of purpose and need
- Applying the decision-based scoping technique
- Integrating the EIS with other planning and regulatory requirements
- Using EIS management tools
- Developing a public involvement strategy
- Preparing public notices and advertisements
- Collecting data

3.2 *Initiating the EIS process*

Mark Twain once quipped, "The secret to getting ahead is getting started." This chapter is devoted to the process of initiating the EIS process. It is important to re-emphasize that NEPA is an environmental *planning* process (Figure 3.1). The principal steps in the EIS process are depicted in Figure 4.2. As just indicated, this chapter introduces the initial prescoping process. While most of these tasks and activities are not required under the Regulations, they can nevertheless promote more effective planning and contribute to a more defensible EIS process. The initial steps described in this section should be considered and adopted where applicable and practical.

Figure 3.1 An environmental impact statement is typically prepared to evaluate impacts of large energy projects. (Courtesy images.google.com.)

3.2.1 Initiating the EIS during the early proposal stage

As witnessed in Chapter 2, NEPA requires that an EIS be prepared for[1]

> ... every recommendation or report on proposals for legislation and other major Federal actions significantly affecting the quality of the human environment....

A *proposal* is defined to exist

> ... at that stage in the development of an action when an agency subject to the Act has a goal and is actively preparing to make a decision on one or more alternative means of accomplishing that goal and the effects can be meaningfully evaluated.... A proposal may exist in fact as well as by agency declaration that one exists. (40 CFR §1508.23)

The EIS is required to be prepared early enough so that it can contribute to decision making and will not be used to rationalize or justify decisions already made (§1502.5). The NEPA regulations also require that preparation of the EIS begin

> ... as close as possible to the time ... in which the agency is developing or is presented with a proposal, so that it can be completed in time to be included in any recommendation or report on the proposal. (§1502.5, §1508.23)

With respect to the aforementioned requirement, a word of caution is in order. A new proposal can be absorbed into election year politics. The ramification of announcing a major or controversial project during an election year should be assessed.

Where an action involves a nonfederal applicant, preparation of the EIS must begin early in the process and "no later than immediately after an application is received" by the agency (§1502.5[b]). If requested by an applicant, the agency must set time limits on the EIS process that also needs to be consistent with the purposes of NEPA (§1501.8[a]).

While the Regulations describe the agency's EIS process in terms of defining a "proposed action," this book actually discourages the common practice of defining a proposed action. Instead, federal officials are encouraged to simply define a "proposal" consisting of a range of reasonable alternatives without defining a "proposed action." This approach is advocated because it can help reduce biases that tend to favor the adoption of the proposed action; thus, this approach advances the goal of promoting a more objective analysis of all reasonable alternatives.

The phrase "EIS process" implies that the preparation of an EIS is not a single isolated event. Webster's dictionary defines a "process" as "A series of actions, changes, or functions that achieve an end or result."[2] Many of the activities described in this chapter are presented as one-time events. In practice, the preparation of an EIS is frequently an evolving and iterative process; consequently, many of the activities described in this book as one-time events will actually be revisited more than once over the course of preparing an EIS.

A well-coordinated planning process typically involves performing certain tasks before initiating the *formal* public scoping process. Informal scoping performed before publication of the EIS notice of intent (NOI) is commonly referred to as *preliminary* or *prescoping*. These initial prescoping tasks are often critical in facilitating the formal public scoping phase that follows. While prescoping can improve the planning process, it cannot substitute for the formal public scoping process that follows the NOI. As detailed in Chapter 4, the formal public scoping process is used to determine the range of alternatives, actions, and impacts that will be investigated.

3.2.2 *Why an EIS provides an early warning sign of trouble ahead*

The EIS process can provide project proponents with a gauge for "testing the waters" of public opinion and opposition. It is frequently the equivalent of a political bellwether, providing an early, sometimes the first, indication that a proposal is headed for trouble, be it an industry, citizens groups, regulators, or other opposing interests. If a project progresses through the EIS process with little controversy, it is much less likely to encounter substantial opposition at the later permitting or "groundbreaking" stage. However, if substantial opposition is encountered during the EIS process, it is frequently the first of many subsequent obstacles that will be encountered. Moreover, as an EIS is often the principal "initiator" in a complex chain of federal requirements, it frequently unveils other unknown problems and obstacles (e.g., regulatory, schedule, budgetary, scientific or engineering, or congressional).

For this reason, the EIS process has sometimes been used as a mechanism for "floating" a proposal to the public and testing the initial reaction. If problems are encountered early on, project proponents may wish to reassess the wisdom of moving forward with the proposed action; should this be the case, the EIS can also serve as an effective tool for identifying other less controversial courses of action (alternatives).

3.2.3 *Identifying the lead and any other cooperating agencies*

The agency having overall responsibility for preparing an EIS is referred to as the *lead agency* (§1501.5, §1508.16). The lead agency is ultimately responsible for the content and accuracy of the EIS. As applicable, the lead agency should solicit cooperation from applicable federal, state, and local agencies as well as from affected Indian tribes in cases relating to their lands or sites, of religious or cultural significance.

In their budget requests, potential lead agencies are expected to request funds sufficient to cover the cost of preparing the EIS (§1501.6). Other lead agency responsibilities include (§1501.7)

- Publishing the NOI in the *Federal Register*
- Determining the scope of the EIS
- Identifying insignificant environmental issues
- Determining the EIS schedule

3.2.3.1 *Cooperating agencies*

Sometimes other agencies assist the lead agency in preparing an EIS. A review is performed to identify any other agencies that may be involved or have an interest in the scope of the EIS. These agencies are referred to as *cooperating agencies*. As funds permit, the lead agency is expected to underwrite the costs of the activities and analyses that it requests cooperating agencies to perform. Cooperating agencies should be contacted at the earliest possible time. Any cooperating agency should identify a point of contact responsible for coordinating the EIS effort with the lead agency. Responsibilities and funding are then allocated to perform data collection, analysis, and other tasks assigned to the cooperating agencies. The author recommends that the lead and cooperating agencies sign a *memorandum of understanding* designating responsibilities between the agencies for performing various tasks.

3.2.3.2 *Identifying and selecting the lead agency*

Potential candidates for the role of lead agency must decide among themselves which one will be appointed. The Regulations specify factors that are used in determining the lead agency designation. These factors, listed in order of descending importance, include (§1501.5[c])

1. Magnitude of the agency's involvement
2. Project approval/disapproval authority
3. Expertise concerning the environmental effects
4. Duration of the agency's involvement
5. Sequence of the agency's involvement

If, after a period of 45 days, the federal agencies cannot agree on who will be designated the lead agency, a request for determination can be filed with the Council on Environmental Quality (CEQ). The CEQ will then make this decision (§1501.5[e]).

3.2.3.2.1 Ensuring qualified agency management and oversight. It is critical that management and the decision maker have qualifications sufficient to allow them to make rational decisions and environmentally sound judgments regarding the preparation of an EIS. Recall the case study presented in Chapter 1 in which the US Nuclear Regulatory Commission prepared EISs for renewing the operating licenses of the nation's fleet of aging nuclear reactors. Many of the problems that plague these EISs can be directly traced back to management problems. While the technical staff responsible for preparing these EISs was generally competent and experienced in NEPA and assessing environmental impacts, the management was perhaps another story. Mr. Brian Holian, the division director, was a nuclear engineer placed in charge of reviewing and approving EISs that involved complex technical and environmental issues. Mr. Bo Pham, who managed these EISs, was a mechanical engineer with marginal environmental expertise; yet he was assigned responsibility for directing, reviewing, approving, and ensuring the accuracy of these scientifically complex EISs. Lack of environmental management qualifications may at least partially explain these problem-plagued EISs. The lesson here is that federal agencies must assemble an interdisciplinary team, including management that understands NEPA and can accurately and objectively assess complex environmental impacts.

3.2.4 Forming and coordinating an interdisciplinary team

Preparation of an EIS may combine the talents of many disciplines (e.g., planning, science, architectural and engineering, socio-economics, scheduling, budgeting, and public relations). The agency is required to prepare the EIS utilizing

> … a systematic, **interdisciplinary** approach which will ensure the integrated use of the natural and social sciences and the environmental design arts in planning and in decisionmaking which may have an impact on man's environment… (§1501.2[a])

The EIS manager is responsible for ensuring that an interdisciplinary team (IDT) is assembled and that a comprehensive planning effort is performed. The disciplines of the IDT staff are to be

> … appropriate to the scope and issues identified in the scoping process. (§1502.6)

The EIS planning and analysis must therefore be performed by professionals with credentials and experience appropriate for investigating complex environmental, planning, and engineering issues. Consistent with NEPA's requirement to perform an "interdisciplinary" analysis, an IDT should be assembled, consisting of professionals representing all principal planning, scientific, and technical disciplines.

3.2.4.1 Interdisciplinary versus multidisciplinary team

It is important to note the distinction between the terms *interdisciplinary* and *multidisciplinary*. The term "multidisciplinary" denotes a process in which specialists representing pertinent disciplines perform their assigned task with little or no interaction. Such practice can lead to problems and disconnects, and fails to meet NEPA's legal requirement for performing an "interdisciplinary" planning process. In contrast, the term "interdisciplinary" denotes a process in which specialists interface and work together in assessing issues.

3.2.4.1.1 Interdisciplinary qualifications. The EIS manager must ensure that the analyses are performed by professional specialists, representing relevant disciplines, who work together to investigate common environmental issues. While the members do not necessarily have to work in the same office or even the same building, they need to coordinate their individual analyses and work together in preparing an integrated planning analysis.

The IDT may require the talents of specialists representing a diverse array of technical disciplines. Such disciplines typically range the gamut from urban and environmental planners, to land use planners, infrastructure specialists, regulatory specialists, toxicologists, health physicists, biologists, geologists, hydrologists, archaeologists, sociologists, and engineers to name just a few. Support staff includes technical editors and graphic artists.

3.2.4.2 Selecting an EIS manager

One of the first items of business is the selection of a qualified EIS manager possessing appropriate environmental and NEPA experience. The EIS manager is responsible for the professional and scientific integrity of the EIS. The manager must ensure that the EIS rigorously explores and objectively evaluates all reasonable alternatives, and discloses and discusses all major points of view. The Regulations also require that the EIS be written in plain English so that it can be readily understand by nonspecialists. The EIS manager must therefore ensure that the statement can be readily understood by the general public.

3.2.4.2.1 Background. Inexperienced management and analysts have been one of the principal reasons for cost overruns, poor planning, faulty analyses, and project delays. Prudence must be exercised in selecting a competently trained manager who can exercise good judgment. Knowledge of the EIS process is essential. The EIS manager should be intimately familiar with the EIS process and environmental issues. To effectively manage and understand complex issues and interrelationships, most EIS managers have a scientific background with experience in diverse environmental and technical disciplines. Depending on the key environmental issues at hand, the EIS manager may need to possess a scientific competency in disciplines as diverse as geology, hydrology and water quality, soils, air quality, biology, cultural resources, hazardous waste, or even air or hydrologic computer modeling.

The following example illustrates what can happen when the EIS manager lacks the appropriate experience and technical skills. The Pacific Northwest National Laboratory (PNNL), located in Richland, Washington, received a contract to prepare a very large and complex EIS. PNNL management assigned an inexperienced lawyer who had majored in political science to manage the EIS. This manager was unable to provide competent direction or coordinate the many technical tasks required. The EIS began to fall significantly behind schedule. Much of the EIS analysis was flawed as a result of incorrect guidance and management direction. This manager was eventually replaced, but not before the EIS had significantly overrun its budget. In the end, it took PNNL nearly three times as long to prepare the EIS as the schedule had originally allowed for. This caused the entire project to overrun its budget and resulted in a significant delay to the overall project schedule. The project ballooned from the original projected estimate of two and a half million dollars to seven million dollars.

Compounding the problem, PNNL was responsible for conducting and managing groundwater measurements at the site. However, the PNNL staff was caught falsifying and mismanaging records. This further delayed the EIS because it cast serious doubts over much of the data being used in the analysis. This shows the importance of developing and validating a quality assurance program that can ensure the integrity of the data used in the analysis.

3.2.4.2.2 Management skills. The EIS manager needs a broad management skill set. These skills include utilizing project management techniques for planning, staffing, cost/schedule control, and teaming skills such as effectively delegating work assignments. Listening and conflict resolution skills are also useful for resolving problems within the team or organization, and for identifying and resolving issues that concern stakeholders.

The EIS manager must seek out and coordinate with other interested and affected organizations, and obtain and consider the views of stakeholders. This involves swift identification of technical issues and establishing means for resolving such issues. The EIS manager is responsible for ensuring that the IDT members

- Understand their assigned roles and responsibilities
- Prepare their respective analyses in compliance with NEPA's regulatory requirements
- Perform ethical and scientifically defensible analyses
- Complete their assigned tasks on schedule and within budget

Among other responsibilities, the EIS manager

- Specifies the final scope of the EIS and significant issues to be investigated
- Assigns responsibilities and resources
- Coordinates various studies and tasks
- Provides ongoing, day-to-day technical direction
- Monitors schedules and budgets
- Resolves problems
- Motivates the IDT

The EIS manager must know how to create an adequate administrative record, in consultation with legal counsel.

3.2.4.2.3 Coordination between the EIS team and the project staff. Preparation of an EIS is sometimes viewed by managers or agency officials as a hindrance to implementing their project objectives. As witnessed in Chapter 1, some agencies still view an EIS as a document preparation process that must be completed as fast as possible so they can implement their predetermined decision. Such mindsets are in sharp contrast to the "systematic and interdisciplinary" process that is to be used in objectively analyzing alternatives and reaching a sound decision. For instance, one EIS manager reported, "Agency officials decided what they wanted to do and then warped the entire process to support this decision." Even worse, managers and officials may delay preparing an EIS until late in the game only to find themselves scrambling to meet a near-impossible deadline.

To help mitigate such problems, project personnel should be educated about the need and importance of the EIS in planning future actions. Coordination meetings may be initiated during the initial planning stage, to educate management about how the EIS can be used as a true planning tool to evaluate and compare the merits of potential courses of action. Such meetings should continue throughout the EIS process. The objective is to ensure sufficient interaction between the EIS team, project managers, and officials.

Table 3.1 Items Discussed at Kickoff Meeting

- Description and scope of the proposal (alternatives/proposed action, if available)
- Roles and responsibilities
- Schedule and budget
- EIS outline (if available)
- Action items
- Description of the upcoming scoping process
- Potentially significant impacts and controversial issues
- Principal planning assumptions (which can affect their analysis)
- Site visit

3.2.4.2.4 Kickoff meeting. Once the IDT has been assembled, a "kickoff" meeting is held to discuss the responsibilities and preparation of the EIS. The IDT is fully briefed on the scope of the proposal, schedule, resources, and key issues. A preliminary EIS outline is provided to each staff member. The schedule and important milestones are fully explained. Each member is assigned specific responsibilities. To the extent information is available, the items listed in Table 3.1 should be discussed.

3.2.4.2.4.1 Tips on conducting effective meetings. Numerous meetings may have to be held between the IDT, EIS contractor, agency officials, engineering staff, public relations specialists, and the like. This is particularly true for large or complex proposals. Many NEPA professionals report that too much time is wasted in nonproductive meetings. It is recommended that an agenda be circulated before the meeting so that attendees know and can prepare for what will be covered; if the agenda is distributed at the last minute, however, attendees may not have sufficient time to consider the issues. It can also provide a valuable technique for refocusing the meeting if it begins to veer off track. Also note that scheduling important meetings on a Friday afternoon can be problematic; attendees may have already left or may be "disengaged," thinking about weekend events.

People tend to think in terms of hour-long blocks of time, and meetings tend to fill the space allotted. A meeting should last no longer than required to get the job done. Table 3.2 lists examples of "meeting norms" that may be prepared and distributed in advance of meetings and explained to attendees at the start of a meeting.[3] If an attendee drones on, all one need do is cut the discussion short by pointing to the item being violated on the group's list of meeting norms. The key is having the courage to intervene to keep the meeting on track. It is important to allow everyone to voice their opinion. A good meeting facilitator observes body language in seeking out what may have been left unsaid and purposely engages those who may be silent.

Table 3.2 Meeting Norms

- Everyone has a right to express their opinion without being ridiculed
- Encourage participation from everyone
- Actively share conflicting ideas and opinions
- Target unresolved issues
- Express disagreement over ideas, not people
- Focus on one issue at a time
- Only one person speaks at a time (no side discussions)
- No one is allowed to dominate the discussion
- No rambling
- No lengthy reiteration of what someone else has already said
- Respect the confidentiality of opinions expressed during the meeting

Informational meetings can be very large. However, one indication of impending problems is when too many people are invited to a meeting where a key decision is to be made. Studies show that if 15 or more people attend a single meeting where a consensus must be reached, do not be surprised to find that the meeting is unproductive and becomes bogged down. Some communication specialists suggest that the ideal size for a decision-making meeting is between 5 and 12 attendees. This provides a cross section of opinions but is less apt to get bogged down with haggling or other nuisances. Carefully consider who needs to attend the meeting (i.e., key decision makers, subject matter experts, managers). Where more than 15 individuals are involved, it may be better to structure the meeting in stages or to consider breakout, submeetings, or even subcommittees.

Be wary of meetings where the organizer cannot explain a simple, precise purpose for the meeting. You may be invited to a meeting that you know will be lengthy or even run over time. For the meetings that you have no direct control over, there may be other options than to attend lengthy meetings. One option is to tell the leader before the meeting starts that you have to leave at a specific time for a prior obligation. That often puts pressure on the leader to keep the meeting on track. Sometimes face-to-face meetings are not essential. You may want to have an assistant attend in your place.

You may need to pick and choose when a meeting is appropriate to physically attend and when it is not. This applies equally to meetings you set and those to which your attendance is requested. You should consider travel time when making a cost/benefit decision regarding attending a face-to-face meeting versus some other venue (i.e., webinars, conference calls, or e-mails). Where a meeting is little more than exchanging information, a conference call or e-mails may be a more productive means of sharing information.

3.2.4.2.5 Standardizing procedures. Ground rules and procedures are established regarding document standards and style. This may involve establishing standards for use of

- Appropriate acronyms
- Methods for referencing material
- Standardized section headings
- Abbreviations
- Punctuation
- Units of measurement to be used

These ground rules and procedures should be discussed at the kick-off meeting or at another forum early in the EIS process. The technical editor may also want to attend and outline editorial guidelines with the entire EIS staff. A memo specifying document standards should be circulated, read, and initialed by all EIS team members to ensure they have read these standards. Direction should also be established for preparing an EIS that can be readily understood by the general public.

3.3 Prescoping

Once the initial activities outlined in Section 3.2 have been completed, the EIS manager may begin the prescoping effort, which will then be followed by the formal public scoping process (described in Chapter 4). Meetings or consultations with other parties and agencies should begin as soon as possible to screen the proposal for potential issues that may need to be addressed.

As part of the prescoping effort, the IDT attempts to identify the preliminary scope of the proposal, including the range of alternatives to be investigated, and key impacts and issues that will need to be evaluated; this task should include a preliminary screening exercise to segregate potentially significant from nonsignificant impacts.

The perception of the potentially significant impacts and issues often changes on visiting the proposed site(s). All team members should personally visit the proposed site(s) and potential affected area(s). A site visit allows analysts to gain an appreciation for the proposal, its affected environment, and potential environmental issues.

3.3.1 Defining the purpose and need

An EIS must "… briefly specify the *underlying purpose* and *need* to which the agency is responding in proposing the alternatives, including the proposed action" (§1502.13). A succinct declaration of the proposal's underlying statement of purpose and need (SPN) for taking action should be prepared as early as possible (see §1502.13). While the EIS team members may unanimously agree on the merits of a particular project, surprisingly, there is sometimes less than unanimous consensus regarding the *underlying* purpose and need. Defining the underlying need is often a nontrivial exercise. Preparing the SPN may appear to be intuitively obvious, yet experience shows that it can be deceptively complicated.

For example, some courses of action may be entirely dismissed, solely on the basis of how the statement of purpose and need is defined. The SPN can sometimes spell the difference between a flawed versus an acceptable EIS. Court rulings have pivoted on how such statements were written.

3.3.1.1 *How the "underlying need" provides a technique for determining the range of alternatives*

Schmidt has shown why a correctly crafted SPN provides a technique for successfully identifying the range of reasonable alternatives for detailed investigation.[4] If the SPN is defined too broadly, the number of alternatives may be virtually unlimited. Conversely, if the SPN is defined too narrowly, it may illegally negate consideration of reasonable or more optimal alternatives. Defined properly, the SPN provides a tool for focusing on and reducing a large set of potential alternatives down to a manageable set of reasonable alternatives.

3.3.1.1.1 Using the SPN to define a set of reasonable alternatives. As just witnessed, the Regulations require that an EIS briefly specify the *underlying purpose* and *need* for taking action. This begs the question: Why did the Regulations use both terms—*purpose* and *need*? Are they redundant terms or do they convey different requirements? As it turns out, there is an important distinction between the two terms. They are related, yet different dictates. As defined in Webster's dictionary, a *need* is "a lack of something useful, required, or desired."[5] A need can simply be viewed as something that is *lacking* or *desired*.

For instance, consider a military action involving long-term storage of high explosives. Rather than state there is a need to "… build an explosives storage bunker at its military training center," the SPN might more properly indicate a need "… to safely maintain and store high explosives." In terms of the potential set of reasonable alternatives, the two statements are actually quite different. The range of reasonable alternatives for building "an explosive storage building at its military training center" is much more limited than a need that involves "safely maintaining and storing high explosives." The first statement of need is limited and largely restricted to different bunker designs at its "military training center." In contrast, the second statement of need could include building an aboveground building, a buried bunker, "disassembling" explosives so they can be stored more safely, locating the facility at various different onsite locations, or even locating the facility at a different site. The second statement of need may provide more opportunities for investigating safer and more cost-effective alternatives.

Consider another example. Imagine an agency proposes to construct a large water well supply system for a military center. Rather than state there is a need to "construct a large water well supply system," the SPN might more properly indicate there is a need to "... increase the quantity of drinking water for the center." Again, the two needs are quite different in terms of their respective set of alternatives. The range of reasonable alternatives for constructing a water well supply system is more restrictive and might be limited to positions where the well would be located, drilling methods used, size of the well, and piping system. In contrast, the set of alternatives for obtaining "... a useable quantity of drinking water" may be much broader, including alternatives such as a water desalination and purification plant, piping the water in from another location, and even a recycling and water conservation program. In this case, the second statement of need provides more opportunities for investigating better environmental and perhaps even more cost-effective alternatives.

3.3.1.1.1.1 The "purpose" for taking action. Now consider the term *purpose*. Webster's dictionary defines *purpose* as a "goal" or "an end or aim to be kept in view in any plan, measure, exertion, or operation."[6] Simply put, a purpose is a *goal* or an *objective(s)* to be met. Thus, the purpose for taking action might involve important environmental, engineering, economic, political, or other goals or objectives.

Reconsider the last example in which an agency has a *need* to "... increase the quantity of drinking water." This was the need, but what about the purpose? The purpose for taking action might involve the following four objectives:

1. Obtaining an economic supply of water
2. Obtaining a supply of drinking water that meets safe drinking water standards
3. Providing a minimum of 275,000 liters per day
4. Could be constructed within a timeframe of 20 months

Such specifications are better viewed as objectives (purposes) than as part of the underlying need (i.e., a lack of something that is required). The importance of this distinction will become clearer in the following sections.

3.3.1.1.1.2 Alternatives to the proposed action versus alternatives to the underlying need. Many EISs routinely view "reasonable alternatives" as alternatives to the proposed action. But this is not correct. As detailed in Chapter 6, the Regulations require an agency to

> … briefly specify the underlying purpose and need to which the agency is responding in proposing the alternatives, **including** the proposed action. (§1502.13)

There is an important distinction between alternatives to the proposed action and alternatives to the underlying need. Note that the provision just cited uses the word "including" in referring to the proposed action. This provision does not direct agencies to define an underlying purpose and need *for the proposed action*. Instead, agencies are directed to specify the purpose and need for proposing "alternatives"—the proposed action simply being viewed as one of the alternatives.

Although this book does not advocate identification of a "proposed action" per se, it is nevertheless a concept in common use. What does the phrase "proposed action" actually imply? As implied in the Regulations, a proposed action is simply the agency's favored approach for satisfying something *lacking or desired* (i.e., underlying need). What, then, are alternatives? Alternatives should *not* be viewed as alternatives to the proposed action. The alternatives are simply options (alternatives) for achieving the underlying need; they are not alternatives to the proposed action.

As we will see, alternatives to the proposed action can be quite different from those crafted to meet the agency's underlying need. Imagine an agency determines there is a need to construct a dam to halt flooding. Notice the difference between alternatives for *constructing dams* versus alternatives to the underlying need of *controlling floods*. The range of alternatives can be quite different. Alternatives to the proposed action (i.e., constructing a concrete dam) are likely to focus on alternative ways of constructing a dam (e.g., concrete, earthen, size, shape, perhaps different locations). In contrast, alternatives to the underlying need of controlling floods are broader and more likely to focus on diverse ways of controlling flooding (e.g., dams, levies, enhancing wetlands, increasing vegetated areas, or diverting water into a nearby reservoir).

Now consider how a change in the definition of underlying need can affect this range of alternatives. Suppose the agency had identified the underlying need as "limiting the damages and impacts from flooding." The range of alternatives might now be expanded to include options such as land use controls that prohibit or limit development in a floodplain.

3.3.1.1.1.3 Difference between a need and the "underlying" need. This leads us to yet another inquiry. What precisely is the difference between a need and the agency's *underlying* need?

Suppose an agency defines a need to "… remove an explosive toxic waste tank." This is a need, but is it really the underlying need? It is instructive to ask why the agency needs to remove the tank in the first place; the underlying need might be better defined as the need "... to protect human health and environmental quality from the consequences of a toxic waste explosion."

Note the difference between the two statements of need. The difference in the alternatives for *protecting the environment and human health* may be quite different from alternative methods for simply *removing an explosive toxic waste tank.*

The first definition of need is more restrictive and may not consider safer or more cost-effective options. For instance, alternatives for simply removing the waste tank might be limited to various engineering methods for excavating and removing the tank. In contrast, there are potentially many more (perhaps better and even less expensive) alternatives for protecting the environment and human health than there are for removing the tank; such alternatives might involve placing a permanent non-entry buffer area around the tank, *in situ* stabilization (e.g., grouting or vitrifying the waste inside the tank), removing the waste for treatment but leaving the tank in place, adding a chemical to the tank that counteracts its explosive potential, or of course removing the tank and waste altogether.

3.3.1.1.1.4 The underlying need defines the range of reasonable alternatives. As indicated earlier, the authors of the Regulations were careful to make a clear distinction between an "underlying purpose and need" versus a simpler "purpose and need." As just witnessed, the set of alternatives meeting the underlying need may be different from that which simply meets the need. Alternatives that meet the need, but not the underlying need, might be viewed as *unreasonable* and, therefore, *eliminated from detailed analysis.* As described in the next section, this distinction provides a valuable technique for screening out what might otherwise be a large or even unbounded range of alternatives.

Graphically, this can be represented as a bulls-eye target. The bulls-eye is the underlying need and arrows represent the alternatives. Any alternative that meets the agency's underlying need is likely to be a reasonable alternative (hits the bulls-eye); conversely, an alternative that does not meet the underlying need may be eliminated from detailed study (see Figure 3.2). NEPA case law supports this scoping technique.

3.3.1.1.1.5 Enlarging or reducing the range of alternatives that need to be evaluated. A "broadly" crafted statement of underlying need can be advantageous as it tends to compel analysts to consider a more diverse range of alternatives for satisfying the underlying need; by forcing them to think outside the box, they may find a cheaper or more environmentally preferable alternative. The disadvantage, of course, is that it can increase

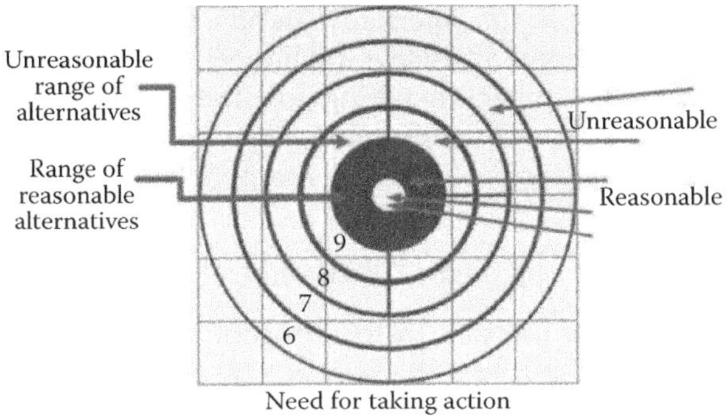

Figure 3.2 Target represents a range of potential alternatives for meeting an underlying need. The arrows on the right side of the figure represent various alternatives (e.g., reasonable vs. unreasonable). The dark bulls-eye represents the *underlying* need for taking action. The outer circles represent the need but not the underlying need for taking action. Alternative actions that meet the underlying need are considered reasonable alternatives. Any alternative that does not meet the underlying need may be deemed unreasonable and can therefore be omitted from detailed investigation.

the complexity of the analysis as a greater number of alternatives may need to be investigated.

Conversely, if the number of reasonable alternatives subject to detailed analysis is unduly large, the range of alternatives may be reduced by writing a more narrowly focused statement of need. In reality, preparation of the statement of underlying need is sometimes an iterative exercise. However, the reader is cautioned that it is inappropriate and potentially illegal to define the statement so narrowly that only an agency's proposed action can meet the statement of need. Although the statement of underlying need provides a valuable tool for defining the range of analyzed alternatives, it must be prepared and used judiciously.

3.3.1.1.1.6 No-action alternative. We will now apply this technique to the no-action alternative using the example involving flood control. The no-action alternative is sometimes viewed as *not taking the proposed action*. Under this interpretation, if the proposed action is to build a dam, the no-action alternative is simply *not to build the dam*. In this case, the impacts of no action may simply be the impacts that would not occur as a result of not constructing the dam (i.e., no construction, recreational, or fishery impacts related to construction of the dam).

However, on the basis of the technique introduced in this section, the no-action alternative should actually be viewed as *not meeting the underlying need*. Thus, if the underlying need is to control floods, the no-action alternative essentially involves *taking no action to control floods*. Under this interpretation, the analysis of the no-action alternative tends to become more sharply focused on the implications of what would happen *if the underlying need is not met*. In this example, the impacts of taking no action would be flooding! Now, consider what the impacts of flooding are. Next, consider what the impacts of constructing a dam to control flooding are, both adverse and beneficial.

3.3.2 *The "purpose" provides a basis for decision making*

This discussion brings us back to the concept of the *underlying purpose* for taking action. Where does the concept of purpose fit into this picture? Recall that "purposes" are *goals* or *objectives* that the agency wishes to achieve. Numerous alternatives may need to be analyzed for meeting the agency's underlying need. But not all of these reasonable alternatives are likely to meet the agency's underlying purpose (objectives). The "purpose" for taking action comes into play during the decision-making process after the reasonable alternatives have been rigorously investigated and the final EIS has been completed. The underlying purpose(s) provide a basis for screening the analyzed alternatives and provides the rationale for selecting the one that best meets the agency's goals or objectives. The decision maker should therefore select the alternative that most closely meets the stated purpose for taking action.

3.3.3 *Identifying potential decisions that may have to be made*

As we have seen, an agency must identify the range of actions, alternatives, and impacts to be investigated as part of its public scoping process. Unfortunately, the staff can perform an extensive scoping process and still find, upon completing the EIS, that the analysis has not adequately addressed future decisions that may need to be made. Despite the fact that they had completed an extensive scoping effort, some agency personnel have reported that they prepared EISs only to find that they "missed the mark" because the analysis did not adequately support decisions that eventually needed to be made. This disconnect is at least partly explained by the fact that agencies sometimes "dive" into the preparation of an EIS, having given little forethought to the actual decisions that eventually need to be considered. The value of identifying potential decisions cannot be overstated.

3.3.4 *Decision-based scoping*

The companion book, *NEPA and Environmental Planning*, presents a technique referred to as decision-based scoping (DBS) for identifying the scope of decisions that may eventually need to be considered. The DBS technique is in marked contrast to the procedure typically used in determining the scope of an EIS; it is particularly useful for complex projects

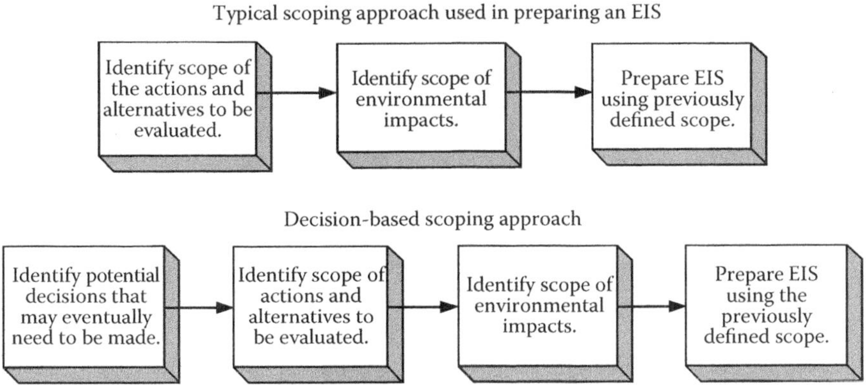

Figure 3.3 Comparison of the decision-based scoping approach with the standard approach typically used in determining the scope of an EIS.

or large programs that involve uncertainties in terms of future decision making.[7] Figure 3.3 compares the DBS technique with the approach typically used in determining the scope of an EIS.

For complex proposals, the DBS in conjunction with a scoping tool known as the decision identification tree (DIT) provides a systematic procedure for rigorously identifying future decisions that may need to be considered. The DIT provides a tool for shaping the bounds of the analysis and ensuring that the scope of the EIS in terms of future decision making is properly defined. Once agreement is obtained on the scope of potential decision making, the scoping effort turns to the standard task of identifying the range of actions, alternatives, and impacts that need to be evaluated. The utility of this approach is particularly valuable for programmatic analyses that involve large or complex planning efforts. This may be particularly true for long-term projects and programs that involve a large degree of uncertainty in terms of future decision making. Thus, this approach can result in long-term cost savings by reducing risks associated with uncertainty during the decision-making process.

3.3.5 Integration with other planning and regulatory requirements

As we have seen, NEPA is an environmental *planning* process. As such, the EIS process is required to be integrated with other related environmental regulatory requirements, permits, agreements, and policies, so that the processes run concurrently rather than consecutively.[8] In preparing an EIS, the Regulations require the lead agency to

> … cooperate with State and local agencies to the fullest extent possible… (§1506.2[b])

To the fullest extent possible, agencies are likewise required to prepare their EISs

> ... concurrently with and integrated with environmental impact analyses and related surveys and studies required by the Fish and Wildlife Coordination Act... the National Historic Preservation Act... Endangered Species Act..., and other environmental review laws and executive orders. (§1502.25[a])

These requirements are applicable to the entire EIS planning process, not simply the preliminary scoping step outlined in this chapter. The author recommends that a detailed schedule of environmental compliance requirements be prepared. The importance of this requirement cannot be overemphasized. Consider the following example involving a high priority waste remediation project that the author worked on. The project was assigned high priority and given a short schedule for completion. The project manager considered the NEPA process to be an unnecessary hindrance. Millions of dollars and nearly two years of design work were invested in preparing the remediation design, which included waste treatment basins for evaporating and concentrating toxic waste. In their haste to speed the project along, the project manager neglected to consider the fact that the evaporation basins were a "treatment" facility subject to a lengthy and rigorous air-permitting process. Fortunately, this permitting requirement was eventually identified during the final stages of the NEPA process. Because this permitting requirement was not identified during the initial planning process, the project was delayed by a year and a half while work began on preparing the air permits. The project manager was removed from the project. This noncompliance could have resulted in the state and Environmental Protection Agency (EPA) denying final project approval, which could have resulted in even greater delays and cost overruns. Worse yet, had the project been approved and implemented without the required permits, the management might have been liable for noncompliance with required permitting requirements. All of this might have been avoided if the project manager had simply taken the time to properly consult and coordinate the project with the NEPA staff and the EPA state regulators.

3.3.6 *Potential environmental statutes and requirements*

As an early planning process, an EIS provides the ideal tool for identifying and coordinating other environmental requirements. For instance, in the example just cited, the air-permitting requirement was finally identified

Table 3.3 Principal Environmental Statutes and Requirements That Need to Be Integrated with NEPA Planning Process

Requirements specifically cited in the Regulations
• National Historic Preservation Act of 1966 (16 U.S.C. 470 et seq.)
• Endangered Species Act of 1973 (16 U.S.C. 1531 et seq.)
• Fish and Wildlife Coordination Act (16 U.S.C. 661 et seq.)
Other environmental laws and executive orders
• Wild and Scenic Rivers Act of 1968 (16 U.S.C. 1271–1287)
• Costal Zone Management Act of 1972 (16 U.S.C. 1451 et seq.)
• Farmland Protection Policy Act of 1981 (7 U.S.C. 4201 et seq.)
• American Indian Religious Freedom Act of 1978 (42 U.S.C. 1996)
• Pollution Prevention Act of 1990 (P.L. 101-508, 6601 et seq.)
• Environmental Justice (Executive Order 12898)
• Protection of Wetlands (Executive Order 11990)
• Floodplain Management (Executive Order 11988)

late in the NEPA process. Table 3.3 denotes three environmental statutes specifically listed in the Regulations as requirements to be integrated with the NEPA planning process (§1502.25[a]). Additional requirements that also commonly need to be integrated with NEPA are likewise listed. The companion book, *NEPA and Environmental Planning*, identifies and describes a host of key environmental regulations and requirements with a focus on how they can be integrated with the EIS process.[7]

3.3.7 *Integrating SEPA, and state and local requirements*

Federal agencies are directed to cooperate with state and local agencies in reducing duplication between NEPA and state and local requirements. More than half of the states in the United States now have some form of state environmental policy acts (SEPA), often requiring preparation of a state equivalent of the NEPA EIS. Where state or local environmental requirements do not conflict with NEPA, federal officials are directed to prepare an EIS that fulfills both requirements.[9] To facilitate efficiency, a SEPA document can also be incorporated by reference into the NEPA EIS. The EIS manager and staff should consider the most effective approaches for coordinating and consolidating federal, state, and local planning/environmental requirements with the NEPA EIS.

3.3.8 *Identifying interim actions*

Actions related to the EIS proposal that need to proceed before the EIS is completed are referred to as *interim actions*. As described in the companion book, *NEPA and Environmental Planning*,[10] NEPA places strict limitations (interim action criteria) on actions that may precede the completion of the EIS (§1506.1). To minimize delays and promote an integrated planning process, the agency should perform a review to ensure that potential interim actions are identified and reviewed. This is a two-stage review process:

1. Identify potential actions related to the proposal that may need to precede the completion of the EIS.
2. Potential actions falling within the scope of the EIS are then examined in terms of the interim action criteria (§1506.1) to verify that they can, in fact, be implemented before the EIS process has been completed.

The reader is referred to §1506.1 of the Regulations, which spells out the interim action criteria. Emphasis is placed on identifying potential problems, such as funding or schedule delays, so that appropriate contingency measures can be developed for any actions that do not qualify for status as an interim action.

3.3.8.1 *Interim action justification memorandum*

The author recommends that an interim action justification memorandum (IAJM) be prepared for individual actions that have been reviewed and found to qualify for interim action status. The IAJM briefly describes the interim actions and provides evidence demonstrating that they meet the eligibility criteria presented in §1506.1 for status as an interim action. This task has many uses, including the fact that it provides evidence that any interim actions were adequately reviewed before being implemented. The memorandum can then be appended to the EIS and any NEPA document that is later prepared for the interim action.

3.4 *EIS management tools*

This section describes some tools and techniques for effectively managing the EIS process. Professional judgment must be exercised (based on the complexity and particular circumstances) in determining the tools and techniques most practical and appropriate for any specific EIS project.

3.4.1 *Management action plan*

A management action plan (MAP) or project management plan provides an initial plan or "road map" for preparing the EIS. Among other things, a MAP can outline roles, responsibilities, and work assignments. This plan describes how deliverables such as the NOI, notice of availability (NOA), and the draft/final EIS will be prepared. Appropriate milestones such as public meetings and resolution of public comments are identified. A

Table 3.4 General Outline for Management Action Plan

- Description of the planning process and how it will be coordinated
- Annotated outline for the EIS
- Roles and responsibilities
- Brief description of the proposal
- Schedule outlining significant milestones
- Work breakdown schedule
- Change request process
- Principal cost and schedule justification assumptions

representative outline of the MAP is provided in Table 3.4. Many of the items noted in Table 3.4 are described in more detail below.

3.4.1.1 Functional roles and responsibilities matrix

An example of a functional roles and responsibilities matrix is illustrated in Table 3.5. This matrix defines personnel responsibilities and helps establish organizational boundaries, especially where more than one entity or organization is involved.

Table 3.5 Example of Functional Roles and Responsibilities Matrix

Name	Title	Organization	Responsibility
Dr. Tripover	EIS manager	Environmental Regulatory Department	Manage the EIS and IDT
Dr. Strangelove	Civil engineer	Program Office	Alternatives description
Mr. Imsoweird	Environmental engineer	ACME Consulting Services	Alternatives description
Ms. Stinkith	Principal scientist	ACME Environmental Consulting	Environmental consequences
Mr. Douglas	Agriculture and hydrologist	Green Acres Consulting Services	Water quality
Mrs. Douglas	Archaeologist	Shady Rest Consulting Services	Cultural resources
Ms. Maybe	Computer modeler	Top Gun Consultants Inc.	Air dispersion modeling
Mr. Ed	Principal geologist	Mr. Ed's Environmental Consulting	Affected environment
Dr. Askmenot	Hydrologist	Top Notch Consulting Inc.	Hydrological issues
Ms. Nottonight	Technical editor	Environmental Division	Document production

3.4.2 *Annotated outline, budget, and schedule*

Mark Twain once shared some of his lifelong experience:

> Never put off until tomorrow, what you can do the day after tomorrow.

This section discusses some of the tasks necessary to keep an EIS on budget and schedule. The EIS manager should prepare an *annotated outline* of the EIS. The outline should be periodically updated to reflect new issues or changing circumstances. The outline typically includes four elements:

1. Annotated outline (including specific issues and topics) to be addressed in the EIS
2. Brief description of how each section of the EIS will be prepared
3. Individual(s) responsible for preparing each section
4. Estimated page limits for each section

Other things that may need to be identified include

- Principal assumptions
- Length of review periods
- Number of reviews
- Estimated number of public scoping and EIS review comments that are expected to be received

The annotated outline provides a basis for preparing the EIS schedule. In coordination with the EIS manager, the IDT members determine the amount of time and effort required to complete their respective tasks and EIS sections. These data are used to generate a schedule showing all significant milestones and deliverable dates.

One reason for EIS schedule delays involves delays as a result of required environmental studies or design work that must be completed. Before the EIS is started, the EIS manager needs to determine how much and what type of information is needed. The EIS manager must understand the full scope of the project, and insist on details. This information will be used to estimate the level of detail the EIS will cover, whether standard methodologies can be employed, and if fieldwork is needed.

3.4.2.1 *Budgeting and the work breakdown structure*

Project delays are occasionally encountered because EIS funding was not adequately factored into early project planning. Agency officials need to estimate and request funding for preparing the EIS as soon as a decision is made to prepare the statement.

3.4.2.1.1 Work breakdown structure. A work breakdown structure (WBS) is a commonly used tool for budgeting and controlling cost. It provides a useful technique for estimating the EIS budget and schedule. The WBS is used to divide the project into its hierarchical phases, deliverables, and work packages. It employs a tree structure showing the subdivision of effort required to achieve each task. The WBS should describe all pertinent EIS tasks and functions, including project management, scoping, public involvement, preparation and review of the draft and final EIS, and issuance of the record of decision (ROD). Figure 3.4 provides a simplified example of how any project (in this case a house) can be broken down into subprojects, work packages, and tasks. The cost, schedule, and level of effort can then be estimated and summed for all project components. Many commercially available software packages can be used in preparing the WBS.

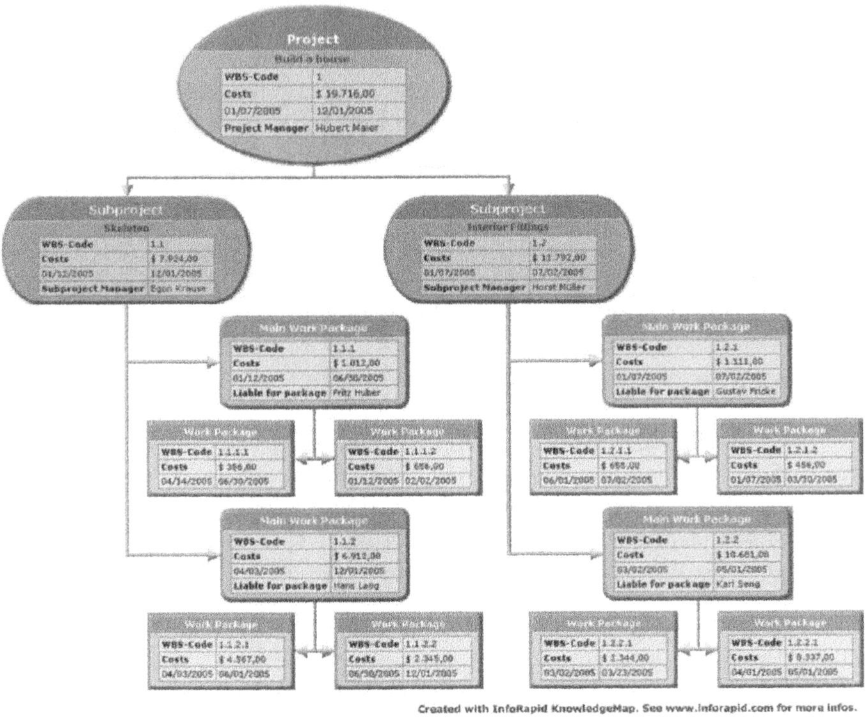

Figure 3.4 A simplified, example of how project cost (in this example of a house) can be estimated by breaking it down into component subprojects, work packages, and tasks. (Courtesy images.google.com.)

3.4.2.2 *Schedule*

The WBS is commonly used to generate the EIS project schedule. The schedule shows all important events, deadlines, and milestones. The schedule lists the start and finish dates of each task. Various types of schedules are in common use. Gantt and PERT charts represent two types of schedules commonly used to manage a project. Both techniques have their own advantages and disadvantages. These schedules can be prepared and updated using many commercially available software packages.

3.4.2.2.1 Managing the schedule and budget. The EIS manager needs to obtain a commitment from each team member to complete their portion of the EIS on schedule. Once the schedule is prepared, indicating who is responsible for each section of the EIS, it is circulated to the staff for review. Maintaining the EIS schedule and controlling the budget requires close attention and coordination. As noted earlier, the EIS manager is responsible for monitoring the team's progress and costs. Progress reports are prepared and circulated on a regular basis, identifying current activities, costs, and problems.

As is often human nature, IDT members sometimes postpone important tasks until the last possible minute. This is particularly true where members are juggling responsibilities among more than one assignment. The EIS manager should not be surprised to find that one or more IDT team members requests additional time to complete an assignment. As a compensation measure, some EIS managers develop a schedule based on a due date that is somewhat ahead of the actual "drop-dead" deliverable date. IDT members are given a completion date shown on the schedule. The EIS manager does not divulge the drop-dead date by which the EIS *must* be finalized; thus, the manager has a built-in buffer for instances where team members require additional time or to cover other delays.

3.4.2.2.1.1 Schedule and budgetary change requests. Schedule and budgetary change requests may result from factors as diverse as a change in scope, new regulatory requirements, or unforeseen environmental issues. A procedure for making change requests should be defined. A change control procedure frequently involves

- Tracking cost deviations from the baseline plan
- Identifying an appropriate course of action
- Obtaining approval for changes

3.4.2.2.2 Gantt chart. A Gantt chart, named after its developer Henry Gantt, is the most commonly used format for presenting an EIS schedule. It is a type of bar chart used to illustrate a project schedule. The

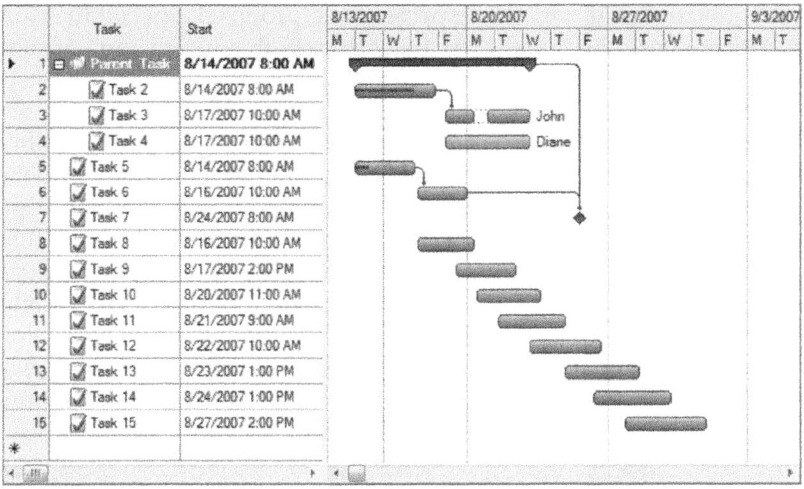

Figure 3.5 Example of a simplified Gantt chart for a project. (Courtesy images.google.com.)

Gantt chart can assist the EIS manager in identifying where additional resources are needed to meet the schedule and ensure that work is performed in a systematic order.

Gantt charts illustrate the start and finish dates of the project elements identified in the WBS. Some Gantt charts show the dependency (i.e., precedence network) relationships between various tasks and activities; for instance, some sections of an EIS cannot be completed until a wildlife survey or water quality testing has been performed. A Gantt chart is particularly useful for small projects that fit on a single sheet or computer screen, but can become unwieldy for larger projects with more than 40–50 activities or so. Figure 3.5 is an example of a Gantt chart.

3.4.3 *Developing a public involvement strategy*

An agency may be open to a legal challenge for failing to make a "diligent effort" to involve the public in the scoping and preparation of the EIS (§1506.6[a]). Technical proposals are often difficult for the general public to understand and comment on. Experience shows that citizens who cannot understand a proposal are often more inclined to be suspicious and to oppose the project. They may also be intimidated and less likely to submit comments that might mitigate controversial issues. The public involvement strategy should be specifically crafted to address such problems. Consistent with NEPA's mandate, the agency needs to be forthright about both the positive and negative aspects of the proposal. Special measures such as those described below can be used to both educate and entice the public to participate in the scoping process.

Preparing an effective public involvement strategy for dealing with stakeholders and their concerns may save an agency endless hours of grief, misdirected effort, and schedule delays, not to mention a tarnished reputation. More innovative methods such as creating an exhibit or perhaps a scale model of the proposed action can be used to stimulate interest. Exhibits of the proposed action may be displayed in an area of public interest such as a mall or town square; agency representatives can even be assigned to the exhibit to explain the proposal and solicit scoping comments from citizens. An agency may likewise consider holding scoping meetings in public malls or at community gatherings such as a county fair.

3.4.3.1 *Managing conflict*

A successful public involvement strategy should focus on identifying and dealing with potential public opposition. Where a proposal is particularly controversial, the agency may want to consider obtaining the services of an impartial public relations facilitator to moderate controversial public scoping meetings.

The public often has little or no experience with the EIS process. A common public misconception (unfortunately sometimes true) is that the agency has made up its mind to implement the proposal and is merely collecting information to support its case. One useful method for defusing suspicion and quelling hostility simply involves continually stressing that the agency is actively *and genuinely* seeking ideas for alternatives, and mitigation and design changes for reducing or eliminating potential impacts. Agency officials should clearly explain that no decision has been made and that the purpose of the scoping process is to provide citizens with the opportunity to influence the outcome of future decisions. The agency should also stress that this is the public's opportunity for raising issues of concern. Often the mere act of education may quell some public distrust.

3.4.3.1.1 *Identifying project opposition.* Sometimes the EIS scoping process is viewed with an "us against them" mentality. Such attitudes are an invitation for trouble. The agency may wish to make an early effort to identify potential project opponents and engage them in the EIS process. The EIS manager can identify points of contention by simply seeking input from the public before beginning the formal scoping process. Such input may allow the agency to modify its proposal in a manner that may minimize later controversy. Providing public interest groups and potential adversaries with an opportunity to voice their opinions and concerns early in the planning stage may prevent significant delays later. To this end, agency officials may want to meet with community leaders and special interest groups before initiating the formal public scoping process. As noted in the case study involving the Nuclear Regulatory Commission in Chapter 1, under no circumstances should agency personnel ever attempt to ignore or "silence" points of view.

3.4.4 Preparing the scoping plan, notices, and advertisements

A well-thought-out scoping plan provides a valuable tool for ensuring that a comprehensive scoping process is performed. The plan should be tailored to the specific proposal on the table. Table 3.6 outlines items such a plan should include.

As shown in Table 3.7, advertisements soliciting comments from the public can take many different forms.

Table 3.6 Items Included in Agency's Scoping Plan

- The entire scoping process including when and where meetings and/or hearings will be held
- The media most appropriate for informing the public about the details of the proposal
- Identifying "how," "when," and "by whom" input will be requested from agencies, organizations, and individuals
- The appropriate format to be followed in managing scoping meetings
- Specific types of information and comments desired from the public
- Procedures for ensuring that all interested parties are represented and afforded an opportunity to participate and provide input
- A specific public involvement strategy (described earlier)

Table 3.7 Means by Which Advertisements Have Been Used to Solicit Comments

- Facebook, Twitter, and YouTube
- Other Internet media
- Television
- Radio
- Local newspapers
- Public meetings and booths at community events
- Direct mail
- Professional journals
- Telephone recordings

Draft media and distribution lists, including news releases, may be prepared as part of the scoping plan. A procedure should also be defined, describing how public comments will be reviewed and recorded.

3.4.4.1 EIS distribution list

The term "EIS distribution list" refers to the list of individuals and organizations that will receive notices and a copy of the EIS (i.e., a list of "contacts"). Consider including the following information in the distribution list:

- Contact information (e.g., name, organization, mailing address, e-mail address)
- Requested EIS documents (e.g., entire EIS, summary only, main body without appendices)
- Preferred format (e.g., printed copy, compact disk [CD], access from a website)
- Additional information (e.g., source for each name on the list, such as from registration at a scoping meeting, request to be placed on the list)

Start compiling the distribution list in the early planning stage and update it throughout the EIS process. Check the accuracy and completeness of the distribution list before printing a draft or final EIS, so that it can be included in the document. Maintain the distribution list in a spreadsheet or database, as these formats allow more flexibility for searching and sorting than a word processing file.

3.4.4.2 Facebook, Twitter, and YouTube and social media

Video and social media often reach a different audience than print media. Public outreach is an integral element in the EIS process. Social media sites such as Twitter, Facebook, and YouTube are increasingly being used to reach the public.[11] YouTube is useful in sharing visual information; Facebook allows picture, video, and text sharing; and Twitter allows short-form "tweets" to convey small amounts of text, such as notices of public comment periods. Many agencies are maintaining active Facebook, Twitter, and YouTube accounts.

Both public participation and agency transparency can be increased by utilizing social media tools. For example, the National Nuclear Security Administration used YouTube to host a video on the preparation of a site-wide EIS. This 4-minute video offered a brief summary of the NEPA process, and provided pertinent information concerning the alternatives, dates of the public comment periods, and details about public hearings. It also provided a live video webcast of its public hearing on its EIS. This allowed those that could not attend the hearing to witness the proceedings.

3.4.5 Establishing an auditable trail and administrative record

With respect to an EIS, *due diligence* is a legal term referring to the level of conscientiousness or standard of care used in preparing the statement. An agency cannot depend solely on environmentally sound decisions as a means of demonstrating due diligence to either the public or courts. If the agency is challenged, a court may order the agency to turn over files and documentation related to the EIS.

3.4.5.1 The agency's administrative record

Agencies are responsible for maintaining an auditable trail, including an "administrative record" demonstrating compliance with NEPA's requirements. It is not easy to define precisely what legally constitutes an agency's administrative record because NEPA does not establish a formal fact-finding process (i.e., a detailed procedure for establishing facts or reaching a final decision). Nevertheless, for the purposes of judicial review, an agency's administrative record (ADREC) can be viewed as[12]

> The body of documents and information considered or relied upon in the process of reaching a final decision.

This body of documents and information includes the entire record that existed at the time the decision was made—not simply the portion of the record read by the decision maker. The ADERC has also been defined to include[13]

> ... all relevant studies or data used or published by the agency complying the statement.

The ADREC may also include informal documentation such as e-mails and memos. Some lawyers draw a distinction between the terms *project file* and *administrative record*. The term project file technically refers to all EIS-related files maintained by the project team, while administrative record refers to the documents that are actually submitted by an agency to the court during an EIS lawsuit. This book simply uses the term "ADREC" and does not make a distinction between these two terms.

It is important to note that while the ADREC is important in demonstrating due diligence with NEPA's requirements, the EIS document must be capable of standing on its own merits. More to the point, the EIS document needs to supply information demonstrating that all key requirements have been satisfied and that the statement provides the basis for reaching the final decision. In one case, for instance, the administrative record contained an important analysis that was not cited in the EIS. When challenged, the agency lost because the EIS analysis was viewed as being incomplete.

3.4.5.2 *A court's review of the agency's ADREC*

A thorough ADREC is crucial in demonstrating due diligence and demonstrating that the agency properly weighed all pertinent factors in reaching its final decision. A plaintiff (party filing the lawsuit) may find it advantageous to scrutinize the agency's ADREC in an attempt to find weaknesses or even lack of evidence that the agency fully discharged its EIS responsibilities; potential adversaries will focus on areas where the agency's ADREC appears weak. In still other instances, a plaintiff's underlying strategy is not even to win a lawsuit but simply to delay the project or to stimulate sizeable public distrust or opposition.

When dealing with questions regarding the adequacy of an EIS, a court's role is to determine if the staff made an adequate and objective effort, judged in light of the "rule of reason" (see Chapter 2, Section 2.4), to investigate the reasonable alternatives and provide relevant information to the decision maker.[13] The court's review of an agency's decision-making process is governed by the Administrative Procedure Act.[14] While agencies routinely deal with complex technical matters, federal judges typically lack such expertise. For this reason, courts generally confine their review to the agency's ADREC.[15] Only in limited circumstances do courts expand their review beyond the ADREC or permit "discovery."[15]

Discovery is the legal process adversarial parties use during NEPA litigation to obtain information they do not have access to and need to support their case. For instance, a plaintiff may petition for documents and information pertinent to its case from a defendant (agency). The defendant would be required to provide such documents and information to the plaintiff according to a schedule set by the court.

3.4.5.2.1 Cases demonstrating importance of maintaining a thorough ADREC. The following two examples illustrate the importance of maintaining a traceable NEPA record of the agency's due diligence, thought process, and decisions. The first example involves issuance of a permit by the US Army Corps of Engineers (Corps) for construction of a dam and reservoir. The plaintiff challenged the Corps, alleging that the EIS prepared for the permit did not address impacts of an existing pipeline that would cross under the proposed reservoir. For its part, the Corps had assumed the pipeline would be relocated, but did not make relocation a condition of the permit. The court concluded that the ADREC did not support the Corps' assumption that the pipeline would be relocated. This assumption should have been documented in the EIS and made a condition of issuing the permit. Moreover, the EIS should have evaluated the impacts of relocating the pipeline.

The second example involves an independent specialist assigned to audit a federal agency's programmatic EIS. He could not understand how certain conclusions were reached. Worse yet, he raised serious questions regarding the origin of some of the data and assumptions used in the analysis. This led to an investigation to determine the basis for the conclusions and the origin of the data. One of the problems was that professional staff turnover had been high and records had not been kept up to date. While the conclusions and data might have been valid, they were nevertheless untraceable. The result was that a significant portion of the analysis had to be regenerated to ensure that it was both accurate and defensible.

3.4.5.2.2 Going beyond the agency's ADREC. The Freedom of Information Act provides a useful tool for gaining access to an agency's internal documents. An agency that appears to be secretive is likely to raise concerns with a judge. Conversely, a well-documented ADREC can reduce the chances that a court will allow discovery. A plaintiff may be granted discovery if it can be shown that there is reason to believe the ADREC is incomplete or that the agency has not been forthcoming. Such reviews, however, tend to be limited to collecting background information or for[16]

> … the limited purposes of ascertaining whether the agency considered all the relevant factors or fully explicated its course of conduct or grounds of decision.… Consideration of the evidence to determine the correctness or wisdom of the agency's decision is not permitted, even if the court has also examined the administrative record.

Allegations that the agency did not investigate reasonable alternatives, overlooked significant impacts, or swept "stubborn problems or serious criticism… under the rug" can raise questions sufficient to justify introducing new evidence outside the ADREC, including expert testimony. Such evidence may be introduced where challenges involve either the adequacy of the analysis or cases involving an agency's finding that an EIS is not required.[13] When it is necessary to establish the reasons for an agency's decision, the court may inquire outside the ADREC, introducing affidavits and testimony. Courts also explore outside the record when it appears that an agency used information outside the ADREC. The courts may likewise inquire outside the ADREC when there is reason to believe that the agency acted improperly or in "bad faith" with respect to its knowledge.[17]

A court may not conduct its own examination if it finds an agency's investigation was inadequate or is not supported by the ADREC. Instead, its duty is to remand the matter back to the agency for further consideration.[16]

Table 3.8 Factors to be Considered in Preparing and Maintaining the Administrative Record

- Maintaining accurate NEPA project files
- Making judgment calls about what documents to include in the record
- Identifying potential records and documents
- Capturing key elements of the NEPA process as part of the administrative record
- If applicable, submitting the administrative record to the court

3.4.5.3 Preparing and maintaining the ADREC

As just witnessed, a well-constructed ADREC enhances an agency's ability to defend its decision to the court and the public. In contrast, a poorly constructed or incomplete record increases the chances that the agency's decision will be overturned by a court. Table 3.8 lists some factors that should be considered in preparing and maintaining a legally sufficient ADREC.

The ADREC should be capable of demonstrating that the agency took a "hard look" at the proposal and was actively seeking alternative views. Many people are involved in shaping plans, assumptions, and internal decisions before the EIS reaches the decision maker; in the event of a legal challenge, this entire process, including all records, might be considered part of the agency's ADREC. Because NEPA is a "full disclosure" act, an outside party's access to the agency's ADREC is usually given considerable deference; that is, they can access all or a substantial part of it.

One should not lose sight of the fact that a judge is trained and works daily with conflicts; an ADREC that reveals a diversity of viewpoints can work in the agency's favor rather than to its detriment. A record that indicates a healthy debate within the agency's planning circles (diverse views, conflicts, evolving ideas) can actually be helpful in demonstrating that the agency was diligently struggling to reach a good decision. Lack of internal conflict may be indicative of a decision that was already made and that the EIS was prepared simply to justify.

3.4.6 The Federal Records Act and maintaining an ADREC

Pursuant to the US Federal Records Act, agencies must retain copies of any draft document showing a substantial change in the agency's thinking. Depending on the complexity of the EIS process, the agency should develop a formal *records management system* for capturing important aspects of the EIS process as part of its ADREC. A formal system can be a mechanism for demonstrating that some of NEPA's most fundamental requirements (i.e., "public," "open," "rigorous," "interdisciplinary," and "systematic" processes) were complied with.[18] The next section describes one such system for managing EIS records. Professional judgment must be exercised in determining which steps are most practically applied on the basis of the complexity and particular circumstances of the EIS process.

3.4.6.1 *Preparing and maintaining a records management system*

As appropriate, a records management system should be capable of reconstructing the investigation of the actions, alternatives, issues, and impacts considered in preparing the EIS. The investigation of both significant and nonsignificant impacts should be documented. At a minimum, a short description of who, when, and what was considered, and important conclusions may need to be recorded.

A point-of-contact, sometimes the EIS manager, should be assigned responsibility for maintaining the ADREC. The ADREC should be reviewed periodically to ensure that it can support and demonstrate a thorough and systematic planning process. All pertinent documents prepared for or used in the EIS planning process should be filed in a centralized storage area (see next section), such as an electronic database or specific filing cabinet(s). The agency should also maintain backup copies.

The EIS manager should ensure that there is a running chronology of events that occurred during the EIS process. Procedures may also need to be established to ensure that the IDT staff maintains records of important events, discussions, assumptions, direction, and conclusions used in their analysis. Public and interagency discussions should be formally documented. Important verbal communications and meeting minutes may likewise need to be recorded. The ADREC may also maintain relevant correspondence, such as official memos, meeting minutes, internal memorandums, public comments, and analysis reports. Discretion should be exercised in documenting meetings and conversations as they may be made publicly available either by a court or by way of the Freedom of Information Act.

Federal agencies are granted considerable discretion in terms of the specific procedures used in maintaining their ADRECs. These procedures vary greatly among agencies. Table 3.9 lists key issues that need to be considered in preparing and maintaining the ADREC.

3.4.6.2 *ARTS and COMTRACK database*

Where the ADREC is voluminous, the agency should consider developing a computerized Administrative Record Tracking System (ARTS) for tracking and managing important ADREC documents, including EIS deliverables, technical reports and studies, personal correspondence, etc. An electronic database has advantages such as providing the capability for quick searches and generating documents. Keyword searches can be performed.

As important issues are reviewed, the results can be documented and entered into this database. The database can maintain a brief summary of each review. The system can be used to demonstrate not only that pertinent impacts and issues were considered but also when, where, and by whom.

Table 3.9 Key Issues to Consider in Developing a Records Management System

- Will an electronic database be used to manage the system?
- Who will be tasked with maintaining the system?
- Are potential ADREC documents identified or segregated in some manner from other project files? If so, how is this being done?
- Are separate files being maintained by different agencies or entities? If so, who is responsible for maintaining key files?
- Who will identify and retain privileged materials?
- Is there a written protocol?
- What method is being used for archiving project-related e-mails? How will other electronic documents and data be stored (e.g., maps, modeling results, engineering drawings)?
- How are attachments handled? How are oversized documents (displays, maps, etc.) archived?
- What "checks" are in place to ensure that proper archiving and filing is maintained?
- What record-keeping requirements or policies must be considered?
- What system is used for "filing" documents?
- Are there any built-in gaps or omissions in the record-keeping system?
- Does an index exist for finding documents?
- Is there a central repository for maintaining e-mails? If not, how will e-mails be located and compiled?
- Where are the documents located (e.g., one central file or multiple files)?
- Will the ADREC records and documents be electronically scanned and incorporated into a database? If so, what technology (e.g., database) will be used? If not, what is the means of structuring the ADREC?
- How will the ADREC be produced for a court or other parties during litigation?

Where a project elicits a voluminous number of scoping or draft EIS comments, the agency may want to consider constructing a Comment Tracking (COMTRACK) database to manage, track, and capture comments and the agency's responses. This is particularly true for large or controversial EISs, where comments may easily run into the thousands. For instance, the Glen Canyon Dam EIS received more than 17,000 scoping comments from citizens in nearly all the 50 states.[19]

3.4.7 *Selecting an EIS contractor*

In a perfect world, an EIS is prepared directly by the responsible federal agency. In practice, this is often not the case; limitations in personnel, resources, and expertise frequently require that an agency contract some or most of their work to a commercial EIS contractor. If the agency obtains the services of an outside contractor, it should designate an *experienced* manager to oversee the effort from start to finish. The agency's EIS manager is responsible for making decisions, providing direction, ensuring the accuracy and adequacy of the analysis, and providing data to the contractor. The EIS manager is also responsible for ensuring that the contractor stays within the assigned scope of work.

3.4.7.1 Statement of work

As necessary, the agency's procurement officer should be contacted for assistance in preparing a statement of work (SOW). The SOW specifies the scope of work, including the specific expertise and resources that the outside contractor is expected to provide. The SOW clearly defines what is expected to be included in the draft and final EIS. An outline of the EIS should be included as part of the SOW. A schedule should also be included, specifying contractor milestones, as well as all important actions that the agency will be responsible for accomplishing.

3.4.7.2 Scheduling

When an EIS contractor will be used, the SOW should lay a foundation for ensuring that schedules are appropriately developed. The agency should provide a basic schedule in the SOW so that the contractor understands what is expected.

The SOW should require the contractor to submit a project management plan early in the process with a detailed schedule showing tasks, durations, specific staff assigned to each task, and potential conflicts. It is important to document these details and identify key assumptions that the schedule is based on. The agency may want to consider making an incentive fee award based in part on the contractor's adherence to the schedule (with exceptions for things that are beyond the control of the contractor).

3.4.7.3 Shopping for a contractor

It is important to obtain a contractor possessing knowledge and expertise with the EIS process and the potential environmental issues. The agency is required to ensure that the analysis is performed using an interdisciplinary approach. As such, the agency must look for a contractor possessing interdisciplinary expertise necessary to adequately evaluate the potential environmental impacts involved with the proposal. The agency therefore needs to verify that the contractor has specialists representing all appropriate environmental disciplines or has access to such expertise. The agency also verifies that the consultant has internal capability or access to any specialized analytical expertise, such as computer modeling, monitoring and laboratory capability, or other technical capabilities that will be required.

3.4.7.3.1 Choosing an EIS contractor. The agency's technical and procurement staff reviews the contractor's proposal in terms of the experience, qualifications, and availability of the professional staff that have been proposed to work on the analysis. Some EIS managers report that a "trick," known in the industry as "Bait-N-Switch," is occasionally used in the consulting world.[20] This occurs where very experienced or recognized experts are proposed in an effort to gain the EIS contract. On obtaining the contract, the company switches these individuals with less experienced staff, so that the experienced personnel can be used on other projects. To counter such tactics, the agency may want to add a stipulation that key individuals identified in the contract may not be substituted without prior written agency concurrence.[21]

3.4.7.3.1.1 Conflict of interest clause. The agency must verify there is no conflict of interest between the potential contractor and the proposal for which the EIS is being prepared. As part of this requirement, the agency must prepare a "disclosure statement" to be signed by the contractor indicating that they have no financial or other interests in the outcome of the project (§1506.5[c]).

Interpreting the conflict of interest provision is not always as straightforward as might first appear. For example, plaintiffs challenged an agency's EIS for a proposed highway interchange. The plaintiffs argued that the agency allowed a private contractor with a conflict of interest (expectation of future work based on the agency's unvarying practice of awarding the final design contract to the company preparing the EIS) to assist in preparing the EIS. They also alleged that the EIS contractor failed to execute the required conflict of interest disclosure statement until after the EIS had been filed. On reviewing the case, the court found that the EIS contractor had no contractual agreement or guarantee of future work at the time the EIS was prepared. It further ruled that the agency's oversight was sufficient to prevent defects that might arise from such a conflict. The court concluded that the ADREC demonstrated that the agency exercised proper managerial oversight, thoroughly reviewed all the contractor's data and analysis, and only used the contractor personnel for technical expertise.[22]

3.4.8 Data collection

It is not uncommon to find that the EIS schedule is not driven by the time required to prepare the statement, but instead by the time needed to compile data and complete field studies. To prevent delays, an early task must often be mounted to collect pertinent data for use in preparing the EIS. It is important to note that data collection is driven by the ultimate scope of the analysis, which may not be completely known until the formal scoping process has been completed. A literature search should be initiated as soon as practical to identify existing scientific studies, engineering reports, related NEPA analyses, and other technical support data.

3.4.8.1 Ensuring data accuracy

An agency is responsible for ensuring the accuracy of data used in the EIS. All data needs to be reviewed and verified to ensure it is accurate and up-to-date. Data should be checked to ensure that it conforms to accepted scientific standards. The following example illustrates the importance of this task. The Sierra Club claimed that the adequacy of data relied on for predicting population levels of various animal species was flawed in a US Forest Service EIS. On examining the case, the court concluded that the Forest Service had failed to respond to criticisms from two respected experts who objected to the use of old data in the analysis. This raised serious questions about the reliability of the environmental impact projections. In ruling against the Forest Service, the court concluded that use of these data was arbitrary and capricious, and therefore unacceptable.[23] This may be one reason the US Forest Service has led the nation in NEPA lawsuits; the Service has lost an excessively large percentage of its cases. Perhaps more telling is the fact that many of these cases involved similar types of deficiencies, errors, and mistakes. According to one NEPA manager with the Service, this is because the agency has expressed an "… arrogant attitude toward the public…" and has "… not learned the lessons of its past mistakes."[24]

3.4.8.2 Incomplete or unavailable data

Sometimes an agency simply lacks the means of acquiring important data. Where key information is *incomplete* or *unavailable*, the agency should consult the Regulations for guidance in dealing with this dilemma (§1502.22). Chapter 6, Section 6.3 of this book summarizes the regulatory requirements for dealing with incomplete or unavailable information.

Because of problems involved in collecting incomplete or unavailable information, agencies may find it helpful to apply a sliding-scale approach (see Chapter 2, Section 2.4) in which the level of effort expended in obtaining key information is commensurate with the potential significance of the impacts. Consequently, more effort, expense, and resources are expended in collecting or generating data where the effects are potentially very significant; conversely, less effort and resources are dedicated to issues deemed to be marginally significant.

An agency has a large degree of discretion in determining the data sources used in preparing the analysis. For example, in one case, a court agreed with an agency. Instead of collecting new data, the agency had used data from an EIS prepared for a previous lease agreement that had examined the direct and indirect impacts of an oil and gas development plan. The court found that the agency had made a reasoned judgment that this prior data was relevant to the new proposal and yielded a useful analysis of the possible cumulative impacts.[25] Thus, while an agency must take a "hard look" at environmental consequences, the EIS "need not be exhaustive to the point of discussing all possible details bearing on the proposed action."[26]

3.4.8.3 *Commonly required types of environmental and engineering data*

Table 3.10 lists types of data commonly needed for developing alternatives and engineering designs. Most data used in an assessment of impacts generally come from established sources (e.g., government agencies, universities, libraries, and other scientific studies). For example, agencies such as the US Fish and Wildlife Service have compiled data such as lists of species within given areas, while the US Geological Survey maintains databases on geological hazards and water resources. Much of these data have only limited public dissemination.

Table 3.11 lists types of data commonly drawn on in preparing the affected environment chapter, while Table 3.12 shows data commonly required in preparing the environmental consequences chapters of the

Table 3.10 Types of Engineering and Technical Data Commonly Needed in Preparing Descriptions of Alternatives

- Footprints (e.g., size of area, amount of disturbed land surface)
- Cost and schedule
- Description, schematics, diagrams, and artist's renderings of the proposal
- Description of any connected or related actions
- Infrastructure modifications or construction
- Potential mitigation measures
- Transportation routes
- Construction and operational personnel requirements
- Construction and operational resource usage for each alternative (e.g., electricity, water, diesel fuel, construction materials)

Table 3.11 Commonly Required Data on the Affected Environment

- Maps (topographic, geological, hydrological, biological)
- Geological surface and subsurface structures, well logs, and potential environmental hazards
- Surface water flow rates, quality, and potential for flooding
- Groundwater flow rates, quality, and aquifer characteristics
- Ecology and habitats
- Sensitive and endangered species
- Socioeconomic data (e.g., housing, infrastructure, traffic rates, minorities)
- Location of historical and archaeological sites
- Meteorological data
- Land use data and maps
- Ambient air quality data
- Ambient noise levels

Table 3.12 Commonly Required Source-Term Data for Assessing Environmental Consequences

- Surface water contaminate releases (e.g., quantities, rates, concentrations, and release points)
- Groundwater contaminate releases (e.g., quantities, rates, concentrations, and release points)
- Air (chemical, radiological, particulate) emissions (e.g., quantities, rates, concentrations, and release points)
- Noise generation
- Chemical or radioactive materials usage (e.g., volume, concentrations, activities)
- Hazardous, nonhazardous, and radioactive waste generation (e.g., types, mass, volume, concentrations)
- Potential accidents (e.g., probabilities, source-term releases)

Table 3.13 Common Types of Environmental Monitoring Studies

- Prevailing wind patterns
- Local air quality
- Stream flow and groundwater flows patterns, quality, and contaminant levels
- Traffic patterns
- Noise levels
- Cataloging indigenous species and habitat

EIS. Actual data requirements, of course, depend on the specific location, context, and nature of the proposal.

3.4.8.4 Collecting data through environmental monitoring

Some types of proposals require premonitoring and environmental sampling to supply baseline information for describing the affected environment and performing the impact analysis. In some cases, the EIS schedule may depend on monitoring environmental attributes such as wind patterns, which may require several years of monitoring. Because environmental monitoring can profoundly affect the EIS schedule, it is an important factor to consider early in the EIS process. Some common types of environmental monitoring are depicted in Table 3.13.

In some cases, environmental monitoring is one of the most expensive components of the EIS process. The EIS manager needs to ensure that sufficient funding is available to support such monitoring. Moreover, monitoring activities can affect the EIS schedule. For instance, some studies such as stream flows or biological studies may need to be performed during specific times of the year such as spring. The traffic volume to a winter recreational area may need to be performed during the winter months. In still other cases, monitoring studies may need to be performed over a

Table 3.14 Examples of Common Field Studies

- Inspections to determine soil characteristics
- Investigations to determine geological hazards
- Surface and groundwater studies to determine hydrologic characteristics
- Wetland studies
- Archaeological and cultural resource surveys

lengthy period, sufficient to account for monthly, seasonal, or even annual variations. The EIS schedule should be planned accordingly.

3.4.8.4.1 Field studies. Field studies may likewise need to be performed to collect data for the analysis. A representative list of studies is shown in Table 3.14.

This chapter has set the stage for Chapter 4, which describes the formal EIS process, beginning with public scoping. As we will see in Chapter 4, the NOI informs the public of the agency's intent to prepare an EIS and invites them to participate in the public scoping process.

3.5 *Summary*

This chapter outlined some of the key steps and tasks that may need to be considered and, if appropriate, performed as part of an agency's

Table 3.15 Checklist of Tasks That Should Be Considered and If Appropriate Performed as Part of an Agency's Prescoping Effort

1. Determine that an EIS must be prepared. If an EIS is required:
   - Designate an EIS manager.
   - Form an interdisciplinary team (IDT).
   - Identify and contact any cooperating agencies.
2. Hold a kickoff meeting:
   - Provide each IDT member with a description of their work assignments and tasks.
3. Begin preparing and circulating progress reports.
4. Initiate technical coordination meetings with other entities that continue throughout the EIS effort.
5. Standardize ground rules and procedures for preparing the EIS:
   - Prepare and circulate a document standards memo, which will be read and signed by all EIS team members.
6. Prepare and maintain up-to-date checklist outlining all important steps that must be conducted during the EIS process.
7. Identify potential decisions that may ultimately need to be considered by the decision maker.
8. Prepare a statement of the underlying purpose and need for taking action.

(continued)

Table 3.15 (Continued) Checklist of Tasks That Should Be Considered and If Appropriate Performed as Part of an Agency's Prescoping Effort

9. Conduct an on-site field trip to familiarize the IDT with the proposal, affected environment, and potential issues.
10. Begin identifying and integrating any applicable state or local environmental requirements with the EIS.
11. Identify and review any interim actions in terms of criteria specified in §1506.1:
    - Prepare an interim action justification for related actions that may need to proceed before the EIS process has been completed.
12. Begin identifying any nonfederal actions that may need to be included within the scope of the EIS.
13. Prepare draft NOI and distribution list.
14. Prepare a management action plan, which includes
    - Function roles and responsibilities matrix.
    - Annotated outline and schedule.
    - Budget, including allocation of funding.
    - Work breakdown structure.
    - Public involvement strategy.
    - Obtain the services of a public relations specialist (if appropriate).
    - Identify potential adversaries.
15. Develop scoping plan:
    - Prepare media distribution list.
    - Prepare notices and advertisements.
    - Develop an administrative record (ADREC) system.
    - Develop an administrative record tracking system (ARTS) and comment tracking (COMTRACK) database (if applicable).
16. Prepare and distribute environmental review forms to the IDT staff (if deemed appropriate).
17. Select an EIS contractor (if applicable):
    - Assign an agency official to oversee the EIS contractor.
    - Prepare statement of work.
    - "Shop" for an EIS contractor (if applicable).
    - Prepare a conflict of interest statement.
    - Select an EIS contractor (if applicable).
18. Data compilation and monitoring:
    - Perform literature search.
    - Establish procedure for ensuring data accuracy.
    - Identify incomplete or unavailable data in terms of requirements specified in §1502.22.
    - Conduct any applicable monitoring.
    - Conduct any applicable field studies.
19. Prepare notice of intent (NOI) (described in Chapter 4).

prescoping process. Professional judgment must be rendered in determining those items most applicable to existing circumstance. Factors as diverse as agency culture, the size and complexity of the planning effort funding and existing data, schedule constraints, and amount of experience the agency has will dictate which steps and tasks are most practical and appropriate.

Table 3.15 summarizes the steps and tasks described in this chapter. This table can assist practitioners in preparing a schedule and budget, and provides a "road map" for planning and managing the EIS prescoping effort. The reader should also note that items indicated in Table 3.15 are not necessarily performed in the same order as indicated.

PROBLEMS AND EXERCISES

1. Why can an EIS provide an early warning sign of trouble ahead?
2. What is an "interim action"?
3. When does a formal "proposal" exist?
4. What is an agency's administrative record?
5. What are three of the responsibilities of the EIS manager?
6. What factors are used in determining the lead agency?
7. An EIS is being prepared for a solar energy farm. The statement of purpose and need states that the underlying purpose and need (SPN) is, "The underlying purpose and need for taking action is to construct a 10-megawatt solar energy electrical generation station to provide power for the neighboring community." Do you believe this SPN is correctly written? If not, how would you rewrite it?

Notes

1. The National Environmental Policy Act of 1969, Section 102(c), as amended (Pub. L. 91-190, 42 U.S.C. 4321-4347, January 1, 1970, as amended by Pub. L. 94-52, July 3, 1975, Pub. L. 94-83, August 9, 1975, and Pub. L. 97-258, § 4(b), September 13, 1982).
2. *Webster's II New Riverside University Dictionary*. The River Publishing Company, Boston (1988).
3. MacLean R. *Meeting Madness: Meetings Consume Far Too Much Time; It's Time to Do Something About It*. Air & Waste Management Association (November 2012).
4. Schmidt O.L. "The statement of underlying need determines the range of alternatives in an environmental document," The Scientific Challenges of NEPA: Future Directions Based on 20 Years of Experience, Session 13– the NEPA Process, Knoxville, TN, October 25–27, 1989; Also "The statement of underlying need defines the range of alternatives in environmental documents," 18 Environmental Law 371-81 (1988).

5. *Webster's New Twentieth Century Dictionary Unabridged* (Second ed.). Simon & Schuster (1983).
6. *Webster's II New Riverside University Dictionary*. The River Publishing Company, Boston, (1988).
7. Eccleston C.H. *NEPA and Environmental Planning: Tools, Techniques, and Approaches for Practitioners*. CRC Press, Boca Raton, FL (2008).
8. 40 Code of Federal Regulations (CFR) 1500.2(c), 1500.4(k) and 1502.25.
9. 40 CFR 1506.2 (c).
10. Eccleston C.H. *NEPA and Environmental Planning: Tools, Techniques, and Approaches for Practitioners*. CRC Press, Boca Raton, FL (2008).
11. Crowley M. Enhanced public participation through social media. *NEPA Lessons Learned* (72) (September 5, 2012).
12. *Haynes v. United States*, 891 F.2d 235 (9th Cir. 1989).
13. *County of Suffolk v. Secretary of the Interior*, 562 F.2d 1368 (2nd Cir. 1977).
14. Administrative Procedure Act, 5 U.S.C. 706.
15. *Crotin v. Department of Agriculture*, 919 F.2d 439 (7th Cir. 1990).
16. *Asarco, Inc. v. EPA*, 616 F.2d 1153 (9th Cir. 1980); *Environmental Defense Fund, Inc. v. Castle*, 657 F.2d 275 (D.C. Cir. 1981).
17. *Animal Defense Council v. Hodel*, 840 F.2d 1432 (9th Cir. 1988).
18. 40 CFR 1501.2 (a) and 1502.14 (a).
19. Randle T. Case study: The Glen Canyon Dam EIS process. *Federal Facilities Environmental Journal* (Winter 1993–1994).
20. Presentation on NEPA consulting given at a CEQ sponsored seminar titled "Implementation of the National Environmental Policy Act on Federal Lands and Facilities," Duke University (April 25–29, 1994).
21. Personal communications, Mr. R.L. Colley and Mr. R.G. Upchurch, Waste Management Federal Services of Hanford Inc., Procurement Department and Contract Administration, respectively.
22. *Associations Working for Aurora's Residential Environment v. Colorado Department of Transportation*, 1998 U.S., App. (10th Cir. 1998), as reported in DOE's *Lessons Learned*, December 98, p13.
23. *Sierra Club v. U.S. Department of Agriculture*, 116 F.3d 1482 (7th Cir. May 28, 1997).
24. NEPA manager, US Forest Service (2011).
25. *Edwardsen v. US Department of the Interior*, 268 F.3d 781 (9th Cir. 2001).
26. *Vermont Public Interest Research Group v. U.S. Fish and Wildlife Service*, 33 ELR 20062 (D. Vt. 2002), citing *County of Suffolk v. Secretary of the Interior*, 562 F.2d 1368, 1375 (2d Cir. 1977).

chapter four

Preparing the EIS

The step-by-step process requirements

This chapter details the step-by-step procedure or process (i.e., process requirements) for preparing a draft and final environmental impact statement (EIS). All pertinent requirements that must be followed in the course of preparing an EIS are detailed, including all regulatory requirements cited in the National Environmental Policy Act (NEPA) implementing regulations (Regulations).[1]

This chapter also draws on lessons learned from the case study presented in Chapter 1; the intent is to help the reader avoid repeating similar mistakes. This chapter provides the basis for Chapter 5, which describes some of the general requirements, processes, and methods commonly employed for analyzing environmental impacts. Chapter 6 describes the actual regulatory requirements (documentation requirements) that the EIS document must meet.

A copy of the NEPA Regulations is provided in Appendix B.* A comprehensive checklist for preparing an EIS is provided in Appendix C.

4.1 *Learning objectives*

- General requirements, direction, and key concepts that underlie the EIS process
- EIS timing requirements and page lengths
- Issuing the notice of intent
- The formal EIS scoping process
- Preparing the draft and final EIS
- Filing and issuing the EIS with the EPA
- Circulating the EIS for public comment
- Postmonitoring and enforcement
- Making a referral to the Council on Environmental Quality
- Preparing a supplemental, legislative, or programmatic EIS

* Specific provisions referenced in the NEPA implementing regulations are abbreviated in this book so as to cite the specific "part" of the Regulations in which they are found. For example, a reference to a provision in "40 Code of Federal Regulations (CFR) 1501.1" is simply cited as "§1501.1."

4.2 General EIS direction and concepts

Mark Twain once advised:

> Plan for the future because that's where you are going to spend the rest of your life.

As we have seen, the EIS process provides a powerful tool for planning future federal actions (Figure 4.1). Chapter 2, Section 2.5 briefly introduced the EIS process. As we have seen, the principal purpose of an EIS is to serve as an "action-forcing" tool

> ... to help public officials make decisions that are based on understanding of environmental consequences and take actions that protect... the environment. (§1500.1[c])

As used in this book, the term "analyzed alternative" refers to an alternative that, in addition to simply being considered, is also evaluated in detail. The term "proposal" is used in referring to the *set* of reasonable alternatives in addition to the proposed action that are evaluated in the EIS. The term "EIS" is a generic expression used to describe a "draft," "final," "supplemental," "legislative," or "programmatic" EIS.

Figure 4.2 illustrates the typical EIS process. Each of these steps is detailed later in this chapter. The EIS process essentially begins at the

Figure 4.1 Logging operations can involve particularly severe impacts requiring preparation of an environmental impact statement. (Courtesy images.google.com.)

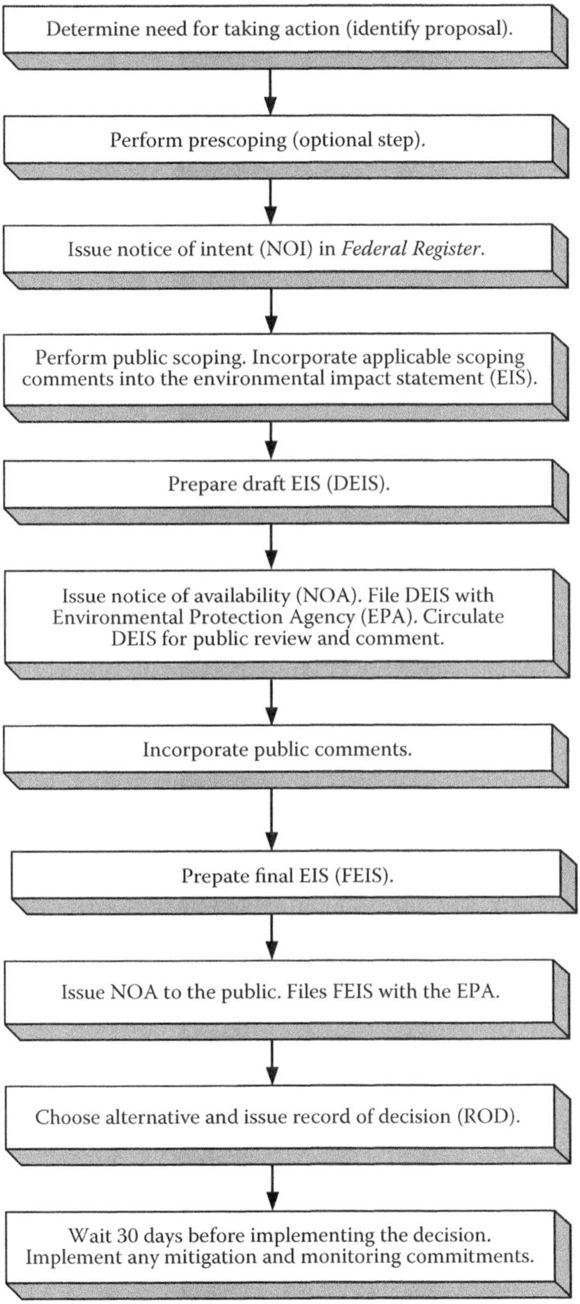

Figure 4.2 Principal steps in EIS process.

time an agency proposes to take action that may significantly affect the environment. The standard EIS process consists of two distinct phases:

1. Draft EIS
2. Final EIS

The following sections address general concepts that are applicable to all EISs.

4.2.1 "Proposal" versus "proposed action"

Project engineers are typically under pressure to ensure that their project is implemented on schedule and within budget. Experience shows that once a proposed action is identified, it often takes on a life of its own. The author refers to this as *project inertia*. As witnessed in the case study presented in Chapter 1, project inertia and instructional factors may bias the EIS analysis in "favor" of the agency's proposed action; this can lead to a situation in which the analysis of alternatives is not afforded treatment equivalent to that of the proposed action. This is a major problem that can lead to a flawed analysis, and worse yet, suboptimal decision making.

To counter project inertia and promote a more objective analysis, the author discourages the common practice of defining a "proposed action." Instead, a "proposal" consisting of a set of reasonable alternatives (analyzed alternatives) is investigated. By not formally defining a proposed action, it is easier to prepare an EIS that affords more equitable treatment to the set of all analyzed alternatives. The goal is to prepare a more objective analysis that identifies an alternative that best meets all relevant planning requirements and agency decision-making factors.* Where this approach is taken, the EIS simply explains that a proposed action was not defined in order to facilitate a more objective analysis of all reasonable alternatives. The following sections present some general time requirements and direction for preparing the EIS.

4.2.2 Timing requirements and page lengths

This section discusses EIS timing requirements, including when to begin the EIS and the duration of the EIS process.

* As explained later in this chapter, the agency must still identify a *preferred alternative*, but this step is done once the proposal (e.g., set of analyzed alternatives) has been thoroughly and impartially investigated.

4.2.2.1 When to begin preparation of the EIS

The Regulations require that preparation of the EIS begin during the early planning stage. Specifically, the preparation of the EIS must

- [Commence] ... as close as possible to the time the agency is developing or is presented with a proposal (§1508.23)
- [Be] ... prepared early enough so that it can serve practically as an important contribution to the decision-making process and will not be used to rationalize or justify decisions already made (§1500.2[c], §1501.2, and §1502.2)
- [Be started early enough so that it] ... can be completed in time for the final statement to be included in any recommendation or report on the proposal (§1502.5)

4.2.2.2 Maximum recommended duration for preparing an EIS

As specified in the Regulations, preparation of an EIS, even for large and complex projects, should normally require 12 months or less to complete; a programmatic EIS may require a somewhat longer period. The Council on Environmental Quality (CEQ) has stated that such a timeframe is within the planning cycle of most large projects.[2] This guidance is questionable and the CEQ offered no basis for this arbitrary period. In practice, the EIS process routinely exceeds this guideline.

4.2.2.3 All EIS timing limits

Table 4.1 presents all time limits prescribed in the Regulations for preparing and issuing an EIS.

4.2.3 Emergency situations and classified proposals

As detailed below, two circumstances require special deliberation:

1. Emergency situations
2. Classified proposals

4.2.3.1 Emergency situations

Occasionally, important circumstances arise in which an agency must respond to an "emergency" situation and does not have time to comply with NEPA's requirements. The NEPA regulations allow for emergency circumstances; the Regulations define specific procedures that are to be followed (§1506.11). The reader should note that this exclusion is interpreted narrowly. Nor does this provision necessarily exempt an agency from all compliance with NEPA. Agencies are expected to consult with the CEQ about making alternative arrangements for complying with NEPA's emergency exclusion clause.

Table 4.1 Compilation of All Time Limits and Periods Prescribed in NEPA Regulations for Preparing and Issuing an EIS

- No decision on a proposed action shall be made or recorded by a federal agency until the later of the following dates (§1506.10[b]):
  1. Ninety days after publication of the notice for a draft EIS
  2. Thirty days after publication of the notice for a final EIS.
- If the final EIS is filed within 90 days after the draft EIS is filed with the EPA, the minimum 30- and 90-day periods may run concurrently. However, subject to §1506.10[d], agencies shall allow not less than 45 days for comments on draft EISs (§1506.10[c]).
- After consultation with the lead agency, the EPA may extend or reduce prescribed regulatory periods. However, if the lead agency does not concur with the extension of time, the EPA may not extend it for more than 30 days (§1506.10[d]).
- If an agency circulates a summary of an EIS and receives a timely request for the entire statement and for additional time to comment, the time for that requestor shall be extended by at least 15 days beyond the minimum period (§1502.19[d]).
- If a draft EIS is to be considered at a public hearing, the agency should make the statement available to the public at least 15 days in advance (unless the purpose of the hearing is to provide information for the draft EIS) (§1506.6[c][2]).
- A legislative EIS may be transmitted to Congress up to 30 days later than its accompanying legislative proposal, to allow time for completion of an accurate statement that can serve as the basis for public and Congressional debate (§1506.8[a]).
- An EIS should normally require less than 1 year to complete.[2]

4.2.3.2 Classified proposals

Sometimes a proposal involves information that might jeopardize national security. Agencies are not required to publicly disclose classified information in an EIS. The Regulations define specific procedures for addressing circumstances involving classified information. These procedures apply only to information that has been properly classified under criteria established by executive order or statute in the interest of national defense or foreign policy. An agency's NEPA implementing procedures should also be consulted for agency-specific direction regarding the handling or exemption of classified information from public disclosure (§1507.3[c]).

It is important to note that this "exemption" does *not* absolve an agency from its responsibility to prepare and use the EIS in the agency's internal decision-making process.[3] Dissemination of any classified portions of the EIS can be restricted to appropriate decision makers and individuals according to established protocols. For example, the EIS may be organized in a way that separates classified from unclassified material. Unclassified portions of the document can be made available for public review while circulation of the classified portion is restricted within an appropriate branch of the agency in accordance with procedures for handling such material (§1507).

4.3 *Issuing the notice of intent*

As "soon as practicable" after deciding to prepare an EIS, the lead agency must publish a notice of intent (NOI) in the US *Federal Register* (§1501.7), which is depicted in Figure 4.3. A brief overview of the *Federal Register* is presented in the next section. Chapter 6 provides a suggested outline for the NOI. This *notice* formally notifies the public of the agency's intent to prepare an EIS and sets the stage for the formal EIS scoping process that follows. The NOI describes both the proposed action and possible alternatives (§1508.22).

In situations where a long lag time may exist between the decision to prepare the EIS and its actual preparation, an agency's NEPA implementing procedures may allow it to postpone publication of the NOI until a date is reached that still provides a reasonable time before work

Figure 4.3 The US *Federal Register.* (Courtesy US *Federal Register.*)

commences on preparation of the EIS (§1507.3[e]). If the proposal is later canceled, the agency should issue a notice of cancellation in the *Federal Register.*

If an agency fails to provide appropriate notice, that lapse may provide sufficient grounds for successfully challenging the EIS. However, a party does not necessarily have sufficient grounds to challenge an agency on this point alone *if* it had received notice by some other means.[4]

Chapter 3, Section 3.4 described the NEPA distribution list. Table 4.2 presents suggested groups to whom notices may be directed, and methods of notification. Where an action may be of national concern, notices must be mailed to national organizations that are expected to have an interest in the matter (§1506.6[a]).

The US Environmental Protection Agency's (EPA's) Office of Wetlands, Oceans, and Watersheds has issued the guidance document, *Community Culture and the Environment: A Guide to Understanding a Sense of Place.*[5] This guide, together with other relevant sources, provides tools for working with community groups to protect the environment.

4.3.1 *Federal Register*

The *Federal Register* is the official daily publication for notifying the public about proposed and adopted government legislation, regulations, rules, executive orders, and other notifications. Information on the availability of government documents, meeting schedules, and decisions is likewise published in the *Federal Register.* It is updated daily by 6 a.m. and is published Monday through Friday, except federal holidays. The *Federal Register* can be accessed at http://www.gpoaccess.gov/fr/index.html.

Table 4.2 Methods for Notifying Parties Where Effects Are Primarily of Local Interest

- State- and area-wide clearinghouses
- Consulting with Indian tribes whose lands or sites are of religious and cultural significance
- Local newspapers and other local media
- Interested community organizations and small business associations
- Newsletters and direct mailing to owners and occupants of nearby or affected properties
- Posting of notice on and off site in the area where the action is to be located

The EPA publishes a list of EISs that they have received from agencies each week and a summary of EIS ratings that have been assigned on the basis of EPA's review of the statement (described later). The easiest way to find NEPA notices is to have access to information such as the agency name, location of the action, and date or date ranges of the publication.

4.4 The formal scoping process

The term "scoping" is an expression used to describe one of the EIS public involvement steps (§1501.7). An agency begins the formal scoping process following publication of the NOI in the *Federal Register.* The concept of a formal public scoping process was one of the features that the public most strongly supported during a review of the draft Regulations in 1977.[6] A well-orchestrated scoping process provides a particularly effective means for focusing the EIS analysis on pertinent issues of true concern. As recommended in Chapter 3, a preliminary scoping process can be performed to gauge the scope of potential actions and impacts as accurately as possible before the formal public scoping process begins.

Each proposed action represents a unique set of circumstances. Not surprisingly, the level of public interest may vary greatly. A sliding-scale approach (see Chapter 2, Section 2.7) is recommended in determining the appropriate degree of public participation. This approach recognizes that the degree of public participation varies with the particular circumstances since some proposed actions, particularly those that are controversial, may necessitate a more extensive public participation effort.

It is important to note that commenting on a proposal is not a "vote" on whether the proposed action should take place. Nonetheless, information provided by the public can profoundly influence the scope of the EIS analysis.

4.4.1 Purpose and goals of scoping

The purpose of scoping is to solicit input from other agencies and the public so that the EIS analysis can be more clearly focused on issues of genuine concern. As noted in Table 4.3, the EIS scoping process is used to determine the range of actions, alternatives, and impacts that will be investigated in the EIS. As indicated in Table 4.3, the range of actions, alternatives, and impacts can each be broken down into three subelements. As shown in Table 4.4, the CEQ has provided supplemental direction listing specific goals that the scoping process should accomplish.[7]

Table 4.3 Range of Actions, Alternatives, and Impacts Determined through Scoping Process

- Actions:
  Connected actions
  Cumulative actions
  Similar actions
- Alternatives:
  No-action alternative
  Other reasonable courses of actions
  Mitigation measures (not in the proposed action)
- Impacts:
  Direct
  Indirect
  Cumulative

Table 4.4 Principal Goals of Scoping

- Ensure that all problems are identified early in the process and are properly studied
- Identify actions that will be examined
- Identify alternatives that will be examined
- Identify significant issues that need to be analyzed
- Identify public concerns
- Identify state and local agency requirements such as permits and land use restrictions
- Eliminate unimportant issues

4.4.1.1 Descoping

Attention is often overly focused on identifying the scope of actions, alternatives, and impact that will be analyzed. Every bit as important, however, is the use of this process to "de-emphasize insignificant issues" so as to "narrow the scope" of the EIS (1500.4[g], 1501.1[d], 1501.7[a][3]). In fact, as part of the scoping process, the lead agency is directed to

> Identify and <u>eliminate</u> from detailed study the issues which are <u>not significant</u> or which have been covered by prior environmental review (Sec. 1506.3), <u>narrowing</u> the discussion of these issues in the statement to a <u>brief presentation of why they will not have a significant effect</u> on the human environment or providing a reference to their coverage elsewhere (1501.7[a][3]).

The author refers to this task as *descoping*. In the author's experience, descoping does not receive sufficient emphasis and is one reason why so many EISs are often improperly focused, excessively lengthy and costly, and take so long to prepare.

4.4.2 *Exemptions to the EIS formal scoping requirement*

It is important to note that any prescoping activities performed before issuing the NOI cannot substitute for the normal scoping process that follows this notice. In the CEQ's opinion, the only exception involves cases where scoping is performed for a NEPA environmental assessment (EA) and *before* a decision has been made to proceed with an EIS. In this instance, the early scoping process may be substituted; this exception only applies if an earlier public notice was issued for the preparation of an EA, clearly indicating that the EA scoping process might be used to substitute for the later EIS scoping process. Once the EIS process begins, however, an NOI must still be issued, and it must state that written comments on the scope of alternatives and impacts will continue to be accepted and considered.[8]

4.4.2.1 *Supplemental and legislative EISs are exempt from formal scoping*

There are two circumstances where the formal public scoping process is not required for the preparation of an EIS (§1502.9[c][4]; §1506.8[b][1]). These circumstances involve the preparation of

- Supplemental EISs
- Legislative EISs

4.4.3 *Initiating the scoping process*

The Regulations do not mandate a specific duration or schedule for performing public scoping, nor do agencies have any obligation to extend a scoping period for public comment beyond the date originally set.[9] The lead agency must publicly identify any EA or other EIS under preparation that is related to the scope of the EIS being prepared. Other environmental review and consultation requirements must also be identified so that these requirements can be integrated with it (§1501.7[a]).

4.4.3.1 *Scoping information package*

Before beginning the formal scoping process, a public information packet should be prepared for public dissemination. The packet may explain the EIS process, emphasizing that the purpose of scoping is to elicit information and that a final decision regarding the course of action to be taken will not be made until the EIS has been completed. The information packet may include items listed in Table 4.5.

Table 4.5 Items Typically Included in Scoping Information Packet

- Description of the proposal, including applicable maps, figures, historical and background material, and other supporting material that will assist the public in understanding and providing comments on the proposal.
- Description of the potential environmental impacts and issues; this discussion should identify the potential environmental resources at risk.
- An invitation requesting interested parties to submit comments and recommendations with any supporting data and evidence.
- How and where comments may be submitted, including a name and contact information for the point-of-contact.
- Brief discussion and purpose of the EIS process with emphasis on:
  - The scoping process
  - Public participation objectives
  - Any other opportunities to provide comments and input, and the process for submitting those comments
- If sufficient information is available at this early stage, a draft schedule and outline for the EIS are also included.

4.4.4 *Performing the scoping process*

Specific steps and measures taken to satisfy the EIS scoping requirement are largely left to the discretion of individual agencies, as are the methods used to seek public input.[10] The agency may choose communication methods it deems best suited to informing the public (whether local, regional, or national) and for obtaining comments and input. Video conferencing, public meetings, conference calls, formal hearings, or informal workshops are all legitimate ways to conduct scoping.

4.4.4.1 *Public scoping meetings*

Scoping meetings provide a popular venue for seeking public ideas, comments, and input on the EIS (Figure 4.4). Under the Regulations, the lead agency is not required to hold public meetings or hearings unless the proposal is highly controversial or specifically requested to do so by another agency (§1506.6[c]). According to the Regulations, the following factors *may* indicate a need to conduct a public meeting or hearings (§1506.6[c]):

- When there is substantial environmental controversy concerning the proposed action
- When there is substantial interest in holding a hearing
- When a request for a hearing is made by another agency with jurisdiction over the action and supported by reasons as to why such a hearing would be useful

Figure 4.4 A public scoping meeting.

While formal scoping meetings may not be required, it is often considered good practice to hold at least one meeting. Many agencies have also adopted NEPA implementing procedures requiring public meetings or hearings. A public announcement regarding a public meeting/hearing should be made *at least* 15 days before any scoping meeting/hearing.

4.4.4.2 *Finalizing the scope of the EIS*

On completing the formal pubic scoping process, the EIS manager, in collaboration with the staff, is responsible for ensuring that all comments are reviewed. All public scoping comments, both written and verbal, must be considered and properly addressed. With respect to preparing the EIS, irrelevant or nonsignificant issues are either de-emphasized or dismissed; however, all public comments must be included in the EIS along with the agency's response (usually in an appendix).

4.4.4.2.1 Preparing EIS implementation plan. The CEQ encourages agencies to publish a "post-scoping document," which is sometimes referred by terms such as the "EIS implementation plan." Its purpose is to notify the public concerning the results of the EIS scoping process. Such a document

> … may be as brief as a list of impacts and alternatives selected for analysis; it may consist of the "scope of work" produced by the lead and cooperating agencies…; or it may be a special document that describes all the issues and explains why they were selected.[11]

If later challenged, the implementation plan can assist agency officials in demonstrating that "due diligence" was exercised in determining the appropriate scope of analysis. If an implementation plan is prepared, it should be issued as soon as possible upon completing the formal scoping process and should be publicly distributed to organizations and interested members of the public. The implementation plan should be prepared to assist the agency in meeting the following three objectives:

- Publicly record the results of the scoping process.
- Provide a compendium of all scoping comments received and the agency's response to these comments.
- Provide a plan, schedule, and outline for preparing the EIS; the plan should assign responsibilities to the lead and any cooperating agencies.

4.4.4.2.1.1 Contents of the implementation plan. On the basis of the results of the scoping process, the implementation plan describes the scope of the analysis, and outlines the proposal and alternatives, as well as mitigation measures, and environmental issues and impacts to be investigated. A proposed outline, page limits, assignment of responsibilities, and a schedule for the EIS may also be included. A generalized outline is suggested in Table 4.6. The actual outline and contents should, of course, be tailored to meet the particular circumstances.

Table 4.6 Suggested Outline for an EIS Implementation Plan

- Outline of the EIS
- Potentially significant impacts and issues to be investigated
- Nonsignificant issues that can either be de-emphasized or entirely eliminated
- Detailed outline of the EIS
- Description of the proposal (including alternatives and mitigation measures)
- Page limits for each section of the EIS
- A schedule and plan integrating the EIS with the agency's time limits for completing various phases of the EIS
- Identification of documents related to the proposal, with emphasis on other NEPA documents that can be used in tiering, or incorporated by reference into the EIS
- Responsibilities of the lead and cooperating agencies, and the EIS contractor (if one is used)
- Compendium of all scoping comments received and the agency's response to the comments

4.4.4.3 Creeping scope syndrome

Once the scope has been defined, the EIS manager is responsible for ensuring that the analysis of actions and alternatives remains within defined bounds. As new information or circumstances arise, the EIS manager is frequently confronted with requests to expand the scope of analysis. One of the most common mistakes that the author has witnessed has simply involved circumstances where an EIS manager allowed the scope of analysis to expand when it was not absolutely necessary. The author refers to this as the *creeping scope syndrome*. It has been responsible for many cost overruns and missed deadlines. What might at first appear to be only a slight increase in scope may result in substantially increased effort.

4.5 Consultation and identifying environmental regulatory requirements

As an environmental planning process, an agency identifies other environmental compliance and consultation requirements. As noted in Chapter 3, Section 3.3, the agency should begin thinking about this step during the initial prescoping phase. As appropriate, regulatory specialists should be consulted in an effort to identify related environmental permitting and compliance requirements that need to be scheduled concurrently with the EIS. As these requirements are identified, the lead agency should contact outside agencies to coordinate these requirements. Table 4.7 lists four items specifically called out in the Regulations as requirements to be integrated with the EIS effort (§1502.25[a]). There are many other consultation and environmental requirements that may also need to be integrated with the EIS process. For more information on consultations and integration of environmental requirements during the EIS process, the reader is directed to the companion book, *NEPA and Environmental Planning*.[12]

One should not fall into the trap of believing that compliance with other environmental laws or regulations will mitigate impacts to the point of nonsignificance or that the impacts will be acceptable. An action that fully complies with all other environmental laws or regulations can still pose a significant environmental toll, particularly from the standpoint of a cumulative impact.

Table 4.7 Environmental Statutes and Requirements That Are to Be Integrated with EIS Effort (§1502.25[a])

- Endangered Species Act of 1973 (16 U.S.C. 1531 et seq.)
- National Historic Preservation Act of 1966 (16 U.S.C. 470 et seq.)
- Fish and Wildlife Coordination Act (16 U.S.C. 661 et seq.)
- Other environmental review laws and executive orders

4.5.1 Endangered Species Act

Following in the footsteps of NEPA, Congress passed the Endangered Species Act of 1973 (ESA). Section 7 of the ESA requires federal agencies to prevent or modify any projects authorized, funded, or carried out by federal agencies that are[13]

> ... likely to jeopardize the continued existence of any endangered species or threatened species, or result in the destruction or adverse modification of critical habitat of such species.

The ESA is administered jointly by the Secretaries of Interior and Commerce:

- The US Fish and Wildlife Service (FWS) is responsible for terrestrial species.
- The US National Marine Fisheries Service (NMFS) is responsible for marine species, including anadromous (fish migrating upriver from the sea to spawn) species of fish.

The principal categories of species and habitats regulated under the ESA are

- *Candidate species*: This category includes plants and animals that have been studied and found to be at risk, and therefore may be proposed for addition to the federal endangered and threatened species list.
- *Threatened species*: An animal or plant species likely to become endangered within the foreseeable future throughout all or a significant portion of its range.
- *Endangered species*: An animal or plant species in danger of extinction throughout all or a significant portion of its range.
- *Critical habitat:* Those areas deemed necessary for the recovery of a species.

The ESA essentially forbids any government agency, corporation, or citizen from "taking" (i.e., harming or killing) endangered species without an *Endangered Species Permit*. In addition to federal actions, the ESA also affects private land use. Penalties for violating the ESA can be as serious as a $50,000 fine and up to 1 year in jail. The ESA also contains a citizen enforcement clause allowing citizens and scientists to sue the government either to obtain listing for a species with dwindling numbers or to comply with the law.

4.5.1.1 Section 7 consultation

Section 7, one of the most important provisions of the ESA, requires all federal agencies to *consult* with the FWS or NMFS if they are proposing an "action" that may affect listed species or their designated habitat.[14] The term "action" is defined broadly to include funding, permitting, and other regulatory actions. This consultation requirement is commonly referred to as the "Section 7 Consultation Process." This consultation process may involve both informal and formal dialogues, as well as preparing a biological assessment and obtaining expert agency opinions. Federal agencies must comply with the following three Section 7 requirements:

1. Perform (if applicable) a formal consultation regarding the potential impacts to species/habitat.
2. Prepare (if warranted) a biological assessment on such proposals (described below).
3. Obtain a permit before monitoring, capturing, killing, or performing other scientific studies on threatened or endangered species.

Federal agencies must review actions they undertake or enable to determine whether they may affect an endangered species or its habitat. If this examination reveals a potential for adverse effects, the federal agency must consult with the FWS or NMFS. Consultation is carried out for the purpose of identifying whether a federal action is likely to jeopardize the continued existence of the threatened or endangered species or adversely affect its critical habitat.

4.5.1.1.1 Informal and formal consultation. Many proposals having potential to adversely affect a listed species can be effectively dealt with through *informal consultation* during the early EIS planning process. Formal consultation may be avoided if project design changes can be made to mitigate the adverse impacts. If the FWS or NMFS determines that a proposed action is unlikely to adversely affect a listed species, further consultation is normally not required.

A *formal consultation* process is normally initiated if it is determined that the proposal could adversely affect a listed species or its critical habitat.

4.5.1.2 The Biological Evaluation and Biological Assessment

If a listed species is unlikely to be adversely affected and formal consultation is not anticipated, a *Biological Evaluation (BE)* is prepared, providing the basis for making a determination during informal consultation. The BE documents an agency's rationale and conclusions regarding the effects of their proposed action.

Where a designated critical habitat, or a threatened or endangered (T&E) species is in the area of the proposed action, a *Biological Assessment (BA)* may be prepared to evaluate potential effects of the project on the species or habitat. The BA is typically coordinated and prepared in conjunction with the agency's EIS process. As appropriate, alternatives and mitigation measures may need to be investigated for avoiding or reducing potential impacts to listed species and their critical habitat. The BA describes the

- Proposed project
- Project area
- Proposed management activities
- Listed species that may occur in the project area (including past surveys for such species)
- How the project may affect listed species or critical habitat (direct, indirect, and cumulative effects)
- Measures for avoiding, reducing, or eliminating adverse effects

A *Biological Opinion* is normally issued within 45 days of concluding the formal consultation period. It states the opinion of the FWS or NMFS as to whether the impacts (including cumulative impacts) of the federal proposal are likely to jeopardize the continued existence of a listed species or result in disruption of critical habitat.

4.5.1.3 Section 9

Under Section 9 of the ESA, it is illegal to "take" any endangered species. The team "take" includes the killing, harming, harassing, or capturing of a threatened or endangered species. This requirement also safeguards critical habitats. An *Endangered Species Permit* can be issued by the FWS or NMFS, allowing an action to go forward that would otherwise be prohibited under Section 9.

4.5.2 National Historic Preservation Act

Similar to NEPA, the National Historic Preservation Act (NHPA) announces a national policy of encouraging preservation of prehistoric and historic resources.[15] While the NHPA does not mandate preservation of such resources, its Section 106 review provision requires federal agencies to take into account the effect of their actions on "historic properties."

The Advisory Council on Historic Preservation (ACHP or "Council") has promulgated regulations for implementing the NHPA, which emphasizes consultation with State and Tribal Historic Preservation Offices (SHPO, THPO), Native American tribes, Native Hawaiian groups, local communities, and other concerned parties about federal decisions that may impact historic properties.[16]

4.5.2.1 The SHPO and THPO

The NHPA encourages each state to designate a State Historic Preservation Officer (SHPO) to administer the state's historic preservation program. In practice, the SHPO actually refers to the staff rather than a single individual. The professional staff of each SHPO office has expertise in history, archaeology, and historic preservation. The THPO is the equivalent of an SHPO in some Indian tribal governments.

4.5.2.2 National Register of Historic Places

The National Register of Historic Places (NRHP) is a list maintained by the National Park Service (NPS) of districts, sites, buildings, structures, and objects deemed by the NPS and SHPO or THPO to be significant in American history, architecture, archaeology, engineering, and culture. Any property listed or eligible for listing in the NRHP is considered to be "historic." Such properties may include archaeological and historical sites, historic buildings or structures, objects (e.g., monuments), and "districts"—properties made up of multiple entities. Landscapes, including wholly natural landscapes, can be eligible for the NRHP if they have historic or cultural significance. It is important to note that an impact on a historic property or other cultural resource is a factor to be considered in assessing the significance of environmental impact (§1508.27b[3]).

4.5.2.3 Section 106 review

As indicated above, Section 106 of the NHPA requires federal agencies to take into account the effect of their actions on "historic properties." The scope of the Section 106 review applies to anything a federal agency plans to do, help someone else do, or permit someone else to do, provided it represents a type of action with the potential to affect historic properties. This does not mean that the agency needs to know that the historic properties would be affected, only that the action *might* affect them (e.g., demolition, earth moving, changes in land use, etc.).

Table 4.8 briefly outlines the Section 106 process. These steps are elaborated in the following section.[17]

The term, "Section 106 review" refers to the review of a project's impacts under Section 106 and the ACHP's regulations. This review is performed on projects planned by federal agencies or involving federal assistance or permits.[18] The Section 106 review considers potential impacts both on places included in the NRHP and on places not listed but that may meet the NRHP eligibility criteria (set forth at 36 CFR §60.4). The ACHP oversees the Section 106 review process with assistance from the relevant SHPO and/or THPO.

Table 4.8 How Section 106 Review Works

1. The federal agency *initiates consultation* with the state and/or Indian tribes, and possibly others interested in the *undertaking* (action) and its possible effects on known or unknown *historic properties;* these entities are called *consulting parties.*
2. With the consulting parties, the agency determines the *scope* of what it needs to do to identify historic properties and determine how they may be affected. One important aspect of this historic scoping process involves determining the area of potential effects (APE)—the area or areas where the undertaking may affect historic properties.
3. The agency undertakes the *identification* of historic places within the APE, usually involving surveys and other kinds of studies, in consultation with the consulting parties.
4. The agency determines whether places in the APE are listed in or eligible for listing in the *National Register of Historic Places,* in consultation with the SHPO/THPO and any other consulting parties.
5. The agency determines whether the proposed action will have *adverse effects* on historic properties, using criteria defined in the ACHP regulations. If not, it proposes a *determination of no adverse effect* for concurrence by the SHPO/THPO and other consulting parties.
6. If there will be an adverse effect upon a historic property, or if the SHPO/THPO does not agree with a determination of no adverse effect, the consulting parties consult to find ways to resolve the adverse effect. This usually leads to a Memorandum of Agreement (MOA), spelling out terms and conditions that the agency is responsible for carrying out.
7. If an MOA is not reached, the ACHP *comments* to the head of the federal agency, who *considers the comments* in deciding whether and how to carry out or approve the action, but need not follow them.

4.5.2.3.1 Consultation. The heart of a Section 106 review is its *consultation* requirement, which requires consultations between the federal agency, the SHPO or THPO, tribes, local governments, and other interested parties. Consultation usually occurs at several points in the process:

1. During the EIS scoping process
2. During the EIS assessment stage when determining whether the project may affect properties either listed or eligible for listing in the NRHP
3. When seeking ways to resolve or mitigate effects that are adverse

Where adverse effects are involved, the end point of Section 106 consultation is usually an agreement among the consulting parties about how the effects will be resolved. If the parties agree, they generally execute a Memorandum of Agreement (MOA), or where an entire program or a complex staged project is involved, a Programmatic Agreement (PA).

An MOA or a PA outlines measures the agency agrees to take to avoid, minimize, or otherwise mitigate adverse effects. In rare instances, agreement is not reached and the ACHP renders a formal comment to the responsible agency head, who must then consider the comment and make a final decision about the project; the agency head need not follow the ACHP's recommendation.

4.5.2.3.2 Integrating Section 106 review with the NEPA and EIS process. While Section 106 review is a completely separate authority from NEPA, the coordination of Section 106 studies with the EIS is strongly encouraged. Integrated coordination tasks generally involve

1. Coordinated scoping
2. Coordinated public involvement
3. Conducting historic property identification and assessing potential effects during preparation of the draft EIS
4. Coordinating consultation on historic property identification, impact assessment, and adverse effect resolution with public review and comment on the draft EIS
5. Including the results of this consultation in the final EIS
6. Referencing any executed MOA, programmatic agreement, or ACHP comment in the record of decision (ROD)

NEPA requires consideration of effects on cultural resources other than historic properties (such as a community's cultural uses of the land and natural resources), as well as effects on historic properties. The interdisciplinary team (IDT) consideration of affected cultural resources should be coordinated with Section 106 review, in consultation with potentially affected Indian tribes and other groups.

4.5.3 Clean Water Act

The Clean Water Act of 1972 extends the definition of waters of the United States to include tributaries to navigable waters, interstate wetlands, wetlands that could affect interstate or foreign commerce, and wetlands adjacent to other waters of the United States.[19]

4.5.3.1 Wetlands

A *wetlands* is defined as an area inundated or saturated by surface water or groundwater at a frequency and duration sufficient to support, and that under normal circumstances does support, a prevalence of vegetation typically adapted to life in saturated soil conditions. Effects on wetlands is an important factor cited in the CEQ regulations for determining the significance of an impact (§1508.27[b][3]).

4.5.3.2 Section 401 water quality certification

Under Section 401 of the Clean Water Act, applicants for a federal license to conduct an activity that might result in a discharge into navigable waters are required to provide the licensing agency with a certification from the state. This certification states that the discharge will comply with applicable Clean Water Act[20] requirements. This may necessitate the application and approval of a National Pollutant Discharge Elimination System (NPDES) permit.

4.5.3.3 Section 404

Section 404 of the Clean Water Act establishes a program to regulate the discharge of *dredged and fills* material into US waters, including wetlands. The Army Corps of Engineers (Corps) and Environmental Protection Agency jointly administer this program. The Corps is responsible for the day-to-day administration and permit review, while the EPA provides program oversight. Under the 404 program, no discharge of dredged or fill material is allowed if

- A practicable alternative exists that is less damaging to the aquatic environment
- The nation's waters would be significantly degraded

4.5.3.3.1 The 404 permitting process. A federal permit is required to discharge dredged or fill material into wetlands and other waters of the United States. As part of the permitting process, the Corps evaluation also includes a review for compliance with NEPA.

No discharge is normally permitted if it would contribute to a significant degradation of wetlands by adversely impacting wildlife, ecosystem integrity, or social amenities such as aesthetics. Neglecting to obtain a permit or comply with the terms of a permit can result in civil and/or criminal penalties. The 404 permitting process involves the following basic steps:

1. The Corps issues a public notice. This notice describes the permit application, including the proposed activity, potential environmental impacts, and location. The public notice invites comments within a specified time.
2. After receiving public comments, the application and comments are reviewed by the Corps and other interested federal and state agencies, organizations, and individuals. The Corps determines whether an EIS is necessary.
3. The Corps evaluates the permit application based on the public comments and its own evaluation. A *statement of finding* is publicly issued explaining how the permit decision was made.

4.5.3.4 Floodplain and wetlands

Executive Orders 11988 and 11990 provide for the protection of floodplains and wetlands, respectively.[20,21] Both executive orders require federal agencies to consider impacts of their actions on floodplains and wetlands through existing review procedures such as NEPA.

Wetlands can be identified by consulting the

- Army Corps of Engineers
- US Fish and Wildlife Service's national wetlands inventory
- Wetlands specialists and federal agency specialists
- State and local wetland inventory databases, land use plans, maps, and inventories
- US Geological Survey topographical maps
- US Department of Agriculture, Natural Resources Conservation Service local soil identification maps and databases

4.5.3.5 Coastal zone management

The Coastal Zone Management Act of 1972[22] established a national program for the beneficial use, protection, and development of the nation's coastal zone.[23] It requires a federal permit to conduct an activity that could affect a state's coastal zone. The applicant must certify to the licensing agency that the proposed activity would be consistent with the state's federally approved coastal zone management program.[24]

4.5.3.5.1 Consistency determination. The National Oceanic and Atmospheric Administration's (NOAA) Coastal Zone Management Act (CZMA) Consistency Regulation[25] regulates activities that can affect a coastal zone. The CZMA requires that all federally conducted or enabled activities affecting coastal zones be undertaken, to the maximum extent practicable, in a manner consistent with approved state coastal management programs.

Any federal agency activity (regardless of location) is subject to the consistency requirement if that activity will affect any natural resources, land uses, or water uses in a coastal zone. The applicant must certify to the licensing agency that the proposed activity would be consistent with the state's federally approved coastal zone management program.[24] If the agency believes there are no reasonably foreseeable effects, it can issue a *negative determination* (i.e., that there are no coastal zone impacts).[26]

Federal agencies may choose, but are not required, to address consistency requirements in an EIS. If a federal agency chooses to include its consistency determination or negative determination in an EIS, the statement must include information necessary to support the determination.

4.6 Preparing the draft EIS

This section describes the step-by-step process for preparing the draft EIS. Many of the techniques useful in managing and preparing the EIS were presented in Chapter 3. It is important to note that there is no standard procedure or approach for preparing an EIS. Regardless of the approach used, the Regulations require that a "systematic" and "interdisciplinary" methodology be used.

Input obtained from the prescoping (Chapter 3) and formal public scoping process is used in determining which staff members will need to participate in the EIS analysis. The respective disciplines will reflect the scope and issues identified during the scoping process (§1502.6). It is important to note that the draft EIS (DEIS) is expected to conform to the scope agreed upon during the agency's formal scoping process.

4.6.1 Preparing the EIS

Before starting actual work on the EIS, agreement is reached regarding the technical approach to be used. A project plan/schedule shows which specialists are assigned to prepare each section of the EIS. The IDT uses input received from the public scoping and consultation process in conjunction with other data in preparing the analysis. Chapter 5 describes the process, including tools, techniques, and methods for performing the analysis.

The specialists are given their assignments and the EIS manager tracks their progress. The specialists submit their respective sections to the EIS manager for review. As necessary, these sections are returned to specialists with comments. The specialists revise and resubmit their sections. This is often an iterative process. The completed sections are incorporated into the EIS document. The EIS manager frequently writes various sections of the EIS that do not specifically fall within the domain or expertise of the IDT staff; for example, the manager may write introductory material and the statement of purpose and need. At a minimum, the completed DEIS will contain the sections shown in Table 4.9 (§1502.10). After the DEIS has been prepared, it is circulated for internal review and approval.

4.6.1.1 Maintaining the EIS schedule

Parkinson's law states, "Work expands so as to fill the time available for its completion." It goes without saying that developing and maintaining a schedule is one of the most important responsibilities of the EIS manager. As detailed in Chapter 3, a schedule should show what work is to be done, who will do the work, and when it will be completed. The status of schedule milestones should be reviewed on a weekly, if not a daily, basis.

Table 4.9 Outline and Format of EIS

1.	Cover sheet
2.	Summary
3.	Table of contents
4.	Purpose of and need for the proposed action
5.	Alternatives including the proposed action
6.	Affected environment
7.	Environmental consequences
8.	List of preparers
9.	List of agencies, organizations, and persons to whom copies of the EIS are sent
10.	Index
11.	Appendices (if any)

An EIS schedule sometimes has to be revised multiple times. An initial schedule may need to be revised as new data needs are identified, environmental issues are discovered, and public comments are received. Some commonly cited reasons for revising the EIS schedule include the following:

- Results of engineering studies and field surveys indicate that an alternative needs to be modified or a new alternative needs to be included in analysis.
- Cooperating agencies are overworked, underfunded, or their priorities and schedule do not coincide with the EIS schedule.
- Public comments have been received that require additional analysis or revision to the EIS.

4.6.1.2 Obtaining data

Obtaining data necessary for performing the analysis is frequently one of the most daunting problems that an agency faces in preparing an EIS. Fortunately, there are a number of modern tools that facilitate data collection.

4.6.1.2.1 Online mapping tools. A geographic information system (GIS) offers rapid and cost-effective analysis of complex data and allows analysts to perform "what-if" scenarios in developing alternatives. Over the last decade, costs have come down while the availability of quality data has gone up. The proliferation of geospatial data on the web makes it simpler than ever to find information and put it to use right away. Sources of publicly available geospatial data useful in preparing an EIS analysis are identified below.

4.6.1.2.2 Data sources. Half a million geospatial datasets can be accessed through https://catalog.data.gov/dataset. This database is part of http://www.data.gov,[27] which provides an alternative method for finding geospatial datasets, as well as many other types of data from federal agencies. For example, energy.data.gov provides information on historic energy use by the federal government and a database of active and pending carbon capture and storage projects worldwide, including technology type, project cost, and schedule.

Files located at geo.data.gov can be freely downloaded for use in GIS software. Each set of files identifies the date of the data and the agency that made the data available. A sample of the useful datasets includes

- Boundaries for federal resource areas
- Soil surveys
- Sole source aquifers
- Parks, refuges, and forests
- Wind speed data
- National wetlands inventory
- Critical habitat for threatened and endangered species
- Census data

Data from this website can be located through keyword search or by browsing lists organized by content type and topic. Also provided on this website are links to applications, such as live map servers that allow one to view the mapped data on the web.

4.6.1.2.3 NEPAssist and EJView. The EPA maintains a web-based GIS tool called NEPAssist. NEPAssist draws information from publicly available federal, state, and local datasets. It allows the IDT, stakeholders, and the public to view information about environmental conditions within a given area of interest quickly and easily at early stages of project development.

One simply logs onto the NEPAssist website and selects a study area. This brings up a map of the selected location. Users can measure distances between points on the map, add custom data (e.g., labels), and generate reports. Users can add various attributes and data layers such as

- Air and water quality
- Water features
- Hazardous waste
- Demographics (schools, hospitals, demographics, administrative and political boundaries)
- Other environmental topics

The EPA also hosts EJView, which was formerly known as the Environmental Justice Geographic Assessment Tool. This GIS allows users to search for a specific facility or area to view a map or report. Its user interface is similar to that for NEPAssist. EJView includes

- Health-related data (e.g., risk of certain health outcomes)
- Neighborhood boundaries
- Information on community-based EPA grants

4.6.1.3 *Keeping the public informed of important changes*

Agency plans may change during the course of preparing an EIS. These changes may involve a decision to suspend, reactivate, cancel, or make major changes in the scope of the EIS. Keeping regulators, cooperating agencies, and the general public informed of the status and schedule is an important objective in maintaining public trust. When an EIS experiences an important change, the public should be notified. Additional public involvement also may be warranted. For example, substantial changes to the proposal may call for additional public scoping.

Sometimes, the EIS schedule may need to be significantly extended. Stakeholders may need to be notified of significant schedule changes. A brief announcement in the *Federal Register* is appropriate to inform the public. Additional announcements may also need to be made in local media, the agency's websites, and via mail or e-mail to interested parties.

4.6.2 *Internal agency review*

Internal review of the preliminary drafts is an important quality assurance measure for maintaining the integrity of the EIS. On its completion, the draft EIS is distributed for management and peer review. This is frequently an iterative process. The EIS manager should therefore not arbitrarily assume that one draft of an EIS will be sufficient. Numerous working versions of the document may need to be circulated for internal review before public release.

It is recommended that copies for internal review be stamped with a label "draft predecisional" or some equivalent designation. Likewise, a date and revision number should be included so that comments can be correctly tied to the appropriate working version of the statement. Line numbers should also be included to facilitate comment review and incorporation. Finally, the EIS manager should be mindful that failure to allow for any public clearance procedures (and appropriate classification review) has been responsible for more than one missed deadline.

The author recommends that personnel not directly involved with preparation of the EIS be included among the reviewers; including reviewers not personally involved with the proposal helps ensure that the final document can be readily understood by non-project personnel.

Environmental impacts and issues are a function of the specific actions that would take place. Yet, reviewers sometimes proceed directly to the section related to their expertise, skipping the affected environment, description of the proposal, and other relevant topics. Lacking this background, reviewers are unlikely to fully comprehend the scope of potential actions or assumptions on which the analysis is based; the reviewers may fail to identify problems or environmental issues that have been overlooked. Accordingly, as practical, reviewers should be instructed to read *all* relevant sections of the EIS.

4.7 *Filing the DEIS with the EPA*

The draft EIS must

> … fulfill and satisfy to the **fullest extent possible** the requirements established for final statements in Section 102(2)(C) of the Act (1502.9[a]).

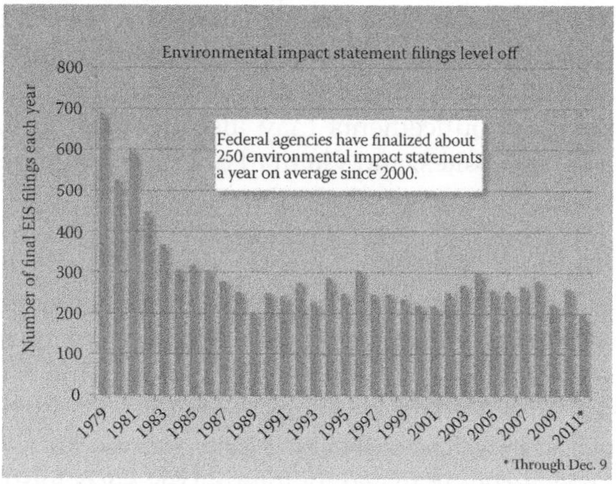

This is an important federal obligation, and most agencies make a diligent effort to comply with its mandate. Unfortunately, some federal officials are more concerned with meeting project schedules than ensuring the completeness and accuracy of their EISs. Recall the case study presented in Chapter 1 involving EISs prepared by the US Nuclear Regulatory Commission for renewing the operating licenses of the nation's fleet of aging nuclear reactors. In one instance, both the project manager and EIS contractor warned that there were deficiencies and inaccuracies in the DEIS and that it was not ready to be publically circulated. Despite these forewarnings, Mr. Pham directed that the DEIS be issued to the public anyway and that the errors would be cleaned up in the final EIS. As indicated in Table 1.1, this is a textbook example of an agency official "sacrificing quality for schedule."

4.7.1 The filing process and public notification

The public must be notified that the DEIS is available for review. Once the DEIS has been reviewed and approved, it needs to be filed with the EPA and circulated for public review and comment. Material referenced in the DEIS also needs to be made available for public review. Where practical, these documents should be provided to the public either free of charge or at a fee that covers only their actual reproduction costs.

4.7.1.1 Public review period requirements

The DEIS is formally filed with the EPA and circulated for public review and comment. The public is notified that the EIS has been circulated for public comment through the EIS filing process. As detailed below, each week, the *Federal Register* lists all EISs filed during the preceding week (§1506.10[a]). All filing and review periods are calculated from the day after EPA's notice of availability (NOA) appears in the *Federal Register.* The EIS filing period and the public comment period begin *after* the DEIS has been transmitted to the EPA and public. To guarantee that the EIS review time meets regulatory requirements, agencies should ensure that any "comment due by" dates in their public notices are based on the date of publication of EPA's NOA in the *Federal Register* (Figure 4.5).

The EPA also should be notified of all situations where an agency has decided to withdraw, delay, or reopen a review period on an EIS. Such notices are published in the *Federal Register.* The following section describes the EPA filing process.

4.7.1.2 EPA's filing responsibilities

In 2011, the EPA issued revised guidelines for filing EISs.[28] This section describes the revised filing process. Once EPA has received the EIS, the statement is given an official filing date and is checked for completeness and compliance with requirements set forth in §1502.10. If the EIS is not "complete" (i.e., if the EIS does not contain or comply with the requirements set forth in §1502.10 of the Regulations), EPA will contact the lead agency to resolve the problem; this is done before publication of the NOA in the *Federal Register.* The EPA maintains a database known as COMDATE, which provides a weekly computerized report, listing all EISs filed during the previous week (§1506.9).

In accordance with requirements set forth in §1506.9 and §1506.10, the EPA is responsible for administering the EIS filing process. The EPA's role in the EIS filing process includes

Federal Register / Vol. 73, No. 8 / Friday, January 11, 2008 / Notices 2023

Commission's Rules of Practice and Procedure.

(G) The Secretary is directed to publish a copy of this order in the **Federal Register**.

(H) The refund effective date in Docket No. EL08–8–000 established pursuant to section 206(b) of the Federal Power Act is 5 months from the date of the filing of the complaint.

By the Commission.

Kimberly D. Bose,
Secretary.

[FR Doc. E8–301 Filed 1–10–08; 8:45 am]

BILLING CODE 6717–01–P

DEPARTMENT OF ENERGY

National Nuclear Security Administration

Draft Complex Transformation Supplemental Programmatic Environmental Impact Statement

AGENCY: National Nuclear Security Administration, U.S. Department of Energy.

ACTION: Notice of Availability and Public Hearings.

SUMMARY: The National Nuclear Security Administration (NNSA), a semi-autonomous agency within the U.S. Department of Energy (DOE), announces the availability of the Draft Complex Transformation Supplemental Programmatic Environmental Impact Statement (Draft Complex Transformation SPEIS, DOE/EIS–0236–S4). The Draft Complex Transformation SPEIS analyzes the potential environmental impacts of reasonable alternatives to continue the transformation of the U.S. nuclear weapons complex to one that is smaller, more efficient, more secure, and better able to respond to changes in national security requirements. While NNSA has revised the document title from that indicated in the Notice of Intent, it remains a supplement to the Stockpile Stewardship and Management Programmatic Environmental Impact Statement. NNSA has prepared this document in accordance with the National Environmental Policy Act (NEPA), the Council on Environmental Quality (CEQ) regulations that implement the procedural provisions of NEPA (40 CFR Parts 1500–1508), and DOE procedures implementing NEPA (10 CFR Part 1021).

DATES: NNSA invites comments on the Draft Complex Transformation SPEIS during the 90-day public comment period, which ends on April 10, 2008. NNSA will consider comments received after this date to the extent practicable as it prepares the Final Complex Transformation SPEIS. NNSA will hold 19 public hearings on the Draft Complex Transformation SPEIS. The locations, dates, and times are listed in the **SUPPLEMENTARY INFORMATION** section.

ADDRESSES: Requests for additional information on the Draft Complex Transformation SPEIS, including requests for copies of the document, should be directed to: Mr. Theodore A. Wyka, Complex Transformation SPEIS Document Manager, Office of Transformation, NA–10.1, Department of Energy/NNSA, 1000 Independence Avenue, SW., Washington, DC 20585, toll free 1–800–832–0885 ext. 63519. Written comments on the Draft Complex Transformation SPEIS should be submitted to the above address, by facsimile to 1–703–931–9222, or by e-mail to *complextransformation@nnsa.doe.gov*. Please mark correspondence "Draft Complex Transformation SPEIS Comments."

For general information regarding the DOE NEPA process contact: Ms. Carol M. Borgstrom, Director, Office of NEPA Policy and Compliance, GC–20, U.S. Department of Energy, 1000 Independence Avenue, SW., Washington, DC 20585, telephone 202–586–4600, or leave a message at 1–800–472–2756. Additional information regarding DOE NEPA activities and access to many of DOE's NEPA documents are available on the Internet through the DOE NEPA Web site at *http://www.eh.doe.gov/nepa.*

SUPPLEMENTARY INFORMATION: *Public Hearings and Invitation to Comment.* NNSA will hold 19 public hearings on the Draft Complex Transformation SPEIS. The hearings will be held at the following locations, dates, and times:

North Augusta, South Carolina, North Augusta Community Center, 495 Brookside Avenue, North Augusta, SC, Thursday, February 21, 2008 (11 a.m.–3 p.m. and 6 p.m.–10 p.m.)
Oak Ridge, Tennessee, New Hope Center, 602 Scarboro Road (Corner of New Hope and Scarboro Roads), Oak Ridge, TN, Tuesday, February 26, 2008 (11 a.m.–3 p.m. and 6 p.m.–10 p.m.)
Amarillo, Texas, Amarillo Globe-News Center, Education Room, 401 S. Buchanan, Amarillo, TX, Thursday, February 28, 2008 (11 a.m.–3 p.m. and 6 p.m.–10 p.m.)
Tonopah, Nevada, Tonopah Convention Center, 301 Brougher Avenue, Tonopah, NV, Tuesday, March 4, 2008 (6 p.m.–10 p.m.)
Las Vegas, Nevada, Atomic Testing Museum, 755 E. Flamingo Road, Las Vegas, NV, Thursday, March 6, 2008 (11 a.m.–3 p.m. and 6 p.m.–10 p.m.)
Socorro, New Mexico, Macey Center (at New Mexico Tech), 801 Leroy Place, Socorro, NM, Monday, March 10, 2008 (6 p.m.–10 p.m.)
Albuquerque, New Mexico, Albuquerque Convention Center, 401 2nd Street NW, Albuquerque, NM, Tuesday, March 11, 2008 (11 a.m.–3 p.m. and 6 p.m.–10 p.m.)
Los Alamos, New Mexico, Hilltop House, 400 Trinity Drive at Central, Los Alamos, NM, Wednesday, March 12, 2008 (6 p.m.–10 p.m.)
Los Alamos, New Mexico, Hilltop House, 400 Trinity Drive at Central, Los Alamos, NM, Thursday, March 13, 2008 (11 a.m.–3 p.m.)
Santa Fe, New Mexico, Genoveva Chavez Community Center, 3221 Rodeo Road, Santa Fe, NM, Thursday, March 13, 2008 (6 p.m.–10 p.m.)
Tracy, California, Holiday Inn Express, 3751 N. Tracy Blvd., Tracy, CA, Tuesday, March 18, 2008 (6 p.m.–10 p.m.)
Livermore, California, Robert Livermore Community Center, 4444 East Avenue, Livermore, CA, Wednesday, March 19, 2008 (11 a.m.–3 p.m. and 6 p.m.–10 p.m.)
Washington, DC, Forrestal Building, 1000 Independence Ave, SW., Washington, DC, Tuesday, March 25, 2008 (11 a.m.–3 p.m.)

Individuals who would like to present comments orally at these hearings must register upon arrival at the hearing. NNSA will allot three to five minutes, depending upon the number of speakers, to each individual wishing to speak so as to ensure that as many people as possible have the opportunity to speak. More time may be allotted by the hearing moderator as circumstances allow. NNSA officials will be available to discuss the Draft Complex Transformation SPEIS and answer questions during the first hour. NNSA will then hold a plenary session at each public hearing in which officials will explain the Draft Complex Transformation SPEIS and the analyses in it. Following the plenary session, the public will have an opportunity to provide oral and written comments. Oral comments from the hearings and written comments submitted during the comment period will be considered by NNSA in preparing the Final Complex Transformation SPEIS.

The Draft Complex Transformation SPEIS and additional information regarding complex transformation are available on the Internet at *http://www.ComplexTransformationSPEIS.com* and *http://www.nnsa.doe.gov*. The Draft

Figure 4.5 Example of a notice of availability for an EIS. (Courtesy US *Federal Register*, Vol. 73, No. 8/Friday, January 11, 2008/notices pg 2023.)

- Receiving and recording EISs
- Documenting the beginning and ending dates for comment and review periods for draft and final EISs, respectively
- Publishing these dates in a weekly NOA in the *Federal Register*
- Retaining the EISs in a central repository
- Determining whether time periods can be lengthened or shortened for "compelling reasons of national policy"

4.7.1.3 Filing EISs electronically

Direction for filing EISs is presented in the Regulations (§1506.9). In 2012, the EPA issued new direction for filing EISs.[29] These changes pertain to EPA's EIS Filing System Guidelines, previously issued in 1989. The EPA no longer accepts paper copies or CDs of EISs for filing purposes. Federal agencies must now file draft and final EISs electronically by submitting the complete EIS, including appendices, to the EPA through the *e-NEPA* electronic filing system. Electronic filing eliminates the need to prepare an EIS filing letter and enables the EPA to host EISs on its website. To sign up for *e-NEPA*, simply register for an account at https://cdx.epa.gov/epa_home.asp.

The draft or final EIS must be submitted in Adobe Acrobat format (.pdf) with the attributes indicated in Table 4.10. Additional instructions can be found on the EPA's website.[30]

The EPA's amended EIS Filing System Guidelines also address existing procedures related to adopting an EIS, and the withdrawal, delay, or reopening of EIS review periods. The following clarifications are provided for filing an EIS electronically:

- Use the EIS title as the file name if submitting the EIS as a single file. Otherwise, name each file using the chapter or subchapter number, followed by its name.

Table 4.10 Requirements for Filing EISs Electronically Using *e-NEPA*

File attributes
• Files are optimized (file size reduced)
• Document text is searchable
• Chapters are bookmarked
• Bookmark view is shown when file is opened
• Metadata are included; use "Document Summary" and enter data into "Subject," "Author," and "Keywords" fields
Required information[a]
• EIS title
• EIS type (i.e., draft EIS, final EIS)
• File size for EIS and appendices (MB)
• Number of pages for each file
• Lead agency(ies)
• Lead agency contact (name, phone number, and e-mail)
• Any cooperating agency(ies) (including federal and other)
• Length of comment period (days)

[a] The reader should check for any changes or new EPA requirements.

- The EIS must be filed using Adobe Acrobat (.pdf) format. An EIS may be divided into multiple files not greater than 50 MB each. Use Acrobat's "Reduce File Size" option to compress the files.
- Files must be searchable. Most Acrobat files, other than scanned documents, are searchable by default. Bookmark EIS chapters and subchapters (bookmark view should be displayed on opening the file).
- Enter metadata into "Document Properties" (title, subject, author, and keywords). Use the EIS title for both the title and subject fields, and the agency name in the author's field.

4.7.2 *Publication of the notice of availability*

Each week the EPA prepares a report listing all EISs filed during the preceding week. This report includes

- Title of the EIS
- EIS status (draft, final, supplemental)
- Date filed with EPA
- An EIS accession number
- State/county of the action that prompted the EIS
- Date comments are due and the agency's contact information
- Agency that filed the EIS

4.7.2.1 *Filing date*

The NOA is published in the *Federal Register* on Friday of the week following EPA's receipt of the draft EIS (§1506.10[a]). The EPA publishes the agency's NOA in the *Federal Register* containing the aforementioned information according to the schedule shown in Figure 4.6. On publication of the NOA, the EPA includes this information in its EIS status report to the CEQ; delivery of the EIS to the CEQ satisfies the requirement of making the EIS available to the president of the United States (§1504.1[c]).

The EIS filing date is defined as the date on which EPA publishes the NOA in the *Federal Register*, not the date that the document is transmitted to the EPA or the date on which it is received by the EPA. Thus, the

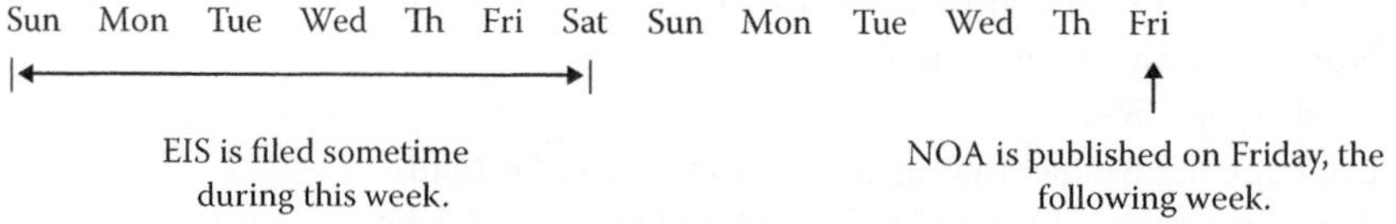

Figure 4.6 Sequence in which NOA is published. This figure shows the publication process relative to date on which EIS is filed with EPA.

minimum EIS commenting and waiting period for the draft and final EIS are both calculated from the date the NOA is published in the *Federal Register* (§1506.10[b], [c], and [d]).

4.7.2.2 Minimum EIS review and waiting periods

The DEIS review period runs for a minimum period of 45 calendar days. Determine the minimum comment period closing date by simply adding 45 days to the date that EPA publishes the NOA.

Of course, the lead agency may set a period longer than the minimum 45-day review period. If a calculated time period ends on a nonworking day, the end date is the next working day (i.e., time periods will not end on weekends or federal holidays). In practice, lead agencies commonly extend comment review and waiting periods beyond the prescribed minimum timing requirements just described.

4.7.2.2.1 Exemptions and minor violations in timing requirements. There is an exception to this timing rule (§1505.10[b]). The EPA has the authority to both extend and reduce the time periods on draft and final EISs based on a demonstration of "compelling reasons of national policy" (§1506.10[d]). If the lead agency does not concur with the extension, the EPA may not extend a prescribed period by more than 30 days. The EPA must notify the CEQ if it reduces or extends any period of time (§1506.10[d]). The reader should note that a party's failure to provide timely comments on review of the DEIS is not considered sufficient reason for extending a prescribed period by the EPA. The CEQ also has the authority to approve alternative procedures for preparing, circulating, and filing supplemental EISs (§1502.9[c][4]).

All EIS regulatory timing requirements are to be strictly adhered to. Nevertheless, an innocent error does not necessarily provide sufficient cause for a successful legal challenge. In one case, an agency was challenged over an irregularity in which an EIS notification was published in the *Federal Register* on the day the EIS was circulated rather than during the following week as specified in the Regulations. A plaintiff challenged this error in court. The court ruled that this minor violation did not affect the ability of parties to review the EIS for the mandatory period following the publication of the NOA; thus, it did not, by itself, constitute sufficient grounds for challenging the EIS.[31]

4.7.3 EPA's EIS repository

Before the new electronic filing system was introduced, EISs filed with the EPA were archived at the Office of Federal Activities for a period of 2 years and made available to office staff only. After 2 years, these EISs were sent to the National Records Center. The EPA Library currently maintains a microfiche collection of final EISs filed between 1970 and 1977, and paper copies of all EISs filed between 1978 and 1990. These microfiches are available through interlibrary loans. The library can be contacted at

Environmental Protection Agency Library
Headquarters Library
EPA West Building
Constitution Avenue and 14th Street, NW, Room 3340
Washington, DC
202-566-0556

One of the largest collections of EISs is available from Northwestern University's Transportation Library. Nearly all of the EISs issued since 1969 are held in both draft and final form.

The Cambridge Scientific Abstracts (CSA) is also a privately owned information company that, for a fee, publishes abstracts and indexes to scientific and technical research literature. It can provide detailed abstracts of EISs published from 1987 to present. Copies of EISs may be obtained by contacting

CSA
Edward J. Reid
Editor, EIS: Digest of Environmental Impact Statements
7200 Wisconsin Avenue—Suite 601
Bethesda, MD 20814
301-961-6742

4.8 Circulating the draft EIS for public comment

At the same time that the completed DEIS is filed with the EPA, it is also circulated to the public, including interested stakeholders and organizations, for comment. An agency may want to notify the public by circulating an advertisement such as the one depicted in Figure 4.7. Most agencies are now circulating their EISs to the public via electronic media. Stakeholders are increasingly indicating that they prefer to receive a web address for accessing the EIS as opposed to receiving a paper copy or compact disk.

As just indicated, the DEIS must be circulated for a minimum comment review period of at least 45 days (§1506.10[c]). Where the DEIS is to be considered in a public hearing or meeting, the EIS manager should make it available to the public at least 15 days before that hearing or meeting; however, this 15-day requirement does not apply in cases where the purpose of the hearing is simply to provide information for the EIS (§1506.6[c][2]).

Figure 4.7 Example of an advertisement notifying the public that an EIS is available for review. (Modified from the World War I Posters Collection, Prints & Photographs Division, Library of Congress, LC-USZC4-3859.)

If the DEIS is found to be "so inadequate as to preclude meaningful analysis" during the comment review period, the agency manager must prepare and recirculate a revised draft (§1506.10); if appropriate, the agency may limit preparation and circulation only to the portion of the EIS that is determined to be inadequate (§1502.9[a]).

4.8.1 Tips for minimizing EIS printing and distribution costs

Agency officials have expressed concerns over the costs of printing and distributing EISs and other NEPA documents. One review found that printing and distribution costs can vary dramatically. For example, a single hard copy of a recent project-specific EIS cost $16; a site-wide EIS, $55; and a very large EIS, several hundred dollars.[32] Because of the expense incurred and the advantages of electronic media, the author recommends that agency officials promote CDs or online distribution of EISs (download EIS via a website) over distributing hard copies. Table 4.11 offers tips for minimizing printing and distribution costs.

Table 4.11 Tips for Minimizing EIS Printing and Distribution Costs

• Prepare an EIS distribution strategy that minimizes the number of printed hard copies of a complete EIS.
• Minimize the use of color maps and figures to the extent feasible; color printing can enhance effective communication but can also have the drawback of significantly increasing printing costs for paper copies.
• Identify stakeholder distribution preferences early by mailing a postcard, sending an e-mail, or providing a form at a public scoping meeting.
• Consider offering stakeholders and the public the following EIS distribution options: a. Printed summary of the EIS b. Printed summary and the complete EIS on CD/DVD c. Complete printed copy of the EIS d. Notification of the EIS's availability online
• Reconfirm stakeholder distribution preferences before distributing a draft and final EIS. In these inquiries, include a statement identifying the default distribution if no response is provided. For example, if stakeholders do not respond to the initial inquiry, then they will receive a subsequent postcard or notification listing the locations of reading rooms that contain a printed copy of the EIS and the website address where the EIS can be downloaded online.
• Factor adequate printing time into the EIS schedule to avoid having to pay higher printing costs for last-minute rush jobs.

4.8.2 *Inviting comments on the DEIS*

As noted earlier, an EIS is a public process and agencies must actively seek public comments on their draft statements. To reduce paperwork, the lead agency should request commentators to be as specific as possible in their remarks (§1503.2, §1503.3).

The EIS manager needs to be mindful that comments can sometimes be voluminous. Consider the experience of the US Forest Service, which prepared a highly controversial national EIS on its Roadless Area Conservation Program for 160 National Forests and Grasslands. The Forest Service received scathing comments and was harshly criticized for what was termed "a destructive and arrogant proposal." Public participation activities for the EIS included about 450 public scoping meetings and hearings. In its scoping process, the Forest Service received more than 517,000 letters, cards, and other submittals containing well over 1 million comments! Form letters and postcard campaigns accounted for about 481,000 of the submitted items.[33] During a 60-day DEIS public comment period, the Forest Service estimated that it received more than 1 million letters, cards, and other items, including about 60,000 individually written letters—6,000 of them from local, state, and federal agencies. The Forest Service assigned 95 full-time staff members to analyze these comments. Such a severe public outcry might have been avoided if the Service had paid more attention to stakeholder's environmental concerns when it first entertained the proposal.

4.8.3 Parties that the agency must seek comments from

Table 4.12 lists public and private parties from which an agency must obtain or at least *request* comments (§1503.1).

Any agency with jurisdiction by law, having special expertise, or authorized to develop and enforce environmental standards is expected to comment on EISs that fall within its jurisdiction, experience, or authority (§1503.2). In some cases, an agency may simply reply that it has no comments.

4.8.4 Circulating a summary

At a minimum, the draft and final EIS should be circulated to the parties indicated in Table 4.13. Where a draft has appendices, it may be circulated to the public without them. Nevertheless, they must be made available upon request (§1502.9). Where a draft or final EIS is "unusually long," the agency may circulate a summary in lieu of the entire EIS (§1500.4[h] and §1502.19); however, the entire EIS must be circulated to the parties indicated in Table 4.13. With the advent of electronic media

Table 4.12 Entities from Which the Agency Is Required to Seek Comments From

- Any federal agency that has jurisdiction by law or special expertise with respect to any environmental impact involved, or that has authorization to develop and enforce environmental standards
- The public
- Any agency that has requested statements on actions of the kind proposed
- Indian tribes, when the effects may be on Indian land or sites of religious or cultural significance to the tribe
- Appropriate state and local agencies authorized to develop and enforce environmental standards
- Applicant (if any)

Table 4.13 Parties to which a Copy of the Entire EIS Must Be Circulated

- The project applicant (if any)
- Any federal agency having jurisdiction by law or special expertise in the environmental impact as well as federal, state, or local agencies that are authorized to develop and enforce environmental standards
- In the case of a final EIS, any person, organization, or agency that submitted substantive comments on the draft
- Any person, organization, or agency requesting the entire EIS

distribution, there is little excuse for not making the entire statement available to the public.

If a summary is circulated and the agency receives a request within a short period of time for the entire statement and for additional time to comment, an extension of at least 15 days beyond the minimum review period must be granted for that requestor (§1502.19[d]).

4.8.5 EPA's Section 309 review

The EPA plays an important role in instilling quality and integrity into the EIS process. Section 309 of the Clean Air Act directs EPA to review and comment on all DEISs. This is commonly referred to as a *Section 309 review*. Fogleman writes[34]:

> Whereas plaintiffs in NEPA lawsuits may be precluded from challenging an agency's substantive decision to proceed with an action, Section 309 expressly grants the EPA administrator the power to comment on the substantive decision, to publish the decision, and to refer the matter to CEQ. Once a matter is referred to the CEQ, agencies tend to accept the CEQ's suggestions or to reach an agreement with the EPA. Thus the EPA has considerable power to ensure that environmentally destructive actions do not proceed. The power is increased substantially if the CEQ agrees with the EPA's comments. This power is largely undeveloped.

The EPA's review comments are made public. If the lead agency fails to make sufficient revisions and the final EIS rating remains "environmentally unsatisfactory," the EPA may refer the matter to the President's Council on Environmental Quality for mediation.

4.8.6 EPA's review

The Section 309 review authority is delegated to EPA's Office of Federal Activities (OFA). This office is required to review and provide comments on the adequacy of the analysis and environment impacts on every draft or final EIS filed with the EPA.[35] As outlined below, the OFA has developed a manual prescribing duties, procedures, and responsibilities for performing the 309 review.[36]

The EPA is responsible for working with the lead agency to resolve comments and any outstanding issues. EPA program offices are responsible for providing technical assistance and policy guidance on reviewing actions related to their areas of responsibility. Each EPA regional office is responsible for carrying out the review process for the proposed federal actions affecting its region. EPA regional offices designate a regional Environmental Review Coordinator who has overall management responsibility for the review process in that region.

4.8.6.1 EPA principal reviewer

An EPA principal reviewer (PR) is designated to coordinate the 309 review for each EIS it receives. The PR prepares a comment letter on the EIS and the proposed federal action. An associate reviewer (AR) may also be assigned to this review. The AR is a person designated to provide technical advice in specific areas and to provide the views of the office in which the AR is located. The EPA tracks this review and the status of the EIS using its publicly available computer database known as COMDATE.

If the agency's preferred alternative is found to be unacceptable, the EPA is required to refer the matter to the CEQ (§1504.1[b]). Other agencies may also make their own review of the EIS. The results of these reviews must be made available to the president, the CEQ, and the public (§1504.1[c]).

4.8.7 EPA's rating system

Once EPA's review of the DEIS has been completed, the PR rates the statement according to the alphanumeric system described below. The designated rating will be cited in a comment letter to the lead agency. Unless an alternate review period is agreed on, the EPA's review comments must be provided to the lead agency within the standard 45-day public review period.

This rating system synthesizes EPA's overall assessment of the EIS and its effect on environmental quality. To the extent possible, assignment of the alphanumeric rating is based on the overall environmental impact of the proposed action, including any impacts that are not adequately addressed in the DEIS. This review rating is normally focused primarily on the agency's preferred alternative identified in the DEIS. The EPA is also expected to comment on specific mitigation measures as well as any actions that may lead to a possible violation of environmental standards. The PR may also rate individual alternatives if

- A preferred alternative is not identified
- There is reason to believe that the preferred alternative may be changed at a later stage
- The preferred alternative has significant problems that could be avoided by selection of another alternative

4.8.7.1 Alphanumeric rating system

EPA uses an alphanumeric rating system to rate the

1. Environmental impact of the action
2. Technical adequacy of the DEIS

As indicated in item 1 above, the DEIS is reviewed to determine the severity of the potential environmental impacts of the agency's preferred alternative. The preferred alternative is rated according to whether the environmental impacts are considered acceptable or unacceptable. As indicated in Table 4.14, the preferred alternative is assigned an environmental rating according to one of four alphabetical categories: LO, EC, EO, or EU.

Each alphabetic rating is also assigned a numeric rating (i.e., 1, 2, or 3) according to the adequacy of the draft document. Together they form an alphanumeric rating system, shown in Table 4.14.

For instance, suppose an EIS receives a rating of "EC-2." This rating indicates that EPA has some "environmental concerns" regarding the action and that the EIS has some shortcomings (e.g., "insufficient information"). It is important to note that most DEISs have some problems that prevent them from receiving a rating of "LO-1."

4.8.7.2 Deficient proposals and EISs

For categories EO, EU, or 3 (Table 4.14), the lead agency will be notified about EPA's concerns before the receipt of EPA's comment letter. For categories EU and 3, the environmental review coordinator must attempt to meet with the lead agency to discuss EPA's concerns before issuing the comment letter. Often the meeting is accomplished through a teleconference. The purpose of this meeting is to

- Discuss EPA's concerns and ways to resolve those concerns
- Become aware of any ongoing lead agency actions that might resolve EPA's concerns
- Ensure that the EPA review has correctly interpreted the proposal and supporting information

Table 4.14 EPA's Alphanumeric System for Rating DEIS

Environmental rating	Adequacy
LO (lack of objections)	1 (Adequate)
EC (environmental concerns)	2 (Insufficient information)
EO (environmental objections)	3 (Inadequate)
EU (environmentally unsatisfactory)	

To ensure objectivity, EPA's comment letter and its alphanumeric rating are not subject to negotiation and will not be changed on the basis of the meeting unless errors are revealed in EPA's understanding of the issues.

4.8.8 EPA's review of the final EIS

The EPA also reviews the final EIS to determine whether the statement adequately resolves any problems identified in EPA's review of the DEIS. A detailed review and submission of comments on the final EIS is performed for proposals that were rated EO, EU, or 3 during the draft stage.

If the EPA administrator determines that the final EIS is still unsatisfactory from an environmental standpoint, EPA is *required* by law to refer (referral) the matter to the CEQ (§1504.1). Any action such as a *referral* to the CEQ is only made after every effort has been taken to resolve the issue(s) with the lead agency. This referral process is described in more detail shortly. Although the EPA does not have authority to halt a project, based on the EIS, the alphanumeric rating may well provide political leverage for bringing it into serious question.

4.8.8.1 Focus of the review

Review of the final EIS is generally focused on major or unresolved issues, and is centered on the impacts rather than on the adequacy of the statement. Normally, EPA's review and comments on the final EIS are limited to issues raised in EPA's comments on the DEIS that have not been adequately resolved, as well as any new and potentially significant impacts that have been identified as a result of information made available after publication of the draft.

It is important to note that the alphanumeric rating denoted in Table 4.14 is not assigned to the final EIS. In its place, EPA prepares narrative comments of its review that generally focus on the impacts and any unresolved issues. As appropriate, the PR will ensure that

1. The EPA receives a copy of the agency's record of decision.
2. The lead agency has incorporated into the record of decision all agreed upon mitigation and other impact reduction measures.
3. As appropriate, the lead agency has included all agreed upon measures as conditions in grants, permits, or other approvals.

4.8.9 EPA monitoring and follow-up

The PR also reviews the record of decision for final EISs on which

- The EPA expressed environmental objections
- The EPA has negotiated mitigation measures or changes in the project design

As appropriate, EPA also performs a follow-up to ensure that

1. Any agreed upon mitigation measures are fully implemented (e.g., permit conditions, operating stipulations).
2. Any agreed upon mitigation measures are identified and spelled out in the record of decision.
3. The EPA participates as fully as possible in any post-EIS efforts to assist agency decision making.

4.9 *Preparing the final EIS*

It is important to note that if the draft EIS is found to be "so inadequate as to preclude meaningful analysis," the agency must revise and recirculate a draft of the appropriate section(s) (§1502.9[a]). The agency is also directed to prepare a supplemental EIS to a draft or final EIS if (§1502.9[c])

- The agency makes substantial changes in the proposed action, relevant to environmental concerns.
- There are significant new circumstances or information relevant to environmental concerns and bearing on the proposed action or its impacts.

The EIS manager must transmit an entire copy of the FEIS to any person, organization, or agency that has submitted substantial comments (§1502.19). Where changes are minor, however, the EIS manager may attach and circulate only changes to the DEIS, in lieu of rewriting and circulating the entire document (§1500.4[m]). The author recommends that discretion be exercised when considering the appropriateness of such an approach.

4.9.1 *Reviewing and responding to public comments on the DEIS*

The EIS manager collects and reviews all comments received from public review of the DEIS.

In preparing the final EIS (FEIS), the EIS manager must consider comments received from public review of the DEIS, both individually and collectively. Frequently referred to as *comment analysis,* the FEIS must review, analyze, and respond to substantive comments on the draft (§1503.4[a] and [b]). Comment analysis can be a complex and daunting effort.

4.9.1.1 *Considering and assessing comments*

The majority of public comments received on the DEIS tend to be received from concerned citizens or groups opposed to the proposal.

It is recommended that public comments be submitted though a single point of contact responsible for coordinating and addressing comments. Similarly, a single point of contact, usually the EIS manager, should be in charge of coordinating agency responses to the public comments.

The IDT considers and responds to comments both individually and collectively. All substantive comments received on the DEIS are attached to the FEIS regardless of whether the comment is considered to merit individual discussion in the statement. Where a comment is exceptionally voluminous, the agency may summarize the comment (§1503.4[b]). Comments received on the DEIS are most often attached in an appendix.

Where comments will be grouped into categories (particularly where a proposal involves complex or voluminous comments), the EIS manager is often challenged in terms of identifying and defining a manageable set of comment categories. Compounding this problem is the fact that lengthy comments frequently transverse a range of diverse issues, problems, and complaints.

4.9.1.1.1 Managing comments. Where comments are voluminous or the EIS is at high risk of a legal challenge, the agency may find it advantageous to capture public comments and responses in a computer database. A computer database, like the COMTRACK database described in Chapter 3, provides the advantage of automating the tracking and management of comments. Parameters such as the commenter's name, date of comment, and either the entire comment or a summary thereof, and the agency's response can be captured in such a database.

4.9.1.2 Responding to comments

In addition to including all public comments received on the DEIS, the FEIS must respond to all substantive comments regardless of whether the comment is deemed to merit individual discussion (§1502.9, §1503.4). If a comment is deemed inappropriate or incorrect, the agency must explain, citing specific sources or reasons, why the comment has not been accepted. For example, if a public comment correctly indicates that impact on an endangered species was not correctly investigated, the agency's response should be that the FEIS analysis has been revised to address this error; if, on the other hand, the agency believes this public comment was in error, the appropriate response would be to explain why the agency believes this comment is incorrect.

An agency is not required to respond at length to vague comments such as a general complaint that "the EIS is inadequate." However, it must provide an adequate response to specific comments. For instance, if a comment correctly indicates that a summary of a groundwater flow model was not included in the EIS, the EIS manager needs to prepare and include a summary of that model as part of the response.[37] The comment response would indicate that this was done.

While comment resolution may appear to be a trivial exercise, experience shows that it can sometimes be a daunting and difficult task. Comments pertinent to the same subject may be aggregated by categories. The task of grouping comments into appropriately defined categories and developing defensible responses can be a challenging effort to say the least. If a number of comments are very similar, the EIS manager may group them together and prepare a single response; when the EIS manager determines that a summary of responses is appropriate, the summary should reflect accurately all substantive comments received on the DEIS.

Public comments and the agency's responses to those comments are typically placed in an appendix to the EIS. However, if the comments are minor, the final EIS might only consist of an errata sheet (i.e., public comments, agency's response to those comments, and corrections to the EIS); in such cases, only the comments, responses, and changes, and not the final EIS, need to be circulated to the public.

4.9.1.2.1 Comments that require a change to the EIS. The FEIS is revised to reflect public comments that merit changes. Potential responses are indicated in Table 4.15 (§1503.4[a]). Where a comment results in a change to the EIS, the agency must explain in its response how and where the EIS has been changed. One common practice is to make initial changes in redline and strikeout mode so that internal reviewers can see the change and its corresponding effect on the document.

An appropriate response to a comment might require that the environmental impact analysis be modified, expanded, or completely redone. For instance, if a comment correctly indicates that the groundwater transport model used an incorrect assumption, the EIS manager is responsible for ensuring that this problem is rectified in the FEIS. In other cases, the agency may need to modify one or more existing alternatives. In still other cases, an alternative not previously investigated may need to be analyzed.

Table 4.15 Potential Agency Responses

- Explaining why comments do not warrant further agency response, citing the source, authority, or reasons that support the agency's position
- Modifying, improving, or expanding the analyses
- Modifying the alternatives
- Making corrections to the text
- Developing and evaluating alternatives that were not seriously considered

4.9.1.2.2 Acknowledging opposing views. The FEIS must address any responsible opposing views that were not adequately discussed in the DEIS (§1502.9[b]). Specifically, the final EIS must (§1502.9[b])

> discuss… any responsible opposing view which was not adequately discussed in the draft statement and shall indicate the agency's response to the issues raised.

The following court case illustrates the importance of complying with this disclosure requirement. The Sierra Club claimed that a US Forest Service EIS used inaccurate data in predicting how a resource management plan would affect population levels of various animal species. This is yet another case in which the Forest Service has lost an EIS lawsuit. In ruling against the Forest Service, the court found that the agency failed to respond to comments from two respected experts who criticized the use of 10-year-old data; use of these data raised serious questions about the accuracy and reliability of the environmental impact projections. The court found that use of these data was arbitrary and capricious.[38]

4.9.2 *Issuing the FEIS*

Once comments on the DEIS have been resolved, incorporated, and approved, the FEIS is ready to be publicly issued. The FEIS is sent to any person, organization, or agency that submitted substantive comments on the draft (§1502.19[d]). The FEIS is publicly circulated using a process similar to that for the DEIS. But in this step, the agency does not actively seek public comments on the FEIS. The purpose of circulating the FEIS is to

1. Allow the public to review the FEIS and understand the nature and impacts of the proposal.
2. Provide other agencies with adequate time to review the FEIS and take any desired action such as referring the matter to the CEQ.
3. Provide potential plaintiffs an opportunity to review the FEIS. This allows potential plaintiffs an opportunity to review the statement and if deemed appropriate, prepare a legal challenge.

4.9.2.1 *Procedures for issuing the final EIS*

As just indicated, once comments have been incorporated and the FEIS is complete, the EIS manager again files the statement with the EPA. While the FEIS must be recirculated, there is no requirement for an agency to request or incorporate comments into it.[39] The EIS manager must transmit an entire copy of the FEIS to any person, organization, or agency that has submitted substantial comments (§1502.19).

As in the case of filing the DEIS, the EIS manager prepares and transmits an NOA to the EPA. The EPA publishes an NOA in the *Federal Register* and is responsible for sending a copy of the FEIS to the CEQ (§1506.9). This filing period cannot occur before the FEIS has been transmitted to commenting agencies and made available to the public (§1506.9).

Rather than a 45-day minimum comment period as in the case of the DEIS, the FEIS has a 30-day minimum *waiting period*. However, during this 30-day waiting period, the public and other agencies may still provide further comments.[40]

4.9.3 *Mandatory 30-day waiting period*

As just noted, the agency must wait for a minimum period of 30 days following the publication of the NOA before making a final decision regarding the proposed action. The 30-day waiting period commences on the date that the NOA is published (§1506.10). As just indicated, this 30-day waiting period is not intended for the purpose of obtaining additional comments but instead to provide other agencies and the public with adequate time to review the FEIS and take action, such as referring the matter to the CEQ or filing a legal suit. No decision regarding the proposal may be made or recorded until the later of the following dates (§1506.10[b]):

1. Ninety (90) days after publication of the NOA for the DEIS
2. Thirty (30) days after publication of the NOA for the FEIS

Where the FEIS is filed within 90 days of the draft, the minimum 30- and 90-day periods may run concurrently. However, agencies may not allow less than 45 days for comments on DEISs (§1506.10[c]). These periods represent the minimum requirements but it is typical for agencies to exceed them, particularly in cases that are highly controversial or involve complex issues.

As an example, imagine that the comment period for a DEIS runs for a period of exactly 45 days starting on the day after the EPA publishes the NOA in the *Federal Register*. The EIS manager promptly incorporates public comments into the draft and completes the FEIS. The EPA publishes the NOA for the FEIS 9 days following completion of the mandatory 45-day comment period. In this case, the 45-day comment period, plus the 9-day comment incorporation period, followed by the mandatory 30-day waiting period, amounts to 84 days, 6 days fewer than the 90-day minimum requirement. The agency must therefore wait out these extra days before recording its final decision in the record of decision. Typically, most EISs exceed the 90-day requirement.

4.9.3.1 Exceptions to the 30-day waiting period

There are a few exceptions to the minimum 30-day waiting period. Agencies involved in rulemaking under the Administrative Procedures Act or other statutes dealing with the protection of health and safety may waive the 30-day waiting period provision and publish the record of decision simultaneously with the FEIS NOA.

Another exception involves cases where an agency has established an appeal process, allowing other agencies or the public to appeal a decision *after* the FEIS has been published. Where this is the case, the EIS 30-day waiting period may not apply (unless the decision is not subject to appeal). In such a case, the EIS must explain the public's right of appeal. Thus, the period for the appeal of this decision and the 30-day waiting period may run concurrently (§1506.10[b]).

4.10 The record of decision

Martin Luther King, Jr. once observed:

> Man is man because he is free to operate within the framework of his destiny. He is free to deliberate, to make decisions, and to choose between alternatives.

The purpose of the ROD is to record the agency's final decision in the form of a concise statement that also discusses the agency's choice from among the various alternatives considered. By requiring an agency to concisely record its decision in the ROD, the decision maker is procedurally forced (at least in theory) to consider the analysis in the EIS.[41] This section examines the process for issuing the ROD.

4.10.1 Choosing a course of action

The decision maker is under no duty to choose an environmentally responsible course of action. However, the EIS *must* be used by agency officials "... in conjunction with other relevant material..." in reaching a final decision (§1502.1). More to the point, the EIS must be used by agency official(s) in making "... decisions that are based on understanding of environmental consequences..." and for "[considering]...actions that protect, restore, and enhance the environment" (§1500.1[c]).

In addition to the environmental consequences, a decision regarding the agency's final course of action can be based on many factors, including *economic* and *technical considerations* as well as the agency's *statutory mission*. The responsible official may select any alternative in the FEIS, provided it has been adequately described and analyzed (§1505.5[e]). A monitoring and enforcement program must be "adopted and summarized where applicable for any mitigation" (§1505.2[c]).

4.10.1.1 Responsible official

A *responsible official* signs the ROD. The responsible official is an agency employee who has the authority to make and implement a decision on a proposed action. The official should coordinate and integrate the EIS review with agency decision making. At a minimum, the author recommends that the decision maker perform the six duties outlined in Table 4.16.

An agency has discretion to change its initial decision regarding the "preferred alternative" cited in the FEIS. Agencies can also change a decision and reissue a ROD that pursues a different course of action, so long as this course of action is adequately analyzed within the FEIS.[42] The EIS manager normally prepares the ROD at the time the final decision is being made, but it may actually be issued at any point after expiration of the minimum waiting period.

4.10.1.2 Decision factors

Recall from the description of Schmidt's model of purpose and need (see Chapter 3) that purposes are *objectives* the agency wishes to achieve. It is during the decision-making process that these purpose(s) comes into play. The purpose (i.e., decision factors) for taking action provides the decision maker with important input (in conjunction with environmental considerations) that is useful in discriminating between the analyzed alternatives. The factors used in reaching a final decision must be disclosed in the ROD (§1505.2[b]). Table 4.17 indicates factors commonly used in reaching a final decision.

Table 4.16 Duties That the Decision Maker Should Perform in Reaching a Final Decision

- Thoroughly review the EIS before making a decision on the proposal. This may include asking the EIS manager, IDT staff, or other specialists to explain any material on which he may have questions.
- Consider the alternatives before rendering a decision on the proposal.
- Consider environmental documents, public comments, and agency responses to those comments.
- Consider results of the consultation process and the conclusions and judgments of other agencies with jurisdiction or expertise.
- Include the decision-making review in the administrative record.
- Make a decision that has been adequately evaluated by the range of alternatives analyzed in the EIS.

Table 4.17 Potential Factors (i.e., Purposes) Commonly Used in Reaching a Final Decision

Examples of decision factors	Explanation
Environmental impact	Selection of an alternative that reduces or eliminates an impact
Mitigation	Selection of a mitigation measure(s) that reduces or eliminates an adverse impact
Cumulative impact	Selection of an alternative that reduces cumulative impacts
Human health risk	Selection of an alternative that reduces risks to human health
Mission	Selection of an alternative that satisfies the agency's underlying need and mission
Technological capability	Selection of an alternative that is scientifically or technologically feasible
Cost	Selection of an alternative that reduces cost
Social economic dislocation	Selection of an alternative that reduces economic dislocation or job losses
Environmental justice	Selection of an alternative that equitably spreads impacts across economic and ethic populations
Consistency	Selection of an alternative that is consistent with other plans or requirements
Regulatory compliance	Selection of an alternative that complies with other regulatory requirements, orders, compliance agreements, and other requirements
Implementation flexibility	Selection of an alternative that provides flexibility in responding to changing circumstances

4.10.1.3 *Bounded alternatives*

In some cases, agencies have chosen a course of action not specifically covered in any one of the alternatives, but has been *appropriately* "bounded" by two or more of the analyzed alternatives. In other cases, agencies have taken a "mix and match" approach, choosing a portion of one alternative and a portion of another alternative, both of which were analyzed in the EIS.

Such decisions should be made judiciously; it is conceivable that the impact of an action "bounded" between two analyzed alternatives might actually exceed projections cited for either one of the analyzed alternatives. Likewise, the impact of mixing and matching aspects of two different alternatives might exceed EIS projections for either one of these alternatives. If such an approach is taken, some type of analysis may need to be performed to demonstrate that the agency's decision does not exceed EIS forecasts.

Finally, lest we forget, an EIS is prepared to provide information sufficient to *discriminate* between various alternative courses of action so as to foster *informed* decision making; a bounding analysis may accurately "bound" potential impacts without providing the decision maker or public with sufficient information for discriminating between alternatives and reaching an "informed" decision. Prudence should be exercised, as an inappropriately bounded alternatives analysis may provide fertile ground for a legal challenge.

4.10.2 *Issuing the ROD and the 30-day waiting period*

As indicated earlier, *no decision* regarding the course of action (alternative) "shall be made or recorded" until the minimum 30-day waiting period following publication of the NOA for the FEIS has expired (§1506.10[b][2]). Unfortunately, the decision has often been finalized well before the waiting period has ended; in some cases, the final decision was made before the EIS was even started.

As we have seen, some agencies have formally established an EIS appeal process. For decisions subject to such an appeal process, the responsible official may be able to sign and date the ROD on the date that the FEIS is transmitted to the EPA and made available to the public; however, for decisions not subject to this appeal process, the responsible official may not sign and date the ROD sooner than 30 days after EPA's NOA for the final EIS is published in the *Federal Register.*

The agency is responsible for preparing the ROD, which may then be "integrated" into any other decision record it has prepared.[43] As a "public" document (§1505.2), the ROD must be made available to the public (§1506.6[b]). Although there is no actual requirement within the Regulations that the ROD be published in the *Federal Register,* virtually all agencies do so.

Although not technically required, copies of the ROD should be distributed to organizations and individuals who received a copy of the final EIS. Once an ROD has been issued, the agency may still change its mind and select a different alternative; this can be done by supplementing or issuing a new ROD.

4.11 *Mitigation, post-EIS monitoring, and enforcement*

As described below, an agency is responsible for ensuring that any mitigation measures, monitoring, and other conditions committed to in the ROD are implemented.

4.11.1 *Mitigation and monitoring transparency*

NEPA requires all agencies of the federal government to make

> available to States, counties, municipalities, institutions, and individuals, advice and information useful in restoring, maintaining, and enhancing the quality of the environment.[44]

It is the responsibility of the lead agency to make the results of relevant monitoring available to the public.[45] NEPA also incorporates the Freedom of Information Act (FOIA) by reference and ensures public access to documents reflecting mitigation, monitoring, and enforcement.[46] Under the FOIA, agencies are required to make available, through "computer telecommunications" (e.g., agency websites), releasable NEPA documents and monitoring results, which, because of the nature of their subject matter, are likely to be the subject of FOIA requests.[47] Mitigation and monitoring reports, access to documents, and responses to public inquiries should be readily available to the public through online or print media, as opposed to being limited to requests made directly to the agency.

4.11.2 *Recent mitigation and monitoring guidance*

In 2011, the CEQ issued guidance to federal agencies regarding the development and implementation of mitigation and monitoring of activities.[48] This guidance is summarized in Table 4.18.

Table 4.18 CEQ Mitigation and Monitoring Guidance

- Agencies should commit to mitigation in decision documents when they have based environmental analysis on such mitigation (by including appropriate conditions on grants, permits, or other agency approvals, and making funding or approvals for implementing the proposed action contingent on implementation of the mitigation commitments).
- Agencies should monitor implementation and effectiveness of mitigation commitments.
- Agencies should make diligent efforts to make information on mitigation and monitoring available to the public, preferably through agency websites.
- Agencies should remedy ineffective mitigation when there is federal action remaining to be taken.

4.11.3 Adaptive management

Mitigation commitments may be structured to include *adaptive management* in order to minimize the possibility of mitigation failure. An example of how adaptive management can be integrated into the EIS process is shown in Figure 4.8. However, if mitigation is not performed or does not mitigate the effects as intended, the responsible agency should, based on its expertise and judgment regarding any remaining

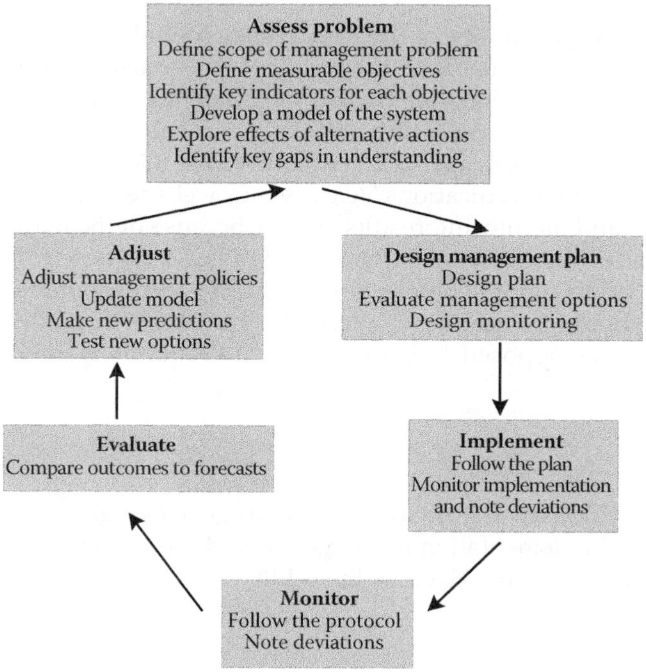

Figure 4.8 Six-stage adaptive management process. Moving clockwise, Step 1 (topmost box) involves assessing the problem. It is sometimes performed in facilitated workshops coordinated with EIS analysis. Participants define the scope of the management problem and investigate potential outcomes of alternative measures. Forecasts of potential outcomes are made, and impacts of alternative adaptive measures are evaluated. Step 2 entails designing and developing a management plan. This stage involves developing a management plan and monitoring program. Step 3 concerns project implementation. The project is implemented and the adaptive management plan is put into practice. Step 4 entails monitoring. Monitored indicators are measured to determine how effective the plan, actions, and any mitigation measures are in meeting objectives. Step 5 concerns project evaluation. This step involves comparing actual outcomes to forecasts and determining reasons for any differences. Step 6 involves adjustment. In this step, new forecasts are made and adjusted to reflect new information. As necessary, this cyclical process repeats. (Courtesy images.google.com.)

federal action and its environmental consequences, consider the need to take supplementary action (1502.9[c]). In cases involving an EA with a mitigated FONSI, an EIS may have to be developed if the unmitigated impact is significant.

4.11.4 Mitigation

The importance of mitigation cannot be overstated. Taking pains to delineate potential impacts of an action certainly serves a purpose, but only if that information is used to find practicable means to prevent, reduce, or offset those impacts through an alternative course of action or by way of mitigation measures. Later generations are not likely to value the yellowing NEPA documents sitting on agency archival shelves, but they may be very appreciative of the mitigation measures that the earlier generation put in place. For instance, as explained in Chapter 1, the NRC routinely evaluates mitigation measures in their nuclear reactor relicensing EISs. However, they rarely, if ever, require an applicant to adopt such measures. Beyond satisfying the CEQ regulatory requirement, one is left to wonder what the purpose of identifying mitigation measures was.

4.11.4.1 Mitigation measures

Mitigation measures include avoiding or minimizing the impacts of an action, repairing the effects of impacts that do occur, and compensating for impacts by replacing or substituting resources that have been damaged (§1508.20). Table 4.19 shows five types of mitigation measures recognized by the CEQ (§1508.20).

Some agencies prepare mitigation action plans (MAPs) as an integral part of the final decision. Regardless of whether they prepare a formal MAP, agencies are expected to successfully carry out the mitigation outlined in the EIS. The EIS is expected to identify reasonable mitigation

Table 4.19 Mitigation Measures

Mitigation methods may include

- Avoiding the impact altogether by not taking a certain action or parts of an action
- Minimizing impacts by limiting the degree or magnitude of the action and its implementation
- Rectifying the impact by repairing, rehabilitating, or restoring the affected environment
- Reducing or eliminating the impact over time by preservation and maintenance operations during the life of the action
- Compensating for the impact by replacing or providing substitute resources or environments

opportunities even if they do not reduce environmental impacts to the point of nonsignificance.

4.11.4.2 Implementing mitigation measures

The lead federal agency should ensure that responsible parties, mitigation requirements, and any appropriate enforcement clauses are included in documents such as authorizations, agreements, permits, or contracts. Such enforcement clauses, including appropriate penalty clauses, should be developed based on a review of the agency's statutory and regulatory authorities. Other agencies that can provide direction and information useful in developing an effective monitoring program include

- The US Fish and Wildlife and National Marine Fisheries Services for evaluating potential impacts to threatened and endangered species
- State Historic Preservation Officers for evaluating potential impacts to historic structures
- US Army Corps of Engineers for evaluating potential wetlands impacts

4.11.5 Monitoring

Developing rigorous or inventive mitigation measures may be highly satisfying, but gains are realized only if the mitigation is actually successful. Postmonitoring is an important step that can be taken to ensure that environmental predictions are not exceeded and that commitments made in the EIS and its ROD are not lost in the haste and confusion of project implementation. Monitoring can be performed as

- A stand-alone element of an agency's NEPA program
- Part of a broader system for monitoring environmental performance

Where applicable, a monitoring and enforcement program is adopted and summarized for any mitigation measures that are adopted (§1505.2[c]). Postmonitoring efforts normally consist of activities such as measuring or sampling air emissions, water discharges, noise levels, and vegetation patterns.

In establishing a monitoring program, it is important to note that specific performance standards against which mitigation measures can be assessed need to be established. Unless those standards are substantially met, the mitigation has not accomplished its intended purpose.

Upon request, the lead agency must inform the public, as well as any cooperating or commenting agencies, regarding the progress and status of any mitigation measures adopted in the ROD (§1505.3[c] and [d]). A supplemental EIS may need to be prepared if monitoring reveals significant new circumstances or information relevant to environmental concerns (§1502.9[c]). The supplemental EIS process is discussed shortly.

4.11.5.1 *Monitoring direction*

A federal agency has a continuing duty to gather and evaluate new information relevant to the environmental impact of its actions.[49] For agency decisions based on an EIS, the regulations require that

> ... a monitoring and enforcement program shall be adopted... where applicable for mitigation. (§1505.2[c])

The regulations go on to state that agencies may

> ... provide for monitoring to assure that their decisions are carried out and should do so in important cases. (§1505.3)

4.11.5.2 *Monitoring objectives*

Postmonitoring is useful in ensuring that

1. Environmental standards are met
2. Mitigation measures are adequately implemented
3. No impacts are encountered that are substantially different from those originally forecast

A monitoring and enforcement plan should be adopted and summarized for any mitigation measures chosen (§1505.2[c]). Agencies are also responsible for making the results of relevant monitoring available to the public.

4.11.5.3 *Monitoring methods*

The form and method of monitoring may be premised on the agency's experience with other monitoring programs that tracked impacts on similar resources. It may also be based on programs used by other agencies or entities. Monitoring methods include

- Agency-specific environmental monitoring
- Compliance assessment
- Auditing systems

4.11.5.4 *Factors considered in prioritizing monitoring activities*

Monitoring plans and programs should be described or incorporated by reference in the agency decision documents. Table 4.20 provides examples of factors that should be considered when prioritizing monitoring activities.

Table 4.20 Examples of Factors That Should Be Considered When Prioritizing Monitoring Activities

- Legal requirements from statutes, regulations, or permits
- Protected resources (e.g., threatened or endangered species, historic site) and the action's impacts on them
- Degree of public interest in the resource or public debate over the effects of the proposed action and any reasonable mitigation alternatives on the resource
- Level of intensity of impacts (§1508.27[b])
- Context in which the impacts occur (§1508.27[a])

4.11.6 *Using an EMS to implement the decision, mitigation, and monitoring*

Sometimes the course of action chosen in the ROD is not properly implemented. A management mechanism can be highly desirable in terms of ensuring that the alternative chosen is correctly executed and that any prescribed mitigation measures are correctly implemented. In 1996, as part of the CEQ's *Improving NEPA Effectiveness Initiative*, the author began a task of investigating commonalities and similarities between NEPA and an International Organization for Standardization (also known as ISO) 14001 Environmental Management Systems (EMS). This investigation resulted in a report summarizing the findings. The author issued this report to the president of the National Association of Environmental Professionals (NAEP) the following year.* In 1998, a strategy for integrating NEPA with an ISO 14001 EMS was published.† In 2000, this report was reviewed and approved by the NAEP Board of Directors and issued to the CEQ as an integrated NEPA management system that should be promoted to all federal agencies. The CEQ later issued guidance on integrating NEPA with an ISO 14001 EMS. This integrated process is summarized below. For more information on the integrated NEPA/EMS process, the reader is referred to the companion book, *NEPA and Environmental Planning*.[50]

4.11.6.1 *Environmental management system*

ISO 14001 provides an internationally accepted specification for an EMS. As used in this chapter, the term "EMS" is interpreted to mean an EMS consistent with the ISO 14001 series of standards. Presidential Executive Order 13148 directed federal agencies to develop EMSs by the year 2005.[51]

* Eccleston C.H., "A conceptual strategy for integrating NEPA with an environmental management system," issued to the President of the National Association of Environmental Professionals, 1997. Issued to CEQ, 1999/2000.

† Eccleston C.H., "A strategy for integrating NEPA with an EMS and ISO-14000," *Environmental Quality Management Journal*, John Wiley & Sons Inc., Spring 1998.

As explained in more detail shortly, strong parallels exist between the goals and requirements of an EIS, and the specifications for implementing an EMS.

As depicted in Figure 4.9, an EMS provides a structured system ("plan–do–check–act") in which a set of management procedures are used to systematically identify, evaluate, manage, and address environmental issues and performance. An EMS provides a mechanism that helps ensure that necessary actions are taken to integrate environmental safeguards and compliance into day-to-day operations. The EMS employs a rigorous monitoring cycle for managing and continually improving a course of action.

4.11.6.2 *Integrating NEPA with an EMS*

When properly integrated with NEPA, an EMS can provide an ideal system for managing, tracking, monitoring, and improving implementation of the agency's chosen course of action, including mitigation measures, monitoring, and other commitments established in the EIS.

Under an ISO 14001 EMS, all personnel whose work could result in a significant environmental impact must receive appropriate training. Such requirements can help ensure that the NEPA decision is correctly and safely implemented.

The NEPA regulations strongly encourage, and in some instances mandate, incorporation of monitoring. Generally, however, the courts have not insisted that agencies incorporate monitoring as part of the NEPA process. In contrast, monitoring is a basic element inherent in an EMS. A properly integrated NEPA/EMS ensures that monitoring is correctly executed.

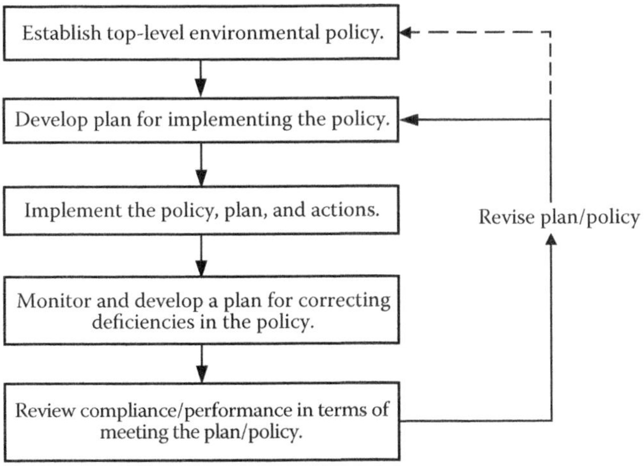

Figure 4.9 ISO 14001 environmental management system.

4.12 Referrals

Interagency disagreements can and sometimes do arise. The Regulations provide a process known as *referral* for resolving disagreements over NEPA. A federal agency can refer any matter to the CEQ, believed to be unsatisfactory from the standpoint of public health, welfare, or environmental quality. The agency that makes the referral is known as a *referring agency* (§1508.24). The procedure for making a referral is detailed below.

As witnessed earlier, the EPA administrator has responsibility under Section 309 of the Clean Air Act to review and publicly comment on EISs. The administrator may refer to the CEQ any EIS deemed "unsatisfactory from the standpoint of public health or environmental quality" (§1504.1[b]).

4.12.1 Referral time periods

Under a provision known as the "predecision referral," any federal agency has authority to refer another agency's FEIS to the CEQ during the mandatory 30-day waiting period if it is deemed to be unsatisfactory; however, the referring agency is expected to make a concerted effort to resolve differences with the lead agency before doing so (§1504.2).

The referring agency has a period of 25 days from the date the FEIS is made publicly available to make a referral in writing to the CEQ (§1504.3[b]). The CEQ will not accept referrals after this time period has expired, except in cases where the lead agency has agreed to extend the referral period. During that same period, other agencies and the public may also submit written comments to the CEQ.

Table 4.21 Actions That CEQ Can Take in Resolving a Referral

- Publish its findings and recommendations
- Conclude that the referral process has successfully resolved the problem
- Initiate discussions to mediate the issue between the referring and lead agencies
- Hold public meetings or hearings to obtain additional views and information
- Determine that the issue is not one of national importance and request the referring and lead agencies to pursue their decision process
- Submit the referral and the response together with the Council's recommendation to the president for action
- Determine that the issue should be further negotiated between the referring and lead agencies until one or more agency heads report to the Council that the agencies' disagreements are irreconcilable

4.12.2 Procedure for making a referral

The referring agency must prepare a letter signed by the head of the agency. This letter notifies the lead agency of the referral and requests that no action be taken with respect to the proposal until the CEQ has acted on it. Once the referral has been made to the CEQ, the lead agency has a period of 25 days to deliver a written response to the CEQ and the referring agency. Table 4.21 shows the actions that the CEQ can take in resolving a referral (§1504.3[f]).

4.13 Supplemental EISs

Sometimes an EIS must be revised or *supplemented*. According to the Regulations, a draft or final EIS must be supplemented if (§1502.9[c])

- There is significant new information or circumstances relevant to environmental concerns that bear on the proposed action or its impacts
- The agency makes substantial changes to the proposed action that is relevant to environmental concerns

An agency may also choose to supplement a draft or final EIS if it is believed that the supplement would further the purpose of NEPA (§1502.9[c][2]). With one exception, preparation of a supplemental EIS follows the same process as a standard EIS. The one exception is that a supplemental EIS does not need to repeat the formal scoping process (§1502.9[c][4]).

While the CEQ does not provide guidance for determining what constitutes new information or circumstances significant enough to necessitate preparation of a supplemental EIS (S-EIS), the courts have provided some factors useful in making such determinations (see Table 4.22).[52]

The author has developed a systematic peer-reviewed tool, referred to as the *Smithsonian Solution* (described shortly) for assisting decision makers in reaching a determination to supplement an EIS. Detail on the use of this tool can be found in Chapter 8, Section 8.7 of the companion book, *NEPA and Environmental Planning*.[53]

Table 4.22 Factors Useful in Determining What Constitutes Significant New Information or Circumstances Sufficient to Require Supplementing an EIS

• Degree of care given in considering new information and determining and evaluating its impacts
• Degree to which the agency supports its decision not to supplement an EIS
• Significance of the new information in terms of environmental impacts
• Probable accuracy of new information

4.13.1 Additional supplementation direction

In one case, a plaintiff brought suit to enjoin (forbid) construction of a dam. The case was partly based on the fact that the Army Corps of Engineers did not prepare a second S-EIS to address concerns raised in new reports regarding adverse impacts on fishing and increased turbidity.[54] The court held that an agency

- Must take a "hard look" at possible new environmental effects. It needs to apply the "rule of reason" (see Chapter 2) when it makes a decision regarding EIS supplementation, even after a proposal has received initial approval. Application of the rule of reason is dependent on the value of the new information to the remaining decision-making process. The decision on whether to prepare an S-EIS is similar to the initial decision to prepare an EIS, i.e., if major federal action remains and if the new information will affect environment quality in a significant manner or to a significant extent not already considered.
- Has a responsibility to continue reviewing impacts of a proposal even after its approval. Nevertheless, new information does not necessarily compel an agency to prepare an S-EIS. In the courts' view, an agency

> ... need not supplement an EIS every time new information comes to light after the EIS is finalized. To require otherwise would render agency decision-making intractable, always awaiting updated information only to find the new information outdated by the time a decision is made.

A new regulation or statute does not necessarily constitute either a change in the proposed action or relevant new information[55]; nor does the mere passage of time compel an EIS to be supplemented.[56] Where an agency decides not to prepare an S-EIS, it should carefully explain its reasoning.

The *arbitrary and capricious* standard of the Administrative Procedure Act provides the criterion used by courts in determining if S-EIS is required.[57] In determining if an agency has acted in an arbitrary and capricious manner, courts consider

- Whether the decision not to prepare an S-EIS was based on a consideration of the relevant factors
- If there has or has not been a clear error of judgment

The court's inquiry must be "searching and careful" but "the ultimate standard of review is a narrow one."[58]

4.14 *Legislative EISs*

Preparation of an EIS may be required for proposals involving congressional "legislation." A legislative EIS (L-EIS) is expected to be integrated with the congressional legislative process (NEPA §102[c], §1506.8). Congress is not a "federal agency" and therefore legislation initiated directly by Congress is not subject to NEPA (§1508.17).

As used above, the term "legislation" includes bills and legislative proposals "developed by or with the *significant cooperation* and support of a federal agency," but does not include requests for appropriations. The term "significant cooperation" means that the legislation has been developed primarily by the federal agency rather than by another source. Drafting a proposal is not sufficient by itself to constitute significant cooperation. An L-EIS is subject to special requirements as described below.

4.14.1 *Preparing a legislative EIS*

The agency having primary responsibility for the subject matter is responsible for preparing a legislative EIS (§1508.17). A draft L-EIS is prepared as part of a formal legislative proposal and transmitted to Congress and the public for review (§1506.8). To allow time for completion of an accurate statement, it may be transmitted up to 30 days following the formal congressional proposal. Because the L-EIS may be used in the congressional debate on the proposal, it must be made available in time for pertinent hearings and deliberations (§1508[a]).

4.14.1.1 *Differences in the L-EIS process*

As described earlier, a typical EIS process is a two-stage procedure: preparation of the draft, followed by the final EIS. A draft L-EIS is prepared in the same manner as that for a nonlegislative EIS, with the following two exceptions:

- No EIS scoping process is required.
- The L-EIS normally ends at the draft stage without preparation of a final L-EIS.

The CEQ notes that legislative proposals are different from other proposed actions normally undertaken by an agency. Important steps (e.g., hearings, votes) are not conducted by Congress in the same controlled and predictable manner as those mandated within an agency. For instance, Congress may vote and hold hearings as it deems appropriate, and may hold hearings or request additional environmental information directly from an agency after it has received the EIS. Consequently, to prevent delays and to support the congressional process, a FEIS for legislative proposals is not required.[41]

4.15 Programmatic EISs

The Regulations state (§1502.4[b])

> Environmental impact statements may be prepared, and are sometimes required, for broad Federal actions such as the adoption of new agency programs or regulations. Agencies shall prepare statements on broad actions so that they are relevant to policy and are timed to coincide with meaningful points in agency planning and decision-making.

Thus, a programmatic EIS (P-EIS) may be prepared, and is sometimes required, for extensive federal actions such as the adoption of new agency programs or regulations. P-EISs should be prepared so that they are relevant to policy decisions and are timed to coincide with meaningful points in agency planning (§1502.4[b]). Furthermore, agencies are encouraged to (§1502.20)

> ... tier their environmental impact statements to eliminate repetitive discussions of the same issues and to focus on the actual issues ripe for decision at each level of environmental review. Whenever a broad environmental impact statement has been prepared (such as a program or policy statement) and a subsequent statement or environmental assessment is then prepared on an action included within the entire program or policy (such as a site specific action) the subsequent statement or environmental assessment need only summarize the issues discussed in the broader statement and incorporate discussions from the broader statement by reference and shall concentrate on the issues specific to the subsequent action.

The following section summarizes decision-making errors that can result when an agency fails to prepare a P-EIS.

4.15.1 The consequences of failing to prepare a P-EIS

Failing to prepare a P-EIS can have significant ramifications in terms of flawed decision making. Recall the case study in Chapter 1, in which the Nuclear Regulatory Commission circumvented its legal mandate to prepare a P-EIS for its program to relicense the nation's fleet of nuclear power reactors. This is a large and pervasive national program affecting the lives of millions of citizens. A P-EIS should have been prepared to question the wisdom of renewing operating licenses for a fleet of aging and antiquated nuclear reactors. It should have investigated programmatic and alternative courses of action such as

- Not renewing the period of operation (no action) for the aging fleet of reactors
- Renewing the operating licenses for the nation's fleet of reactors (proposed action)
- Renewing some types or classes of reactors, but not other more hazardous designs
- Pursuing alternative energy sources, including renewable energy
- Replacing the existing antiquated plants with a generation of modern and safer plants

Such a programmatic examination would have allowed the decision maker, public, and Congress to understand the optimum path forward, as well as the ramifications and risks associated with relicensing an entire generation of aging nuclear reactors. The Commission completely failed its responsibility to examine broad and programmatic directions, or alternatives to the relicensing program. In doing this, the Commission may have exposed citizens to a much higher degree of risk from a catastrophic nuclear accident.

In circumventing its responsibility to prepare a P-EIS, the Commission erred from the "get go." The relicensing program should never have begun until the P-EIS had investigated and determined a course of action with respect to this national program. This is a clear violation of direction provided in the NEPA regulations and NEPA case law (§1508.18[b][3]; 40 CFR §1502.4[b]). The lesson to be learned is that a P-EIS needs to be prepared for new and broad-based programs.

4.15.2 *Programmatic EISs and tiering*

As noted above, a P-EIS provides a critical and valuable tool for establishing a high-level direction for subsequent lower-level NEPA analyses (lower-level, project-specific EISs or environmental assessments); the lower-level NEPA analyses evaluate the details and specific actions for implementing the programmatic alternative adopted in the P-EIS ROD.

Subsequent NEPA analyses need only summarize the issues discussed in the P-EIS; such discussions can be incorporated from the P-EIS by reference, so that decision makers may concentrate on issues specific to these lower-level decisions (§1502.20).

4.15.3 *Determining appropriate scope of a P-EIS*

The Regulations do not provide direction for determining the scope of issues or the amount of detail, discussion, and analysis appropriate for examination within a P-EIS. Lack of definitive direction can also result in inconsistencies in the treatment of NEPA documents, and increased risk that a project may be challenged as a result of inappropriate coverage or treatment of an issue. Lacking systematic guidance, such determinations tend to be made on an ad hoc basis. The companion book, *NEPA and Environmental Planning*, provides a systematic and peer-reviewed tool for determining the scope, including whether a particular topic or issue should be addressed in a P-EIS.

The reader should note yet another advantage that a P-EIS offers. The assessment of cumulative impacts is an issue receiving increased attention and scrutiny. Failure to adequately address cumulative impacts is now one of the most widely litigated issues in NEPA. A P-EIS provides perhaps the most valuable tool for identifying, evaluating, and making decisions based on the cumulative effects.

PROBLEMS AND EXERCISES

1. Where is the NOI for an environmental impact statement published?
2. When should preparation of an EIS begin?
3. Assume that an EIS was prepared 4 years ago for the construction of a small air field. Owing to budgetary limitations, the field was never built. Funding is now available. But a public opposition group maintains that the EIS is too old and must be redone. No significant changes have been made to the proposal and no new environmental issues have surfaced in the interim. Is their argument correct?
4. Briefly explain EPA's alphabetic system for rating environmental impact statements.
5. What is the process for referring an EIS to the CEQ?
6. What are the principal differences between a normal EIS and a legislative EIS?
7. An agency's security officer has just written a memo indicating that NEPA is a public process that requires open review of the proposal in an EIS. Because the proposal involves classified information, it cannot be openly disseminated. Therefore, an EIS is not required for this classified project. Is this conclusion correct? If not, what can be done to comply with NEPA without compromising classified information?
8. Suppose that the comment period for a DEIS runs for a period of exactly 45 days starting on the day that the EPA publishes the NOA in the *Federal Register*. The agency promptly incorporates the comments into the draft and the EPA publishes the NOA for the FEIS,

10 days following completion of the mandatory 45-day comment period. How long must the agency wait before a final decision can be rendered?

Notes

1. 40 Code of Federal Regulations [CFR] Parts 1500–1508.
2. CEQ. *Forty Most Asked Questions Concerning CEQ's National Environmental Policy Act Regulations*, Question No. 35, 46 *Fed. Reg*. 18026 (March 23, 1981), as amended.
3. *Weinberger v. Catholic Action of Hawaii/Peace Education Project*, 454 U.S. 139 (1981).
4. *Northwest Coalition for Alternative to Pesticides v. Lying*, 844 F.2d 588 (9th Cir. 1988).
5. EPA 842-B-01-003 (November 2002).
6. CEQ. *Preamble to Final CEQ NEPA Regulations*, 43 *Federal Register* 55978, Section 1 (November 29, 1978).
7. CEQ. *Guidance Regarding NEPA Regulations* (48 FR 34263) (1983).
8. CEQ. *Forty Most Asked Questions Concerning CEQ's National Environmental Policy Act Regulations*, Question No. 13, 46 *Fed. Reg*. 18026 (March 23, 1981), as amended.
9. *Citizens Against the Collider Here—Illinois, Inc. v. DOE*, 1988 WL 94142 (N.D. Ill. 1988).
10. CEQ. Memorandum: Guidance Regarding NEPA Regulations, 48 *Federal Register*, 34263 (July 28, 1983).
11. CEQ. Memorandum for General Counsels, NEPA liaisons, and Participants in Scoping, Section II, Part A, number 6 (April 30, 1981); Also 46 FR 25461, Notice of Availability of Memorandum to Agencies Containing Scoping Guidance (May 7, 1981).
12. Eccleston C.H. *NEPA and Environmental Planning: Tools, Techniques, and Approaches for Practitioners*. Chapter 4, CRC Press, Boca Raton, FL (2008).
13. 16 USC 1531 et seq.
14. 16 U.S.C. Section 1536(a)(2).
15. 16 U.S.C. §470 et seq.
16. 36 CFR Part 800.
17. This section was prepared with the assistance of Dr. Tom King, a cultural resource specialist with more than 40 years experience dealing with such issues.
18. 36 CFR Part 800, Protection of Historic Properties (2001).
19. 33 USC 1341.
20. Executives Order 11988, Floodplain Management (May 24, 1977).
21. EO 11990, Protection of Wetlands (May 24, 1977).
22. 16 USC 1451 et seq.
23. 16 U.S.C. 1451-1464.
24. 16 USC 1456(c)(3)(A).
25. 65 FR 77123-77175 (December 8, 2000).
26. 15 CFR 930.35.
27. USA.gov. http://www.data.gov/catalog/geodata.

28. Environmental Protection Agency. *Amended Environmental Impact Statement Filing System Guidance for Implementing 40 CFR 1506.9 and 1506.10 of the Council on Environmental Quality's Regulations Implementing the National Environmental Policy Act* (January 14, 2011), https://www.federalregister.gov/articles/2011/01/14/2011-758/amended-environmental-impact-statement-filing-system-guidance-for-implementing-40-cfr-15069-and#h-6.
29. Environmental Protection Agency. *EIS Filing System Guidelines*, 77 FR 51530 (August 24, 2012), http://energy.gov/sites/prod/files/EPA_Amended_EIS_Filing_Guidance_08-24-12.pdf.
30. United States Environmental Protection Agency. http://www.epa.gov/.
31. *Del Norte County v. United States*, 732 F.2d 1462 (9th Cir. 1984), Cert. Denied, 469 U.S., 1189 (1985).
32. Department of Energy. *NEPA Lessons Learned* (74): 6 (March 2013).
33. DOE. National Environmental Policy Act, *Learned Lessons* (24): (September 1, 2000).
34. Fogleman V.M. *Guide to the National Environmental Policy Act Interpretations, Applications, and Compliance*. Quorum Books (1990).
35. Clean Air Act, 42 U.S.C. §7609 (2000).
36. EPA. *Policy and Procedures for the Review of Federal Actions Impacting the Environment* (October 3, 1984).
37. CEQ. *Forty Most Asked Questions Concerning CEQ's National Environmental Policy Act Regulations*, Question No. 29a. 46 *Fed. Reg*. 18026 (March 23, 1981), as amended.
38. *Sierra Club v. US Department of Agriculture*, 116 F.3d 1482 (7th Cir May 28, 1997).
39. *Stop H-3 Asso. v Lewis*, 538 F. Supp 149 (D. Hawaii 1982).
40. CEQ. *Forty Most Asked Questions Concerning CEQ's National Environmental Policy Act Regulations*, Question No. 34b, 46 *Fed. Reg*. 18026 (March 23, 1981), as amended.
41. CEQ. *Preamble to Final CEQ NEPA Regulations*, 43 *Federal Register* 55978, Section 4 (November 29, 1978).
42. CEQ. *Forty Most Asked Questions Concerning CEQ's National Environmental Policy Act Regulations*, Question No. 5a, 46 *Fed. Reg*. 18026 (March 23, 1981), as amended.
43. CEQ. *Forty Most Asked Questions Concerning CEQ's National Environmental Policy Act Regulations*, Question No. 34a, 46 *Fed. Reg*. 18026 (March 23, 1981), as amended.
44. 42 U.S.C. Section 102(2)(G).
45. 40 C.F.R. §1505.3(d).
46. 42 U.S.C. Section 102(2)(C).
47. 5 U.S.C. §552(a)(2); 40 C.F.R. §1506.6(f).
48. CEQ. Memorandum for Heads of Federal Departments and Agencies: Appropriate Use of Mitigated Findings of No Significant Impact (January 14, 2011).
49. 42 U.S.C. §4332[2][A].
50. Eccleston C.H. *NEPA and Environmental Planning: Tools, Techniques, and Approaches for Practitioners*. Section 2.7, CRC Press, Boca Raton, FL (2008).
51. Executive Order 13148, Greening the Government through Leadership in Environmental Management (April 21, 2000).
52. *Warm Springs Dam Task Force v. Gribble*, 621 F.2d 1017, 1024 (9th Cir. 1980).

53. Eccleston C.H. *NEPA and Environmental Planning: Tools, Techniques, and Approaches for Practitioners*. CRC Press, Boca Raton, FL (2008).
54. *Marsh v. Oregon Natural Resources Council*, 490 U.S. 360, 109 S.Ct. 1851 (1989).
55. *National Indian Youth Council v. Watt*, 664 F.2d 220 (10th Cir. 1981), *citing Concerned Citizens v. Secretary of Transportation*, 641 F.2d 1, 6 (1st Cir. 1981).
56. *Coker v. Skidmore*, 941 F.2d 1306 (5th Cir. 1991); *Sierra Club v. United States Army Corps of Engineers*, 701 F.2d 1011, 1036 (2d Cir. 1983).
57. Administrative Procedure Act, §706(2)(A).
58. *Marsh v. Oregon Natural Resources Council*, 490 U.S. 360, 109 S.Ct. 1851 (1989); *Citizens to Preserve Overton Park, Inc. v. Volpe*, 401 U.S. 402, 416 (1971) Id. at 375, 376, 378.

chapter five

Performing the EIS analysis

Chapter 4 described the step-by-step procedural process for preparing and issuing an environmental impact statement (EIS). A thorough scientific analysis of potential environmental impacts provides the foundation upon which a decision to pursue a given course of action will later be made. This chapter details four features essential to the goal of performing a rigorous and sufficient environmental impact assessment (EIA). It describes:

1. All pertinent requirements that must be followed in preparing the EIS analysis, including all applicable regulatory requirements cited in the National Environmental Policy Act implementing regulations[1]
2. A systematic, general purpose process for performing the EIA
3. Tools, techniques, and methods for performing an EIA
4. Leading methodologies used in performing EIAs

Special consideration is devoted to describing the affected environment and reasonable alternatives (Figure 5.1). Timely topics such as cumulative impact and greenhouse gas (GHG) assessments are likewise described in detail. This material sets the stage for Chapter 6, which describes all applicable EIS regulatory and document requirements (a.k.a. documentation requirements) that the draft and final EIS document must satisfy.

A more detailed examination of impact assessment and methodologies, including evaluating cumulative impacts and GHG emissions, is provided in the companion book, *Environmental Impact Assessment: A Guide to Best Professional Practices*.[2] A copy of the NEPA implementing regulations (Regulations) is provided in Appendix B.* A comprehensive checklist for preparing the EIS is provided in Appendix C.

* Specific provisions referenced in the NEPA implementing regulations are abbreviated in this book so as to cite the specific "part" of the NEPA implementing regulations in which it is found. For example, a reference to a provision in "40 Code of Federal Regulations (CFR) 1501.1" is simply cited as "§1501.1."

Figure 5.1 Even "clean" energy projects can involve severe impacts requiring preparation of an environmental impact statement. (Courtesy images.google.com.)

5.1 Learning objectives

- General NEPA requirements for performing the environmental analysis
- The action–impact model used in performing the environmental analysis
- Six key environmental impact assessment methodologies
- Assessing and describing the affected environment
- Investigating the range of reasonable alternatives
- Performing a human health impact assessment
- Performing a cumulative impact assessment
- Resolving Eccleston's Cumulative Impact Paradox
- Performing an assessment of GHGs and climate change
- A five-step procedure for assessing GHG emissions
- Analyzing the risk and consequences of potential accidents

5.2 Requirements governing the EIS analysis

Mark Twain once mused:

> Get your facts first, and then you can distort them as much as you please.

This chapter is about the first half of Twain's entertaining statement: "... getting the facts straight." The National Environmental Policy Act (NEPA) implementing regulations (Regulations) specify requirements intended to promote the scientific accuracy and integrity of the EIS analysis (Figure 5.2). These requirements are examined in this section and throughout this chapter.

5.2.1 Rule of reason and sliding-scale approach

An overly strict or unreasonable application of a regulatory requirement may lead to decisions, a course of action, or a level of effort that is wasteful, ridiculous, or absurd. A *rule of reason* is an important mechanism used by the courts for injecting reason into the analysis of impacts in an EIS. Reasonable or common sense is to be exercised in determining the scope and detail accorded to issues, alternatives, and impacts considered in the analysis.

5.2.1.1 Sliding-scale approach

Consistent with the rule of reason, the author recommends application of the sliding-scale approach in determining the scope and level of effort devoted to the assessment of issues and impacts. As detailed in Chapter 2, Section 2.6, a sliding-scale approach recognizes that the amount of effort expended on investigating a specific issue or addressing a particular regulatory requirement should vary with the significance of the potential impact and its importance to the decision-making process. Thus, environmental impacts are investigated with a degree of effort commensurate with their importance in the decision-making process. Impacts with potentially severe consequences may warrant considerably more investigation, time, and resources than do those with marginally significant effects.

Figure 5.2 One of the 80,000 dams in the United States. Dams have important economic benefits as well as environmental issues. (Courtesy images.google.com.)

5.2.2 *Conducting a fair and objective analysis*

A fair, objective, and impartial analysis is central to NEPA's congressional mandate of providing decision makers and the public with information that will facilitate informed decision making. Table 5.1 provides a list of regulatory citations that underscore the importance of performing a fair and objective analysis. The case study cited in Chapter 1 involving the Nuclear Regulatory Commission is a vivid example of how public safety and environmental quality can be jeopardized when a federal agency disregards its legal responsibility to prepare an open, fair, and objective analysis. Failure to perform a fair and objective analysis can cast doubt on an agency's integrity and may provide project foes with a legitimate legal basis for challenging an agency's EIS and decision-making process. An agency must strive to avoid even the perception that the EIS analysis has been conducted in an unfair or biased manner.

5.2.3 *Requirements for performing a scientific analysis*

T.H. Thompson once mused:

> If you tell a man there are three billion stars in the universe, he'll believe you. But if you tell him a bench has just been painted, he has to touch it to be sure.

Thompsons's observation is as applicable to an environmental analysis as it is to painted benches. Evidence is the key to defensibility, both in the minds of the public and in the courthouse. Conclusions must be

Table 5.1 NEPA's Regulatory Direction for Performing Fair and Objective Analysis

The EIS shall… provide full and fair discussion of significant environmental impacts and [inform] … decisionmakers and the public of the reasonable alternatives… (§1502.1).
Agencies shall not commit resources prejudicing selection of alternatives before making a final decision (§1502.2[f]).
Environmental impact statements shall serve as the means of assessing the environmental impact of proposed agency actions, rather than justifying decisions already made (§1502.2[f]).
The EIS shall… not be used to rationalize or justify decisions already made (§1502.5).
The EIS shall… objectively evaluate all reasonable alternatives…. (§1502.14[a]).
The agency shall… disclose and discuss… all major points of view on the environmental impacts of the alternatives including the proposed action. (§1502.9[a]).

supported by a solid, scientifically grounded, and defensible analysis. NEPA requires, and the courts demand, that a thorough, comprehensive, and accurate analysis be performed. The accuracy and integrity of an analysis may be greatly influenced by the methods and procedures used. The architects of NEPA recognized and addressed this problem by providing specific direction for ensuring accuracy and confidence in the scientific analysis. This direction is provided in the following sections.

5.2.4 Requirement for developing methods and procedures

As specified in the NEPA statute, agencies are tasked with responsibility for developing methods and procedures that facilitate analysis of environmental amenities and values. Specifically, agencies are instructed to

> … identify and develop **methods** and **procedures**… which will insure that presently unquantified environmental amenities and values be given appropriate consideration in decision making along with economic and technical considerations. (§1502[2][b] of NEPA)

This requirement is also paraphrased in the Regulations (§1507.2[b]). The Regulations do not provide any insight as to what "methods" and "procedures" should be developed. They do, however, specify certain attributes that such methods and procedures should possess. Other requirements related to the goal of performing an accurate environmental analysis are also presented in the following subsections. Consistent with this requirement, the author describes a number of tools, techniques, and methods in Chapters 3 through 6 of this book for performing the analysis and preparing the EIS.

5.2.5 Rigorous analysis

Emphasis must be placed on ensuring that a thorough and professionally credible analysis has been performed. The analysis must be accurate, of high quality, and scientifically credible. For example, the talents of a professionally qualified interdisciplinary team are to be employed in performing the analysis. Members should have experience with the issues and impacts identified in the scoping process. Assumptions and models need to be justified on the basis of their technical merits. Completion of the preliminary analysis should be followed with a rigorous peer review. Table 5.2 provides a summary of selected Regulatory citations underscoring these requirements and constraints.

Particular emphasis is placed on performing a rigorous analysis that thoroughly investigates the reasonable alternatives and their respective

Table 5.2 Requirements for Performing Accurate and Thorough Scientific Analysis

Accurate scientific analysis... is essential to implementing NEPA (§1500.1[b]).
Agencies shall ensure the professional integrity, including scientific integrity.... They shall identify any methodologies used.... (§1502.24).
Identify methods and procedures required by Section 102(2)(b) of NEPA to ensure that presently unquantified environmental amenities and values may be given appropriate consideration (§1507.2[b]).
The information must be of high quality. Accurate scientific analysis,... [is] essential to implementing NEPA (§1500.1[b]).
... the analysis is supported by credible scientific evidence (§1502.22 [b][4]).

impacts. Selected regulatory citations regarding this requirement are provided in Table 5.3. Direction regarding other important aspects of performing the analysis is provided in Table 5.4.

While the Regulations require a rigorous scientific analysis, they equally stress the importance of performing an assessment in an efficient manner. Direction for limiting the analysis and reducing the size of the EIS is presented in Chapter 6. This guidance is illustrated in the following regulatory citations:

Table 5.3 Performing Rigorous Analysis of Reasonable Alternatives and Impacts

Rigorously explore and objectively evaluate all reasonable alternatives (§1502.14[a]).
Devote substantial treatment to each alternative considered in detail (§1502.14[b]).
[Prepare] analytic rather than encyclopedic environmental impact statements (§1500.4[b]).
[The EIS shall be]... supported by evidence that the agency has made the necessary environmental analyses (§1502.1).

Table 5.4 Other Related Direction for Performing an Analysis

Providing adequate detail
Identify environmental effects and values in adequate detail so that they can be compared to economic and technical analysis (§1501.2[b]).
Cost–benefit analysis
If a cost–benefit analysis... is being considered... discuss the relationship between that analysis and any analysis of unquantified environmental impact, values and amenities (§1502.23).
Analysis and supporting data
Agencies shall employ writers of clear prose... to write statements which will be based upon the analysis and supporting data from the natural and social sciences and the environmental design arts (§1502.8).

> Agencies shall focus on significant environmental issues and alternatives and shall reduce paperwork and the accumulation of extraneous background data. (§1502.1)

5.3 Six-step technique for analyzing impacts

Figure 5.3 provides a general purpose six-step technique for performing the EIA. The author refers to this procedure as the *action–impact model* (AIM). The AIM is described in more detail in the companion book, *NEPA and Environmental Planning.*[3] This model lays the foundation for much of the direction presented later in this chapter.

The AIM consists of the following six steps described in this section:

1. Identifying actions
2. Identifying and quantifying environmental disturbances
3. Identifying and describing potentially affected receptors and resources
4. Performing the environmental impact analysis
5. Assessing significance
6. Investigating mitigation measures and performing monitoring

5.3.1 Actions

Analysts must understand what action(s) will take place before the EIA can commence. The Regulations have this to say about a proposal:

> [A] proposal exists at that stage in the development of an action when an agency subject to the Act [NEPA] has a goal and is actively preparing to

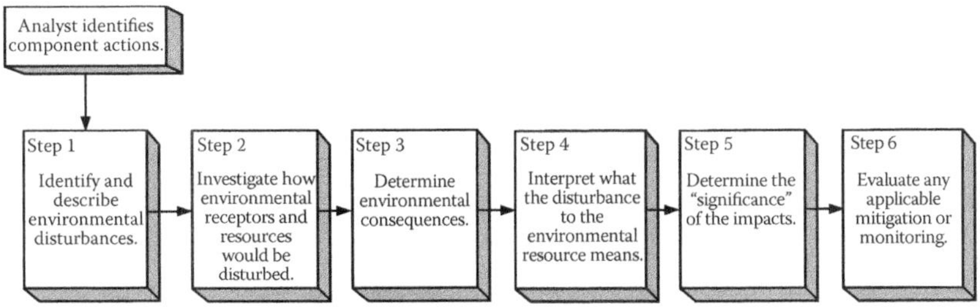

Figure 5.3 Action–impact model. This figure shows a general purpose six-step approach for performing environmental impact assessment.

> make a decision on one or more alternative means of accomplishing that goal and the effects can be meaningfully evaluated. (§1508.23)

It is imperative that the proposal be accurately defined (see top block in Figure 5.3) before beginning the EIA. Occasionally, an agency begins the mechanical process of preparing an EIS, such as scoping and writing the analysis, before they have established a firm grasp of what it is they wish to accomplish. In one instance, an agency was in the midst of preparing an EIS for an ill-defined proposal whose value was suddenly diminished by evolving political considerations. In response, the EIS had to be altered to address how the agency would *not* perform the action rather than perform the action. The contractor writing the EIS described the situation as "the proposed action is no action."

Once the need for taking action has been clearly identified, the agency begins developing one or more actions (alternatives) to accomplish the need for the proposal. For the purposes of NEPA, agencies typically identify one action (proposed action) they feel best accomplishes the need for taking action. Other reasonable approaches for accomplishing the need are commonly termed *reasonable alternatives* (strictly speaking, the proposed action is simply one alternative within the range of reasonable alternative actions). The assessment of alternatives is of such importance that the Council on Environmental Quality (CEQ) has called it the "heart of the environmental impact statement" (§1502.14).

5.3.1.1 *Component actions*

For the purposes of NEPA, most proposals or proposed actions actually consist of a set of "component actions." For example, a proposal to open a new forest area to logging might involve a group of component actions: surveying, access road construction, construction of a bridge, construction of a supply shed and field building, application of herbicides, the actual logging operations, revegetation, and monitoring and collection of field data. Thus, a proposal typically consists of a set of discrete component actions, including any "connected actions" related to the proposal.

All component actions must be identified and adequately evaluated as part of the EIA. Agencies must be careful not to illegally segment interrelated activities into separate actions that are evaluated in independent NEPA analyses (i.e., segmentation).

5.3.2 *Environmental disturbances*

An action or a set of component actions produces *environmental disturbances* (e.g., air emissions, effluents, noise, ground disturbances, extraction of groundwater or surface water, water products). These environmental disturbances must be identified and described in detail sufficient to support a subsequent analysis of their effect on environmental resources (see Step 1, Figure 5.3). It is important to note that such "disturbances" are not normally, in and of themselves, environmental impacts.

5.3.3 Receptors and resources

Each environmental disturbance changes or perturbs one or more *receptors* or *environmental resources* (i.e., "human environment"). Examples of receptors or environmental resources include air, water resources, cultural and archaeological resources, wildlife, habitats, and of course humans. For example, air emissions affect the air resources while effluents affect water and ground resources (receptors). Likewise, environmental disturbances such as air emissions or effluents can affect human receptors. The analysis must determine what receptors (resources) will be affected by an environmental disturbance (see Step 2, Figure 5.3).

5.3.4 Impact analysis (consequences)

An EIS must evaluate and describe the "impact" of the disturbances on receptors or environmental resources. To this end, the analysis is directed at determining how environmental disturbances would change or alter (i.e., environmental impact or consequences) the receptors/resources (see Step 3, Figure 5.3).

This analysis is conducted on a resource-by-resource basis in the environmental consequences section of the EIS (see Figure 5.3). The result is a set of consequences (i.e., environmental effects or impacts). The terms "consequences," "impacts," and "effects" are essentially synonymous (§1508.8[b]). The CEQ recognizes three distinct types of impacts (§1508.25[c]): (1) direct, (2) indirect, and (3) cumulative.

The term "effects" is defined to include

> … ecological (such as the effects on natural resources and on the components, structures, and functioning of affected ecosystems), aesthetic, historic, cultural, economic, social, or health, whether direct, indirect, or cumulative. (§1508.8)

5.3.5 Interpreting the impact

The EIS analysis must interpret what the disturbance to the receptor/resource *means* (see Step 4, Figure 5.3). For example, extracting water from a well (disturbance) may affect the water table, by causing it to drop by

3 meters. But what does this drop really mean? Perhaps it means that the amount of water available for cattle ranching is decreased by 5.5% and agricultural production in the area is reduced by 15%.

5.3.6 *Significance*

One of the most important steps in the EIA involves determining the "significance" of an impact (see Step 5, Figure 5.3). That is, the analysis must determine the *importance* of the impact. Specific factors are presented in the Regulations (§1508.27) for assessing the significance of an impact (see Table 5.5). Additional information on significance and its interpretation can be found in the companion book, *Preparing NEPA Environmental Assessments*.[4]

Significance is a relative concept, often requiring a substantial degree of professional judgment to reach a rationale and defensible conclusion

Table 5.5 Factors Considered in Evaluating Significance in Terms of Intensity

1. Impacts that may be both beneficial and adverse. A significant effect may exist even if the federal agency believes that on balance the effect will be beneficial.
2. The degree to which the proposed action affects public health or safety.
3. Unique characteristics of the geographic area such as proximity to historic or cultural resources, park lands, prime farmlands, wetlands, wild and scenic rivers, or ecologically critical areas.
4. The degree to which the effects on the quality of the human environment are likely to be highly controversial.
5. The degree to which the possible effects on the human environment are highly uncertain or involve unique or unknown risks.
6. The degree to which the action may establish a precedent for future actions with significant effects or represents a decision in principle about a future consideration.
7. Whether the action is related to other actions with individually insignificant but cumulatively significant impacts. Significance exists if it is reasonable to anticipate a cumulatively significant impact on the environment. Significance cannot be avoided by terming an action temporary or by breaking it down into small component parts.
8. The degree to which the action may adversely affect districts, sites, highways, structures, or objects listed in or eligible for listing in the National Register of Historic Places or may cause loss or destruction of significant scientific, cultural, or historical resources.
9. The degree to which the action may adversely affect an endangered or threatened species or its habitat that has been determined to be critical under the Endangered Species Act of 1973.
10. Whether the action threatens a violation of federal, state, or local law or requirements imposed for the protection of the environment.

concerning the significance of an impact. In Chapter 1, we learned how the Nuclear Regulatory Commission's management routinely reaches the indefensible, if not ridiculous, conclusion in its nuclear power plant EISs that the risk of a severe accident, such as a full-scale nuclear meltdown, is "small." When an agency attempts to deceive the scientific community and public about the significance of an event like a full-scale nuclear meltdown, its officials should not be surprised to find that they have lost credibility with stakeholders and much of the public. As outlined earlier in this chapter, it is absolutely essential that an agency perform a fair, objective, and rigorous assessment of the significance of its actions.

5.3.6.1 Assessing significance

The Regulations provide practical direction for assessing significance of environmental impacts. The assessment of significance is based on

1. The context in which the impact occurs (§1508.27[a])
2. Ten intensity factors (§1508.27[b]) (see Table 5.5)

Despite this guidance, the assessment of significance can be a daunting problem; perhaps the key reason behind this problem is the inherent subjectivity involved in evaluating these 10 significance factors as well as the context in which impacts occur. Adding to this difficulty is that the EIA must cite evidence used in reaching conclusions regarding the significance of the impacts.

5.3.6.2 Context

As just noted, analysts must gauge significance in terms of

> … several **contexts** such as society as a whole (human, national), the affected region, the affected interest, and the locality. (§1508.27[a])

In other words, they must consider the context in which the impact occurs in terms of society as a whole (human, national), affected region, affected interest, and the locality in which the impact would occur. As shown in Table 5.5, 10 intensity factors are also to be considered in assessing significance.

5.3.7 Mitigation and monitoring

As described in Chapter 4, Section 4.11, the EIS must identify and evaluate means for mitigating potential impacts (see Step 6, Figure 5.3). With respect to the requirements of NEPA, however, the agency is not normally obligated to adopt any of these mitigation measures in its record of decision.

A monitoring program is also an integral element of a well-planned environmental process, particularly where there is a chance that the impact projections could be exceeded. The reader is referred to the companion text, *NEPA and Environmental Planning*, for a more detailed treatment of mitigation and monitoring.[5]

5.4 Impact assessment methodologies

Impact assessment methodologies have always been an area of interest among NEPA practitioners. Numerous impact assessment methodologies have been developed since the 1970s, and they continue to be improved upon. Some widely used assessment tools and methods include but are certainly not limited to

- Geographic information systems
- Matrices
- Checklists
- Networks
- Carrying capacity
- Ecosystem analysis

5.4.1 Geographic information system

A geographic information system (GIS) is a computer-based tool for collecting, storing, analyzing, and displaying spatial and geographic data. It was briefly introduced in Chapter 4, Section 4.6. As illustrated in Figure 5.4, a GIS provides NEPA practitioners with a powerful tool for analyzing impacts. Potential applications include

- Using US Environmental Protection Agency and other data to identify information and databases such as water quality reports for particular water bodies or facility reports for hazardous waste sites
- Identifying specific features on a map such as cities and infrastructure, forested areas, water bodies, deserts, wetlands, etc.
- Identifying environmental attributes such as air and water quality zones, soil types, and demographics and socioeconomic considerations
- Visualizing changes resulting from development and growth over time
- Identifying species distributions and habitats
- Identifying and distinguishing developed areas such as urban and rural locales, and undeveloped regions
- Delineating zoning and land use

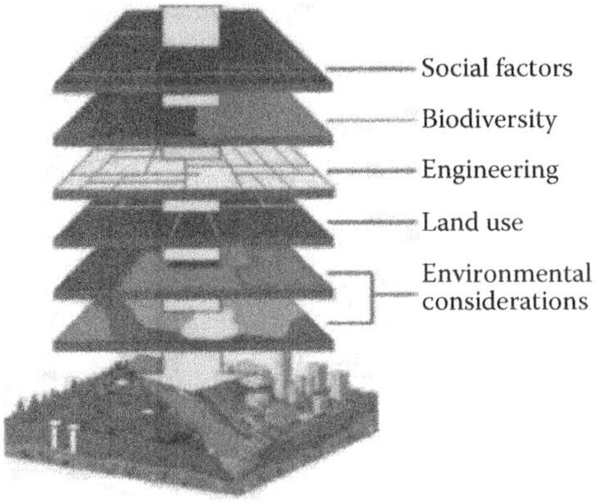

Figure 5.4 Various layers can be combined to form an image of affected environment. (Courtesy of US Department of Energy.)

ArcGIS Explorer Desktop is a free GIS software package. The CEQ has assembled an inventory of more than 150 government data services that can be used with most GIS software applications. Table 5.6 shows examples of federal agency GIS database resources from CEQ's inventory.

A GIS can identify the presence of resources that could be significantly affected.[6] By identifying these resources, the GIS offers planners a powerful tool for determining how environmental, socioeconomic, and cultural resources could be affected by potential actions.

5.4.1.1 *How a GIS can be used in preparing EIS*

The GIS provides a particularly powerful tool for analyzing difficult effects such as indirect and cumulative impacts, including modeling past and future development (e.g., induced development, enabled actions, and demands on infrastructure support). For instance, it can be used in analyzing an area's socioeconomic conditions, the capacity and reliability of local waste treatment, water and energy consumption, transportation infrastructure and its ability to meet present and future demands, and changes in resources or land use over time.

The GIS can help identify particular resources or features that may trigger the need for other review or permitting requirements. For example, it can identify whether the proposed action would occur in or near a wetlands, floodplains, prime or unique farmland, wilderness areas, critical habitat, airspace or military operations, national parks, impaired waters, air quality nonattainment areas, and places on or eligible for listing on the National Register of Historic Places.

Table 5.6 Examples of GIS Resources

Environmental Protection Agency
- National Hydrology Dataset
- WATERS (Watershed Assessment, Tracking and Environmental Results)
- EPA Cleanup Sites

Federal Emergency Management Agency
- National Flood Hazard Layer

National Oceanic and Atmospheric Administration
- Habitat Areas of Particular Concern and Critical Habitat Designations
- The Multipurpose Marine Cadastre

National Park Service
- National Register of Historic Places
- NPScape

US Fish and Wildlife Service
- National Wetlands Map
- Critical Habitat

US Geological Survey
- The National Map
- The National Atlas
- Protected Areas Database of the United States
- National Land Cover
- The Historical Natural Hazards Database

The GIS can also assist analysts in developing mitigation measures (§1508.20). For instance, it can be useful in modeling sedimentation, erosion, discharges, and emissions. It can further support a monitoring program by providing a means of visualizing large amounts of information obtained from field surveys, monitoring stations, and other sources.

5.4.2 *Matrices*

In 1971, Luna Leopold developed a widely used method for assessing environmental impacts. Matrices are among the most popular and widely used impact identification methods.[7] The Leopold Matrix, or a modified version thereof, is perhaps the most widely used matrix method. A *Leopold Matrix* is actually an extension of the *environmental checklist* (described below), in as much it can be viewed as a two-dimensional checklist, with project activities listed on one axis and potentially affected environmental resources or similar attributes on the other; it can allow analysts to assess the importance of individual interactions between activities and resources. Measures of potential magnitude and significance are used to score each impact component. There are many variations of the Leopold Matrix. An example of a simplified Leopold Matrix is shown in Figure 5.5.[8] An "X" indicates a direct impact, while an "O" depicts an indirect effect.

Leopold's original matrix comprised a large grid of 100 potential project actions along the horizontal axis versus 88 environmental attributes on the vertical axis. The matrix does not indicate factors such as the timing or duration of impacts.

Analysts may elect to simply note the presence or absence of an impact in the matrix cells. A more powerful approach, however, is to score impacts based on factors such as importance, magnitude, or probability of occurrence. For example, the values entered into each cell may represent quantified air emissions or number of disturbed acres. Because of the difficulty in quantifying many environmental attributes, matrices often use a subjective or relative score. Weighting schemes are also common.

Matrices are generally superior to checklists because they relate actions to environmental attributes. They also have the advantage of depicting environmental impact data in a simple format. The Leopold Matrix attempts to quantify impacts; however, in the absence of a systematic scoring system, determining the significance of an impact is ultimately subjective. Another disadvantage is that impacts can be double counted.

Where numerical data have been acquired, a matrix can provide a mathematical tool for (multiplying values in individual cells using techniques from Matrix algebra) evaluating combined or synergistic effects, as well as cumulative impacts of multiple actions on individual environmental resources. While principally used in socioeconomic assessments, they are increasingly being used to assess physical impacts on environmental resources.

Affected habitat	Impoundment	Channelization	Grazing	Logging	Surface mining	Row crop agriculture	Groundwater mining	Flow diversion	Flow augmentation	Urbanization	Irrigation	Hydropeaking
Sediment yield	x		x	x	x	x				x	x	o
Water yield	o		x	x	x	x	x	x	x	x	x	
Channel morphology	x	x	x	x	x	x			o	x	x	o
Substrate characteristics	x	x	x	x	x	x			o	x	x	x
Cover	o	x	x	x	x	x				x	x	o
Timing of flows	x						o	o	o	o	x	x
Magnitude of peak flows	x	o	x	x	x	x				x		x
Magnitude of low flows	x		o	o	o	o	x	x	x	x	x	x
Thermal regime	x		o	o	o	x	o	x	o	x	x	
Water quality	o		x	x	x	x		o	o	x	x	

***Figure* 5.5** Example of simplified Leopold Matrix.

5.4.2.1 Evaluating cumulative impacts

Figure 5.6 shows the Cluster Impact Assessment Procedure developed by the Federal Energy Regulatory Commission to assess the cumulative impacts of small hydroelectric facilities within a single watershed. This method uses a matrix for each resource consisting of relative effect ratings based on a scale of 1 to 5, arranged by resource component (salmon, migration, spawning habitat). Each resource matrix contains a summary column that represents the total cumulative impact score across various components for each project.[9] In theory, this approach provides a robust tool for assessing impacts. However, it presents a challenging task in terms of constructing a correct model and obtaining reliable data that yield accurate results.

5.4.3 Environmental checklists

An environmental checklist consists of a single-column overview of a proposed action, with only a coarse characterization of the type and magnitude of potential environmental impacts. There are many variations of the checklist method. A simplified example is shown in Table 5.7.

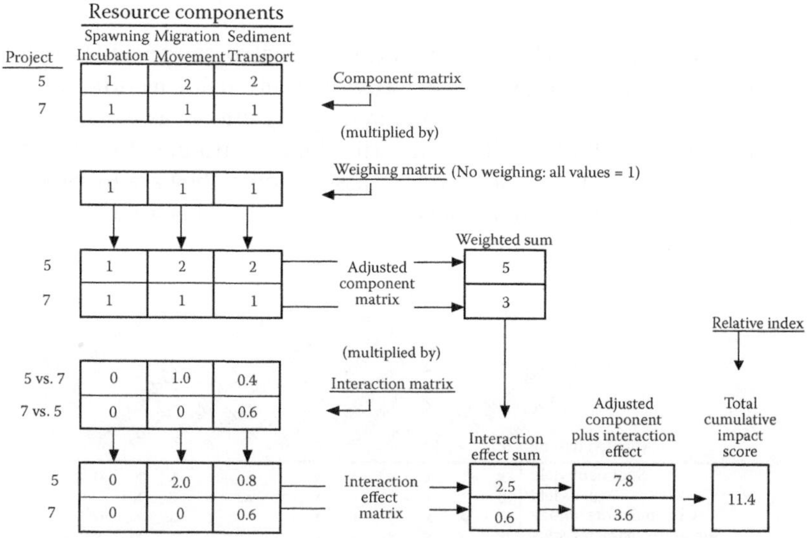

Figure 5.6 Example of Cluster Impact Assessment Procedure matrix for evaluating cumulative impact using matrix algebra techniques to link three resource components and two projects. A component matrix is multiplied by a weighting matrix to produce an adjusted component matrix, which is then multiplied by an interaction matrix to produce an interaction effects matrix. The final result is a cumulative impact score. (Courtesy of CEQ.)

Table 5.7 Simple Checklist for Assessing Impacts on Geological Resources

Item	Short-term	Long-term
Soil disturbances		×
Land excavations		×
Surface water hydrology	×	
Surface water quality		×
Groundwater hydrology		
Groundwater quality		
Aesthetics		×

Although checklists can provide helpful tools, they cannot take into consideration all specific resources or disturbances, or other environmental attributes that may be encountered, particularly on large or complex projects. Their strength lies in the fact that they can be easily used to systematically identify an array of impacts; they can provide analysts with a simple tool for reducing the likelihood that important impacts will be overlooked.

Nonetheless, the simplicity of a checklist can also be its principal disadvantage. Their use frequently discourages thinking (i.e., "tunnel vision") and may provide a false sense of a complete assessment; an incomplete checklist can result in a flawed analysis in which important impacts are overlooked. Thus, checklists are frequently ineffective because of their incompleteness or because they contain so many irrelevant impacts that they essentially become too large and unwieldy to be of practical use. These disadvantages can sometimes be compensated for by developing checklists tailored to specialized types of projects (constructing bridges, roads, or power plants) where specific actions and impacts tend to be encountered.

A diverse set of environmental checklists have been developed to provide analysts with a systematic framework so that they do not overlook important environmental considerations. Beyond the standard checklist (see Table 5.7), three additional types of checklists are in wide use:

- **Questionnaire checklists:** Provide a series of questions relating to the impact of a project; these checklists are particularly useful for less experienced practitioners.
- **Descriptive checklists:** Provide lists of environmental parameters, including information on impact identification and assessment that can assist analysts in identifying relevant impacts.
- **Weighting (scaling) checklists:** The most complex type of checklist, uses weighting factors to assess unquantifiable and intangible impacts using a common scoring and weighting scale.

In addressing cumulative impacts, the checklist needs to incorporate all of the activities associated with the proposal, as well as past, present, and reasonably foreseeable future actions. Figure 5.7 shows a hypothetical environmental checklist for identifying potential cumulative effects of a highway project.[10]

5.4.4 *Networks*

Another analysis method consists of the *network* and the *system diagram.* These methods provide a systematic and rigorous technique for identifying and tracing potential impacts through sequential cause–effect linkages. The concept of using a network diagram to progressively trace cause-and-effect relationships was pioneered by Sorensen in 1971.[11] Network and system diagrams provide one of the best tools for identifying cause-and-effect relationships.

Computerized expert systems can facilitate network and system analyses. There are many variations of the network and system analysis methods. The network analysis proceeds in only one direction (forward). In contrast, a system diagram allows loops or feedback from one part of the system to another. System diagrams can also provide a superior means of illustrating interrelationships.

Figure 5.8 illustrates the use of a simplified network diagram used by the European Commission in assessing the impacts of widening a channel. Network and system diagrams can also provide a robust method for assessing cumulative impacts. Figure 5.9 illustrates how a system diagram is used to assess the indirect effects of a single activity resulting in a cumulative impact on a single resource.[12] Specifically, this system diagram shows how fish spawning has been degraded as a result of aerial application of herbicides through five different pathways resulting in low-dissolved oxygen concentration and high-sediment stress. In this example, the low oxygen level is caused by decreased plankton growth and increased oxygen consumption from debris pollution and erosion. Increased sediment is also caused by increased erosion and debris pollution following loss of riparian vegetation.

5.4.5 *Carrying capacity analysis*

A method known as *carrying capacity analysis* is based on the fact that many environmental and socioeconomic systems have inherent limits or threshold levels. Carrying capacity is the maximum population of a species that an area or ecosystem can sustain. A system is considered to be unsustainable if the carrying capacity is exceeded. For example, a reservoir can only supply a finite amount of water to users, a road system can only efficiently accommodate a certain level of traffic, and a rangeland can only sustain a certain number of deer.

Potential impact area	Proposed action (three stages)			Other past actions	Other present actions	Future actions	Cumulative impact
	Construction	Operation	Mitigation				
Topography and soils	**	*		*			**
Water quality	**	**	+	*		*	***
Air quality		**		*			**
Aquatic resources	**	**	+	*		*	**
Terrestrial resources	*	*		*			**
Land use	*	***		*		*	***
Aesthetics	**	***	+	*			**
Public services	*	+				+	+
Community structure		*			*		*
Others							

*, Low adverse effect; **, moderate adverse effect; ***, high adverse effect; +, beneficial effect.

Figure 5.7 Example of environmental checklist for identifying cumulative effects of a highway project. (Courtesy of CEQ.)

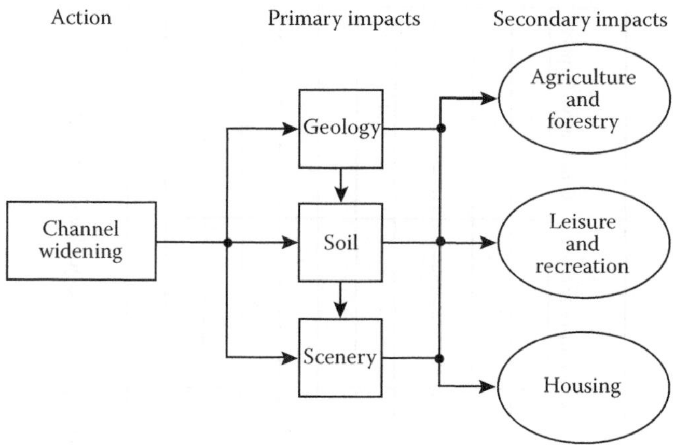

Figure 5.8 Network diagram used by the European Commission (1999).

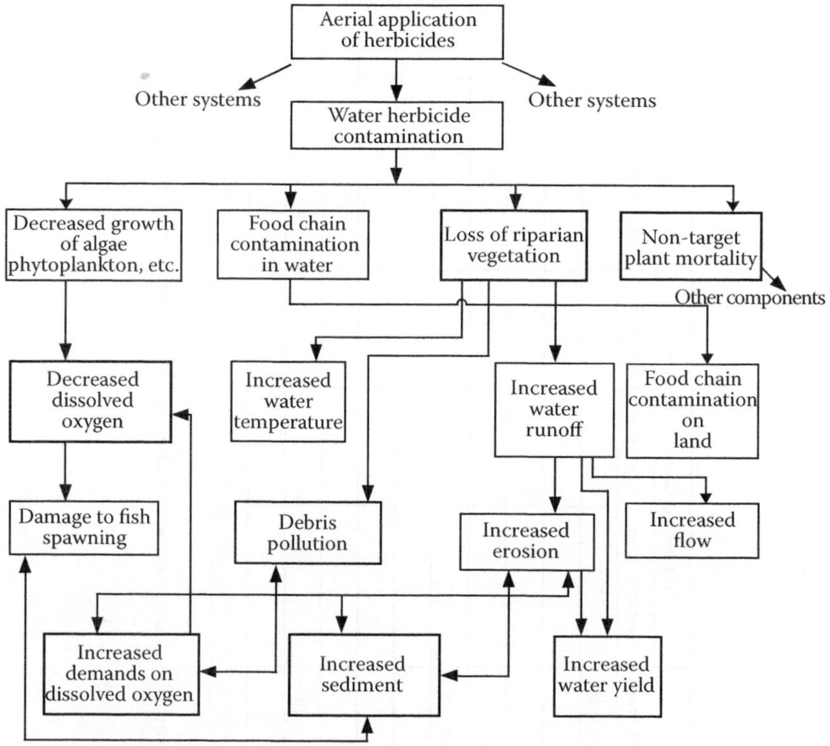

Figure 5.9 A system diagram can provide an excellent tool for assessing cause-and-effect relationships. In this example, a system diagram is used to evaluate the indirect cumulative impact of aerial herbicide application on an aquatic system. (Courtesy of CEQ.)

Figure 5.10 illustrates an example of carrying capacity. This figure shows what may happen once a population exceeds the carrying capacity. The species population overshoots the natural carrying capacity of the ecosystem and then crashes. The population then recovers and increases to the point where it again exceeds the carrying capacity. This oscillating cycle repeats indefinitely.

In terms of assessing cumulative impacts, the carrying capacity method can be used to identify thresholds for resources and systems, and can provide metrics for monitoring increases in incremental resource usage. The analysis begins with the identification of potential limiting factors (e.g., grazing land in a pasture). Mathematical equations may be developed for quantifying the capacity of the resource in terms of these limiting factors. This approach can be used to estimate the effect that a given project would have on the remaining resource capacity.

5.4.6 Ecosystem analyses

Traditionally, EIAs have tended to provide independent analyses of stand-alone resources such as air quality, hydrology, wildlife, and habitats. Although separate or segmented assessments tend to be more straightforward and easier to perform, they also tend to obscure interdependencies and interrelationships, particularly with respect to investigating cumulative impacts. Recognition of the interconnectedness of these resources analyses

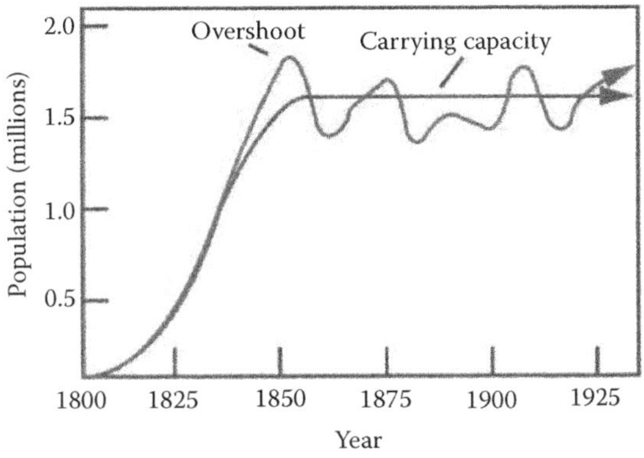

Figure 5.10 Population of a species generally has some given carrying capacity—point above which population growth cannot be sustained. Carrying capacity inhibits higher population because the amount of resources required to support population size has been exceeded. This figure shows what happens once a population exceeds carrying capacity. Species population overshoots the natural carrying capacity of the ecosystem and crashes. Population then recovers and increases to the point where it again exceeds carrying capacity. This oscillating cycle may repeat indefinitely. (Courtesy of images.google.com.)

has facilitated development of more comprehensive methodologies such as *ecosystem* and *watershed* management approaches. An ecosystem approach provides an important tool for assessing cumulative impacts because it considers the full scope of ecological resources that could be affected and their interrelationships. The ecosystem approach involves three basic principles:

1. Address interactions among ecological components that are needed to maintain the functioning of the ecosystem
2. Take a "big picture" approach when assessing the ecosystem
3. Use a diverse suite of biological indicators in assessing impacts

The CEQ has issued guidance on using an ecosystems approach in performing NEPA analyses.[13,14] Figure 5.11 shows a simplified schematic diagram illustrating the use of ecosystem management in performing a comprehensive assessment of the flow of nitrogen across land, water, and air resources.

Some basic biodiversity principles for performing a comprehensive ecosystem analysis are indicated in Table 5.8.[13] An ecosystem approach can provide a broad regional perspective, which addresses the following three cumulative impact principles:

- **Use natural boundaries:** Ecosystem analysis uses ecological boundaries or regions (watersheds, basins) in considering ecosystem functioning and in addressing issues such as habitat segmentation.

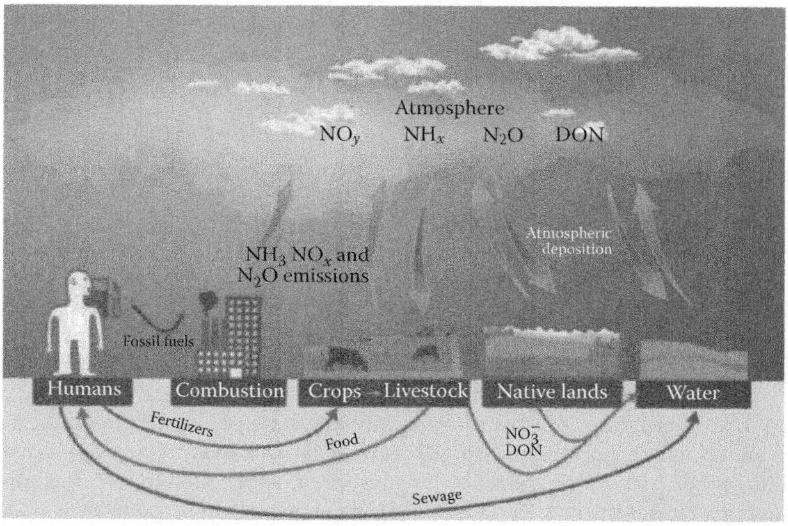

Figure 5.11 Simplified schematic illustrating application of an ecosystem management analysis approach in evaluating the flow of nitrogen across land, water, and air resources. (Courtesy of images.google.com.)

Table 5.8 Principles of Biodiversity Conservation

1. Take a "big picture" or ecosystem view.
2. Protect communities and ecosystems.
3. Minimize fragmentation. Promote the natural pattern and connectivity of habitats.
4. Promote native species. Avoid introducing non-native species.
5. Protect rare and ecologically important species.
6. Protect unique or sensitive environments.
7. Maintain or mimic natural ecosystem processes.
8. Maintain or mimic naturally occurring structural diversity.
9. Protect genetic diversity.
10. Restore ecosystems, communities, and species.
11. Monitor biodiversity impacts. Acknowledge uncertainty. Be flexible.

Source: CEQ 1993.

- **Focus on the resource or ecosystem:** Ecosystem analyses address biodiversity considerations and incorporate use of ecological condition indicators in assessing impacts.
- **Address resource or ecosystem sustainability:** Ecosystem management approaches specifically address the interactions and processes necessary to sustain the composition and function of an ecosystem.

5.5 *Investigating and describing the "affected environment" and "alternatives"*

This section provides direction for describing and assessing the "affected environment" and "reasonable alternatives." The documentation requirements for describing the affected environment and reasonable alternatives are described in Chapter 6.

5.5.1 *Describing the affected environment*

The Affected Environment section of an EIS describes the environmental baseline (affected environment) for investigating the impacts of a proposal. The description includes the existing physical, ecological, cultural, and socioeconomic resources *before* they are altered by the proposal (e.g., air and water quality, hydrology, sensitive species, historic structures, socioeconomics). The potentially affected environment is frequently referred to as the region of influence (ROI).

The EIS manager provides direction and monitors the effort to describe the affected environment, ensuring that this section is prepared in a cost-effective manner. Experience has shown that this is an area frequently subject to abuse in terms of excessive effort, unwarranted detail, and misspent resources. A tool (Sufficiency-Test Tool) detailed in the companion book, *NEPA and Environmental Planning*, provides a practical and defensible method for determining the description and level of detail that is sufficient to ensure that resources have been adequately described.[15] The description of the affected environment requires consideration of two domains:

- Spatial bounds
- Temporal bounds

5.5.1.1 *Determining spatial boundaries*

The geographical extent of the ROI may vary with the respective environmental resources. As part of the EIS scoping effort, the IDT establishes the geographic bounds of the study area (i.e., perimeter of the affected environment), recognizing that the size of the study area varies as a function of the specific impact and resource under investigation. To this end, the Affected Environment section of the EIS delineates boundaries of the ROI for each individual environmental resource and describes why these boundaries were chosen.

Depending on the nature of the activity and resources involved, some spatial dimensions may be very limited (e.g., project noise on humans), while in other cases they may be global in extent (e.g., ozone depletion or greenhouse emissions). Physical boundaries often provide a reasonable and defensible basis for delineating the borders of environmental resources. For example, a watershed may be defined as the geographical bounds for a hydrological or biological investigation.

In some cases, the affected environment may have no physically definitive boundaries. Frequently, the spatial boundaries are simply defined as the distance to which an environmental disturbance can be felt (i.e., ROI). It is reasonable to delineate this boundary as the maximum reasonable distance to which significant effects can be expected.

These boundaries are subject to change as new information becomes available. The EIS should include a rationale justifying how the spatial bounds were chosen. Maps should be used in delineating boundaries.

5.5.1.2 *Determining temporal boundaries*

Practitioners must consider both the timeframe (i.e., temporal boundary) over which specific actions will occur as well as the period in which impacts will continue if the action ceases. As with spatial boundaries, the temporal boundary may likewise vary with each impact to an environmental resource. For example, determining the temporal bounds for a noise analysis may be much shorter than the long-term effect on a groundwater aquifer. The temporal boundaries and any supporting assumptions are identified in the EIS. The EIS should include a rationale justifying how the temporal bounds for each resource were chosen.

5.5.2 Investigating reasonable alternatives

The analysis of alternatives is the "heart" of the EIS (§1502.14). A thorough investigation of a reasonable range of alternatives is a requisite for achieving NEPA's ultimate goal of excellent decision making. Regulatory citations pertinent to this requirement are presented in Table 5.9.

5.5.2.1 Identification and assessment of alternatives

As depicted in Figure 5.12, there are many potential courses of action that may need to be investigated for reducing impacts. By definition, the NEPA regulations recognize three types of alternatives (§1508.25[b]):

1. No action alternative
2. Other reasonable courses of action (including a proposed action)
3. Mitigation measures (not included as part of the proposal)

The range of alternatives may vary with the level of discretion the decision maker can exercise in reaching a final decision. Specifically

> The range of alternatives discussed in environmental impact statements shall encompass those to be considered by the ultimate agency decisionmaker. (§1502.2 [e])

Table 5.9 Direction for Analyzing Alternatives

Study, develop, and describe appropriate alternatives to recommended courses of action in any proposal which involves unresolved conflicts concerning alternative uses of available resources as provided by section 102(2)(E) of the Act (§1501.2[c]).
Use the NEPA process to identify and assess the reasonable alternatives to proposed actions that will avoid or minimize adverse effects of these actions upon the quality of the human environment (§1500.2[e]).
The... environmental impact statement shall.... inform decisionmakers and the public of the reasonable alternatives.... (§1502.1).
Agencies shall focus on significant environmental issues and alternatives (§1502.1).
The range of alternatives discussed in environmental impact statements shall encompass those to be considered by the ultimate agency decisionmaker (§1502.2[e]).
[The alternatives] section is the heart of the environmental impact statement (§1502.14).
Rigorously explore and objectively evaluate all reasonable alternatives.... (§1502.14[a]).
Develop and evaluate alternatives not previously given serious consideration by the agency (§1503.4[a][2]).

Figure 5.12 There are many potential courses of action that may need to be evaluated for reducing impacts. (Courtesy images.google.com.)

5.5.2.2 *Identifying alternatives*

An interdisciplinary effort should be used in identifying *all* feasible and practical approaches, for meeting the agency's underlying need for taking action. This effort should exhaustively pursue all potential avenues such that "no stone is left unturned."[16] Particular emphasis may be placed on identifying "best value" solutions within the defined constraints.

The identification and assessment of alternatives can be a dynamic process that evolves over time. While the steps described in this section are generally presented as a one-time event, in practice, the synthesis of alternatives is frequently an iterative process that may be revisited multiple times throughout the EIS process; this is particularly true for complex proposals. The EIS manager may need to consider additional or fewer alternatives as the options and their respective impacts become better understood. For example, the investigation of alternatives may need to be broadened to include new options identified on the basis of public comments received from circulation of the draft EIS (§1503.4[a][1] and [2]).

5.5.2.2.1 *Set of "reasonable" versus "analyzed" alternatives.* The *rule of reason* was briefly introduced in Chapter 2, Section 2.7.2 and Section 5.2.1. As viewed by the courts, NEPA is governed by the rule of reason. Properly exercised, this rule provides a defensible basis for defining the scope of reasonable alternatives. Consistent with the rule of reason, only *reasonable alternatives* are required to be analyzed. Reasonable alternatives are interpreted to mean those alternatives that are "practical or feasible" from a "technical and economic standpoint"; "common sense" is expected to be exercised in discriminating reasonable alternatives from those that are not.[17]

Chapter 3, Section 3.3 described how the statement of purpose and need (SPN) can provide a useful and defensible tool for identifying a range of reasonable alternatives for detailed investigation.[18] If the SPN is defined too broadly, the number of alternatives may be virtually unlimited. If, on the other hand, it is defined too narrowly, the SPN may eliminate consideration of potentially superior alternatives. Properly defined, the SPN provides a powerful tool for focusing on and reducing a large set of potential alternatives to a manageable set of reasonable alternatives for detailed examination.

5.5.2.2.1.1 Analyzed alternatives. The set of alternatives that are actually analyzed in detail is often referred to as the *analyzed alternatives*. The number of analyzed alternatives is frequently smaller than the set of identified alternatives; however, the set of analyzed alternatives must represent a full range of reasonable options. Table 5.10 lists criteria that may be helpful in screening a set of all possible alternatives down to a more manageable set of reasonable range for detailed analysis (e.g., analyzed alternatives). The Sufficiency-Test Tool presented in the companion book, *NEPA and Environmental Planning*, provides a practical and defensible tool for determining the description and level of detail that is sufficient to ensure that each alternative has been adequately described.[16]

5.5.2.2.2 Dismissing alternatives. Prudence must be exercised in eliminating alternatives. An alternative cannot necessarily be dismissed simply because it lies outside the legal jurisdiction of an agency or conflicts with a local or federal law (§1502.14[c]).[18] Alternatives can be rationally dismissed for being

- Technically or economically infeasible
- Speculative in nature
- Premised on unproven techniques and methods
- Too exotic

Table 5.10 Criteria That May Be Useful in Screening and Identifying a Manageable Set of Reasonable Alternatives for Detailed Analysis

• Is it economically reasonable?
• Is it technically practical or feasible?
• Is it reasonable and practical from the standpoint of "common sense"?
• Does it satisfy the agency's statement of underlying need?
• Could it show promise for avoiding or reducing environmental impacts?
• Does it meet reasonably defined schedule requirements?
• Does it meet functional requirements or constraints?

A defensible rationale should be cited for each alternative considered and then dismissed from detailed investigation. In one case, a court did not agree with an agency's rejection of an alternative as being economically unfeasible; the agency's administrative record failed to *demonstrate* why the alternative was considered economically unfeasible; such a demonstration may require analytical evidence (e.g., preparation of economic analysis, profit-margin studies, etc.).

5.6 *Assessing direct and indirect impacts, and significance*

As described in Section 5.3, each "component action" may produce *environmental disturbances* (e.g., emissions, effluents, sound waves, ground disturbances, or biological disruptions). These disturbances must be identified and described in detail sufficient to allow analysts to evaluate how they could affect their respective environmental resources. Pathways and cause-and-effect relationships are identified. As described in Section 5.4, tools such as network and system diagrams may be particularly useful in performing this task. Once cause-and-effect relationships are understood, analysts can begin assessing *how* the environmental resources or systems would respond to the actual disturbances.

As we have seen, environmental impacts are analyzed on a resource-by-resource basis. The focus of this step is on determining how environmental disturbances would affect or change (impact) the baseline environmental conditions. The ultimate goal is to prepare an analysis that clearly conveys to the reader what the disturbance to a given resource actually means. The EIS must evaluate three distinct types of environmental impacts (§1508.25[c]):

1. Direct
2. Indirect
3. Cumulative

5.6.1 *Describing impacts*

The assessment of environmental impacts attempts to gauge how a change to an environmental resource would affect humans and environment quality. The severity or degree of the potentially significant impacts dictates the appropriate level of analysis to be performed. Still, there is often considerable disagreement over the level of detail that is sufficient to adequately describe the impacts. Here again, the Sufficiency-Test Tool presented in the companion book, *NEPA and Environmental Planning,* provides a practical and defensible tool for determining the description and level of detail that is sufficient to ensure that impacts have been adequately described.[16]

Instead of simply stating that a resource "would be impacted," the analysis must indicate *how* environmental resources would be affected. For instance, it is much more instructive to describe how and to what degree a habitat would be perturbed than to simply state (as is sometimes the case) that the "… habitat would be impacted." To the extent feasible and practical, an environmental effect should be quantified using a technically appropriate unit of measurement.

Affected *resources or populations* should be clearly delineated and described. Where applicable, the *time* and *period* over which the impact would occur should be indicated, as should its *likelihood or probability*. As a minimum, analysts should strive to convey the following information to the reader:

- **Magnitude:** Where practical impacts should be quantified. If quantifiable measures are not possible or practical, effects may be described using qualitative descriptors such as "immeasurable," "minor," "large," or "substantial."
- **Duration and timing:** To the extent feasible, the EIS should denote the duration over which the effects would persist. The analysis should indicate when the impact would begin and end.
- **Beneficial or adverse:** The analysis should clearly describe the effect such that there is no misunderstanding as to whether an impact is beneficial or adverse. It is not uncommon to find that an action can result in both beneficial *and* adverse impacts.

5.6.2 "Reasonably foreseeable" versus "remote or speculative" impacts

As we have seen, the rule of reason is used by the courts in determining the degree to which impacts are investigated. Environmental effects deemed to be remote and speculative are generally considered to be unreasonable and therefore need not be investigated in detail. In contrast, reasonably foreseeable significant impacts must be evaluated. With respect to NEPA, there are no generally agreed upon definitions of the terms "reasonably foreseeable," and "remote or speculative." The author defines a *reasonably foreseeable impact* as

> Having a discernible cause-and-effect relationship such that there is an ability to reasonably predict or to reasonably anticipate that an action or impact will occur.

Impacts are generally considered to be reasonably foreseeable if there is a logical connection between the action and its resulting effect. In other words, an impact normally needs to be investigated if there is a reasonably discernible cause-and-effect relationship between the action and its subsequent impact.[19] With respect to cumulative effects, reasonably foreseeable impacts are interpreted broadly, to include impacts of actions even if an action has not been *formally* proposed.[20]

5.6.2.1 Remote or speculative

Fogleman has identified factors used by the courts for determining when an impact should be deemed remote or speculative[21]:

1. Level or degree of confidence that the agency has in predicting the impact
2. Amount of information available to the agency that provides a basis for describing the impacts in a manner meaningful to the decision maker

According to Fogleman, an action is likely to be deemed reasonably foreseeable if it is a logical "stepping stone" to potential local or regional development or accelerates such development. Conversely, the degree of speculation increases as a projected impact becomes removed or dissociated from the precipitating action. Adding an *additional step in the causal chain of events* tends to increase the degree of speculation, even if the incremental step (by itself) is considered reasonably foreseeable.

Table 5.11 lists common characteristics useful in discriminating between actions or impacts that are deemed to be virtually certain, "reasonably foreseeable," or "remote or speculative." This guidance is based on case law, regulatory inferences, and professional experience.

5.6.3 Indirect impacts

Indirect impacts are defined as those effects that are

> … caused by the action and are **later in time** or farther **removed in distance**, but are still reasonably foreseeable. Indirect effects may include **growth inducing effects** and other effects related to induced changes in the pattern of land use, population density or growth rate, and related effects on air and water and other natural systems, including ecosystems. (§1508.8[b], emphasis added)

Table 5.11 Common Characteristics of "Reasonably Foreseeable," versus "Remote or Speculative" Actions or Impacts

	Virtually certain	"Reasonably foreseeable"	"Remote and speculative"
Interpretation of these terms	Clear or present causation; one would have to be "dumb and blind" not to see or notice it.	Does not require a Nobel Laureate or PhD to identify the potential impact.	Very difficult to prove; pure conjecture; "grasping at straws."
Stage of planning	A formal proposal or plan has been prepared.	Preparation of a proposal or plan is under active consideration or development.	No effort has been made to formulate a tangible proposal or plan.
Degree of control over the action or impacts	The agency has a considerable degree of control over the potential action or impact (e.g., alternatives, approval or authorization, mitigation measures).	The agency can probably exercise some degree of control over the action or its impact.	The agency has little or no control over the action or impact; the temporal and spatial domain may allow intervening actors.
Confidence or degree of concern	High degree of confidence in the environmental forecasts; often involves recognized and definite concerns about the action or impact.	Some level of confidence in the environmental forecasts; may involve some real concerns with respect to the action or impact.	Little or no confidence in the projections; often involves little concern over the actual action or impact; disagreements often tend to be centered on personal values or concerns rather than scientifically established concerns.
Debate	The action or impact tends to involve a high degree of certainty or degree of confidence; however, there are often technical debates over the specifics.	There may be some important disagreements or conflicts among experts.	Substantial conflict and virtually no agreement among experts.

(*continued*)

Table 5.11 Common Characteristics of "Reasonably Foreseeable," versus "Remote or Speculative" Actions or Impacts (Continued)

	Virtually certain	"Reasonably foreseeable"	"Remote and speculative"
Outside entities	Often involves clear concerns among outside citizens groups or other entities.	Often involves some concern or interest among outside entities.	Little concern or interest among outside entities.
Benefit/impairment	Some entities will probably either benefit or be impaired by the action/impact; entities can be clearly identified.	Some entities may benefit or be impaired by the action/impact; however it may be uncertain as too exactly who.	Specific entities cannot be clearly identified as either benefiting or being impaired.
Evidence	The agency has direct and documented or scientific knowledge; specific past examples or clear precedents provide a basis for making projections.	General studies of similar examples indicate a reasonable likelihood that the projections will occur.	Vague or implausible examples or precedents.
Documentation	Substantial evidence and documentation to support conclusions.	Some evidence or documentation to substantiate conclusions.	Little credible data; a theoretical worst-case analysis is questionable or very conjectural.

Table 5.12 Comparison of Traits for Direct versus Indirect Impacts

	Direct impact	Indirect impact
Time domain	"Now"	At some time in the future
Space domain	"Here"	Removed in distance
Is it a growth-induced effect?	Possibly	Frequently

There can sometimes be a fine line in the distinction between direct and indirect impacts. Table 5.12 shows key characteristics useful in differentiating between a direct and an indirect impact. As indicated by the table, an indirect effect is removed from the agency's proposal in the time and/or spatial domain.

5.6.4 Interpreting significance

As noted in Section 5.3, once the analysis of impacts has been completed, the IDT must interpret and explain the significance of the impacts. The goal is to convey to the decision maker and public the implications of the impact. The EIS needs to clearly convey to the reader the importance (e.g., large or small impact) of the effects. This discussion should interpret how, why, and to what degree environmental impacts are significant.

For instance, what would a contaminate release into a stream used by a community for its water supply mean in terms of the incidence of human cancer rates? What would the number of fish killed really mean in terms of environmental quality and long-term sustainability? The companion book *NEPA and Environmental Planning* provides more extensive direction on assessing and interpreting significance.[22]

5.7 Performing a health impact assessment in an EIS

Even when human health is not the primary focus of an agency proposal, an action may still have significant health impacts that should be factored into the decision-making process. Such an inquiry may involve preparation of a Health Impact Analysis (HIA). The National Research Council of the National Academies has prepared guidance for preparing an HIA (HIA Report).[23] This section summarizes this guidance.

An HIA is a systematic process for assessing the potential effects of a proposed policy, plan, program, or project on the health of individuals, a population, and the distribution of those effects within the population. The HIA can serve as a basis for recommendations on monitoring health effects and mitigating adverse effects.

The analysis

- Uses a host of data sources and analytic methods
- Considers input from stakeholders to identify effects
- Communicates effects to decision makers and the public

5.7.1 General guidance

Both the NEPA statute and its Regulations, as well as executive orders establish the foundation for including an analysis of health impacts in an EIS. In fact, one of the factors cited in the Regulations for determining significance involves (§1508.27[b][2])

> The degree to which the proposed action affects public health or safety.

An appendix to the National Research Council's report provides general NEPA guidance on the issues summarized below.

5.7.1.1 Determining when to analyze health impacts

The NEPA regulations require that health impacts be analyzed in detail only when there is reason to conclude that they may be significant (§1501.7[a][3]). The HIA report identifies potential factors to be considered in determining the significant health impacts:

- Scoping comments
- Whether health concerns are controversial (§1501.7 and §1508.27[b][4])
- Whether the proposal could result in significant changes to factors known to affect health, such as changes in
  - Emissions of hazardous substances
  - Community demographics
  - Industry actions or practices
  - Employment, government revenues, or land use patterns
  - Modes or safety of transportation; access to natural resources
  - Food and agricultural resources

5.7.1.2 Determining the appropriate scope of analysis

As appropriate, the EIS considers potential direct, indirect, or cumulative health impacts of the proposal (§1508.8). Determinants of health can include factors as diverse as

- Quality and affordability of housing
- Access to employment and government revenues

- Quality and accessibility of parks, schools, and transportation services
- Neighborhood safety
- Exposure to environmental hazards
- Quality and affordability of food resources
- Extent and strength of social networks

5.7.1.3 Identifying affected populations

The description of the affected environment provides the baseline against which health impacts of various alternatives can be compared. The HIA report advises that the baseline include a concise description of public health status and health determinants relevant to the health impacts that will be analyzed. Consultation with relevant health agencies may be desirable.

5.7.1.4 Performing the assessment and mitigation measures

While the NEPA regulations do not provide specific guidance on methods for assessing health impacts, they establish basic standards and expectations regarding the interdisciplinary and scientific approach to be used.[1] When faced with uncertainty, this may include making informed judgments about reasonably foreseeable impacts.

The EIS must consider health-based mitigation measures. Mitigation may be implemented by the lead agency and through actions taken by a cooperating agency, another government entity, or a local, state, or tribal health department, or through voluntary actions taken by a project proponent.

5.8 Performing the cumulative impact assessment

As we have seen, the NEPA regulations defines cumulative impacts as

> … the impact on the environment that results from the incremental impact of the action when added to other past, present, and reasonably foreseeable future actions, regardless of which agency (federal or non-federal) or person undertakes such other actions. Cumulative impacts can result from individually minor but collectively significant actions taking place over a period of time. (§1508.7)

Thus, a cumulative impact assessment (CIA) must evaluate an action's incremental (i.e., direct and indirect) impacts combined with the effects of other past, present, and reasonably foreseeable future actions, to provide the decision maker and the public with a full understanding of the potential significance. One of the principal reasons for performing a CIA is to determine if a project's small incremental impact, when added to other impacts, could breach a significance threshold (i.e., the straw that breaks the camel's back). The results of the CIA should be incorporated into the agency's overall environmental planning.

This section examines requirements and best professional practices for performing a CIA. The companion book, *Environmental Impact Assessment*, provides a detailed examination of best professional practices for evaluating cumulative impact.[2]

5.8.1 Avoiding legally deficient analyses

One court set down widely cited requirements for performing an adequate CIA. This direction is indicated in Table 5.13 and has been reinforced by other courts.[24]

The assessment of cumulative impacts is a convoluted requirement that has been the subject of endless disputes and confusion, to say nothing of litigation. One study of court cases involving CIA issues concluded that the most common reason agencies lost in court was an inadequate analysis of past, present, and reasonably foreseeable future actions; agencies also lost numerous cases because their CIAs lacked supporting data or rationale.[25]

5.8.1.1 Examples of flawed cumulative impact assessment

As witnessed in the case study presented in Chapter 1, the Nuclear Regulatory Commission prepares EISs for renewing the operating licenses of nuclear power plants. It stretches the imagination to believe that any meaningful CIA can be provided in a terse one or two sentence statement. Yet, consider how a recent and typical relicensing EIS addresses the cumulative impacts of operating a nuclear plant. The EIS looked at cumulative impacts across 10 different environmental resources (air, water, terrestrial ecology, etc.). The CIA consisted of woefully inadequate "assessments" as short as one or two sentences in length. Not only do their

Table 5.13 Direction Provided by Courts for Performing Adequate Cumulative Impact Assessment

The specific area in which effects of the proposed project would be felt
Impacts that are expected in that area from the proposal
Other past, present, and reasonably foreseeable future actions
Expected impacts from these past, present, and reasonably foreseeable future actions
Overall cumulative impact if the individual impacts were allowed to accumulate

CIAs fail to meet NEPA's regulatory direction, they also do not comply with direction provided by the courts as shown in Table 5.13. Consider four of the Commission's one-sentence CIA assessments in light of the court direction provided in Table 5.13 and the definition of a cumulative impact just presented (§1508.7):

1. **Water quality effects:** "Cumulative impacts to water quality would not be expected because the small amounts of chemicals released by these low-volume discharges are readily dissipated in the receiving waterbody."
2. **Land use effects:** "The significance of any impacts is so minor and localized that cumulative impacts are not an issue."
3. **Terrestrial ecological effects:** "No mitigation measures beyond those implemented during the current term license would be warranted and little potential for cumulative impacts is indicated."
4. **Groundwater effects:** "Hence, the contribution of plant operations (during the license renewal period) to the cumulative impacts of major activities on groundwater quality would be relatively small."

This was the total verbiage devoted to the subject of cumulative impacts for each of the four disciplines. Now consider how each of these one-sentence assessments suffers from the following six common errors:

1. **Dismissing the cumulative impact because the project impact is deemed small:** The four "assessments" dismiss the potential for a cumulative impact based on the flawed supposition that because the impact of the proposal is small, there is essentially no cumulative impact. This defeats the very purpose for performing a CIA. The principal purpose of a CIA is to combine impacts, regardless of how small, to determine if the sum total of the impact threatens to breach the threshold of significance (§1508.7). A significant cumulative impact can occur *even though the project impact is small.* Mr. Bo Pham, the Commission's public official responsible for managing and preparing these EISs, approved these statements without even understanding the underlying purpose for performing a CIA.
2. **Neglecting to define spatial bounds:** Because the CIA involves other past, present, and future activities, the CIA geographic (spatial) boundaries are typically different and often much larger than those used in the analysis of direct and indirect impacts. The courts have been adamant in providing direction that the spatial bounds be explicitly defined. Yet, not a single CIA in these EISs defined its spatial bounds. Without defining the spatial bounds for each environmental resource, it is next to impossible to even determine what

is being considered and analyzed in terms of cumulative actions and cumulative impacts. A defensible CIA *must* define the spatial bounds of the assessment.

3. **Neglecting to define temporal bounds:** As with the problem involving spatial bounds, the CIA temporal boundaries may be different from those used in the analysis of direct and indirect impacts. The courts have been adamant in demanding that temporal bounds be defined. Despite such direction, not a single CIA in these EISs defined its temporal bounds. Without defining the temporal bounds for each environmental resource, it is difficult, if not impossible, to determine the extent to which cumulative actions and impacts have been adequately addressed. A defensible CIA *must* define the temporal bounds of the assessment.
4. **Neglecting to identify impacts of other actions:** As just noted, the adequacy of a CIA depends on how accurately the analysis considers impacts of other past, present, and reasonably foreseeable actions. The courts have stated that the impacts of other past, present, and reasonably foreseeable actions be identified and investigated. This requires identifying other past, present, and reasonably foreseeable future actions. Yet, as illustrated in the four examples above, this was not done in even a single instance. It is impossible to perform a comprehensive CIA without considering the impacts of other past, present, and reasonably foreseeable actions.
5. **Failing to rigorously add the impact of the proposal to other actions:** An accurate and defensible CIA can only be performed by "adding" or combining the impact of the proposal to that of other past, present, and reasonably foreseeable actions. In most instances, this was either not done or was performed in a less than rigorous manner that lacked defensible or convincing evidence.
6. **Failing to consider cumulative significance:** The central purpose for performing a CIA is to determine if a proposal's impact contribution (however large or small) is sufficient to breach the threshold of significance; the CIA provides the means of determining if a project would result in a significantly cumulative effect upon a given resource. As witnessed by the four examples above, no serious attempt was made to assess cumulative significance by combining the project impacts with other projects and activities.

5.8.1.2 Concealing cumulative risk

The problems just cited pale in comparison to the way NRC conceals cumulative risk of a major nuclear accident from the public. That the cumulative impacts from a major accident would be felt by millions and could sweep across many states if not a sizable portion of the North American continent is undeniable; radiation released from a single accident could threaten tens of thousands of citizens; then there are the potentially paralyzing socioeconomic and relocation impacts of a major accident.

As we have seen, the NRC looks at the risk of an accident and reaches the confounding conclusion that the risk of a "serious accident" such as a full-scale meltdown of a nuclear reactor is "small." But even more troubling is the fact that nowhere within these relicensing EISs has any consideration been given to evaluating *cumulative risk*. As noted in the case study in Chapter 1, the NRC typically presents the following terse 48-word "canned" statement concerning the risks posed by a "severe" nuclear accident:

> The probability weighted consequences of atmospheric releases, fallout onto open bodies of water, releases to ground water, and societal and economic impacts from **severe accidents** [such as a nuclear meltdown] are **small** for all plants. However, alternatives to mitigate severe accidents must be considered for all plants that have not considered such alternatives.

As we have seen, this bewildering conclusion is scientifically indefensible in light of recent experiences and circumstances. But there is an even more troubling flaw that has been concealed from the American public. The NRC's analysis of severe nuclear plant accidents only considers the probability and consequences of an accident from a single nuclear operating station. But this is not the case at all. There are actually 104 commercial nuclear reactors in the United States located in 31 states. The actual "cumulative risk" to the American citizens posed by an entire fleet of operating reactors is much greater than that posed by a single lone nuclear reactor. Consider this analogy; a driver has a much higher (cumulative) risk of being struck by another automobile if there are 100 other vehicles traveling down the highway compared with a situation where there is only one automobile.

Yet, the Commission's EISs do not even acknowledge, let along compute or disclose, this total or cumulative risk. Nor do they describe (as required by the Regulations) the actual cumulative consequences (impact) from multiple catastrophic nuclear accidents. Nevertheless, the Commission grants renewed operating licenses based on scientifically unsound assessments approved by management which is responsible for ensuring the accuracy of these analyses. Unfortunately, such egregious practices raise an even larger question. If the Commission approved CIAs that did not even meet rudimentary regulatory and legal requirements, why should the scientific community and public have confidence in the Commission's conclusion that these nuclear reactors can be safely operated for an additional 20-year period? This example clearly illustrates the responsibility that agencies bear in computing and disclosing cumulative risk and impacts to the public.

5.8.2 *Defining the cumulative impact baseline*

As noted in Section 5.8.1, a defensible CIA requires that the agency properly identify and define the bounds of the analysis and the impacts of other past, present, and reasonably foreseeable actions.

5.8.2.1 *Defining spatial and temporal boundaries*

As noted earlier, one of the reasons that the CIA tends to be more challenging than the corresponding evaluation of direct and indirect effects is simply due to the difficulty of defining the spatial (geographic) and temporal (timeframe) boundaries. If these boundaries are defined too broadly, the analysis can become exhausting and unwieldy. If defined too narrowly, the analysis may be insufficient to inform decision makers of potentially significant cumulative impacts. Just as it is used in identifying other key issues or effects for analysis, a well-orchestrated scoping process plays an integral role in identifying the spatial boundaries and timeframes.

5.8.2.2 *Identifying other past, present, and future activities*

Past, present, and reasonably foreseeable future activities can be identified once the temporal and spatial boundaries are defined. Identifying the impacts of other past and present actions is sometimes relatively straightforward. Identifying the impacts of reasonably foreseeable activities can be more daunting.

5.8.2.2.1 Reasonably foreseeable actions. Reasonably foreseeable actions include those projects and activities that are ongoing and are likely to change or expand, as well as those that don't yet exist but can be reasonably anticipated. For instance, urban sprawl may be a future growth-induced impact that could result from a proposed action to construct a large plant employing several thousand workers or perhaps a highway interchange.

Municipal planning and zoning offices are often a good source of information on projects that are contemplated or under review. Common sources of information concerning reasonably foreseeable future activities are noted in Table 5.14.

5.8.3 *Five-step procedure for assessing cumulative impacts*

The importance of preparing an accurate CIA is demonstrated in a case where the Sierra Club sued the US Forest Service, claiming that its EIS had improperly investigated cumulative effects of various land uses for a resource management plan. The Sierra Club argued that the EIS simply cited a laundry list of individual effects. The court sharply rebuked the Service for failing to include all effects of the various activities that could occur and that it had not evaluated impacts of various activities in combination with one another.[26]

Table 5.14 Sources of Information on Reasonably Foreseeable Future Activities That May Need to Be Included in Cumulative Impact Assessment

Projects directly related to or associated with the proposal
Projects or activities not directly related to or associated with the proposal, but which would likely be induced as a result of the project's approval (support facilities, stores, malls, housing)
Projects identified in a development plan (such as a comprehensive plan or master plan) for the area
Projects officially announced by developers or project proponents
Projects currently undergoing regulatory review with a reasonable possibility of approval
Projects that have been formally approved

A simplified five-step procedure for performing a CIA is outlined in Table 5.15 and described in detail in the companion book, *Environmental Impact Assessment*[2]; this book also details a more rigorous and systematic 15-step procedure for performing a CIA.

As just noted in the case study, the CIAs prepared by the NRC do not even comply with even this simplified five-step approach. The *proximate cause test* (described in the next section) noted in Step 1 can help prevent unnecessary analysis and wasted effort.

5.8.3.1 Proximate cause: defining limits of the analysis

In placing limits on the extent of the analysis, the US Supreme Court appears to have provided direction indicating that NEPA requires "a reasonably close causal relationship" between an impact and its cause.[27,28] This ruling has its roots in the *Doctrine of Proximate Cause* from tort law.[29] It can be defined as follows:

> "Proximate cause" is merely the limitation which the courts have placed upon the actor's responsibility for the consequences of the actor's conduct. In a philosophical sense, the consequences of an act go forward

Table 5.15 Simplified Five-Step Procedure for Performing Cumulative Impact Assessment

Step 1: Apply the proximate cause (outlined below) screening test
Step 2: Reference or describe the environmental resources to be reviewed
Step 3: Determine spatial and temporal bounds
Step 4: Determine no-action baseline
Step 5: Determine proposal's incremental impact by combining the impacts of the proposal with the effects of all other past, present (no-action baseline described in Step 4), and reasonably foreseeable future actions

> to eternity, and the causes of an event go back to the dawn of human events, and beyond.... As a practical matter, legal responsibility must be limited to those causes which are so closely connected with the result and of such significance that the law is justified in imposing liability. Some boundary must be set to liability for the consequences of any act, upon the basis of some social idea of justice or policy.[30]

The word "proximate" means "close in space and time, or close in relationship." In other words, the Supreme Court interpreted the CIA provision in terms of restricting an agency requirement to only consider the incremental impact *proximately* caused (closely related) by the proposed action in the context of the existing conditions, together with other present and future actions affecting the same resource. For a detailed discussion of how the proximate cause test can be used in evaluating cumulative impacts, the reader is referred to the companion book, *Environmental Impact Assessment*.[2]

5.8.3.1.1 Limiting the scope of analysis. Consistent with the proximate cause test, the Supreme Court has emphasized that agencies may properly limit the scope of their cumulative effects analysis based on practical considerations. In this case, the court wrote[31]:

> Even if environmental interrelationships could be shown conclusively to extend across basins and drainage areas, practical considerations of feasibility might well necessitate restricting the scope of comprehensive statements.

5.8.4 Performing the CIA

Performing a CIA for additive effects can be relatively straightforward. In some instances, the EIS may only need to sum the magnitudes of the effects to assess the combined effect. Consider the daily cooling water use from a reservoir based on an actual CIA for a proposed power plant (Table 5.16). The table shows the daily water consumption from existing users and compares it to the available storage capacity of the reservoir. It then adds the incremental use from a future water user and the proposed action to obtain the cumulative use (impact).

The reader should note that the CIA would have been more accurately performed by adding the existing use to the future water user, *All-in-One Power*. The consumption of the proposed power plant would then be added to determine what incremental effect the proposed action would have in terms of all other existing and future users.

Table 5.16 Cumulative Daily Water Use from a Reservoir That Would Provide Water for a Proposed Power Plant

Facility	Water withdrawn (mgd)	Water returned (mgd)	Net water use (mgd)	Reservoir storage required (acre-ft)	Percent of total reservoir storage
Existing water users					
Regional water system	6.00	5.40	0.60	108	0.01
Nearby manufacturing plant	6.30	5.70	0.60	108	0.01
City of Turkeyville	60.00	0.00	60.00	10,200	0.99
Town of Bestcity	0.50	0.45	0.05	9	<0.01
Town of Neatestcity	0.06	0.00	0.06	23	<0.01
Subtotal	72.86	11.55	61.31	10,448	1.01
Future water users					
All-in-one Power	25.20	2.10	23.10	4158	0.40
The proposed action	3.14	0.75	2.39	430	0.04
Subtotal	28.34	2.85	25.49	4588	0.45
Total reservoir use	101.20	14.40	86.80	15,036	1.46

Table 5.17 compares a narrative versus a quantitative description of the cumulative effects associated with an increase in nitrogen oxide (NO_x) concentrations.

In reality, many CIAs are not as straightforward as illustrated by these two examples. To complicate matters, a cumulative effect may result from simple *additive* disturbances or from complex *interactive* phenomena that can be much more complex. As explained in the next section, the very definition of a "cumulative impact" can result in a paradox.

5.8.5 Eccleston's Cumulative Impact Paradox

The importance of assessing cumulative impacts is underscored by one of the factors required to be considered in reaching a determination regarding potential significance:

> … whether the action is related to other actions with individually insignificant but cumulatively significant impacts. Significance exists if it is reasonable to anticipate a **cumulatively significant impact** on the environment. (§1508.27 [b][7])

Table 5.17 Example of How a Table Can Be Used to Summarize a Cumulative Impact Analysis in Terms of Narrative versus Quantitative Description

	Past action	Present action	Proposed action	Future action	Cumulative effect
Narrative description	No discernible effect on NO_x levels	Notable deterioration in visibility during spring, but standards are still met	Visibility further affected by the project, but standards are still met	Increased vehicle emissions are expected	Standards are likely to be exceeded
Quantitative assessment		5% increase in NO_x concentration, but standards are met	10% increase in NO_2x concentration, but standards are met	5% increase in NO_x concentration	20% increase in NO_x concentration will exceed regulatory standards

As described below, this significance criterion can result in a paradox known as *Eccleston's Cumulative Impact Paradox* (*Eccleston's Paradox*).[2]

5.8.5.1 Eccleston's Paradox

By definition, a finding of no significant impact (FONSI) means an action that

> 1. ... Will **not** have a **significant** effect on the human environment (§1508.13)

Moreover, a "Categorical Exclusion" [CATX] means

> 2. ... A category of actions which do not individually or **cumulatively** have a **significant effect** on the human environment and which have been found to have no such effect... and for which, therefore, neither an environmental assessment nor an environmental impact statement is required.... (§1508.4)

A puzzling paradox arises when one considers these provisions in terms of considering the effects of other potentially significant past, present, and reasonably foreseeable future activities (cumulative impact baseline). Consider the following example. A proposal is made to construct a federal building in a crowded downtown business area of a large city. The area has already sustained significant cumulative impacts across multiple environmental and socioeconomic resources. For instance, natural vegetation and wildlife habitat originally present in the area have been destroyed, and the downtown area is now paved with concrete, buildings, and skyscrapers. The underground aquifer from which the city derives its drinking water has been contaminated, and the water table has sustained a significant drawdown. Ambient air quality has been significantly degraded. Fish and other aquatic species in nearby rivers and streams have experienced a substantial decline. Destruction of wetlands and construction of impermeable pavement has increased the risk of flooding within the city and downstream of the city. Streets are noisy and congested with traffic. The visual quality of the once rustic setting has been significantly degraded. As a result of the impacts of past and present actions, a number of environmental resources have already been significantly affected, from a cumulative standpoint. The proposed project in conjunction with other reasonably foreseeable future actions will only worsen these problems.

A strict interpretation of regulatory citations no. 1 and no. 2 (above) leads to the conclusion that a project is *not* eligible for either a CATX or a FONSI if that proposal adds *any* contribution to a cumulative impact that has already breached the threshold of significance. Moreover, the impact of an action in an EIS need not be investigated in detail as long as the effect is deemed to be nonsignificant; but if the resource has already sustained a significant cumulative impact, then any incremental contribution from the proposed action can be deemed significant, requiring a detailed examination of the impact regardless of its magnitude.

This leads to a paradox. If an environmental resource has already sustained a cumulatively significant impact, how can a decision maker declare that any action contributing *any* incremental impact (however small) is eligible for a CATX, FONSI, or does not require detailed examination in the EIS? Paradoxically, this leads to the conclusion that many (if not most) mundane activities should actually be ineligible for a CATX or FONSI, thus requiring preparation of an EIS; moreover, many impacts routinely dismissed in the EIS as nonsignificant should instead be classified as cumulatively significant and therefore subject to detailed investigation, including analysis of applicable alternatives and mitigation measures for reducing the significance of their impacts.

Thus, strict compliance with the aforementioned regulatory provisions (nos. 1 and 2) results in an unreasonable and voluminous increase in both the number of required EISs (even where the incremental impacts would be small) as well as substantially increasing the level of effort within an EIS to investigate cumulatively significant impacts. This regulatory interpretation might even eliminate the use of many if not most CATXs and FONSIs. For instance, in the example of the downtown area just described, a strict interpretation of cumulative significance leads to the conclusion that a federal agency would have to prepare an EIS to construct something as mundane as a walkway or a small parking lot. Clearly, a strict interpretation of NEPA's regulatory requirements can lead to absurd and unreasonable results. This violates court direction that NEPA is to be governed by the *rule of reason described in Section 2.7.*

5.8.5.1.1 Importance of resolving the paradox. As we have seen, NEPA is governed by the rule of reason (see Sections 2.7 and 5.2). That is, "reason" should prevail when a regulatory requirement results in an absurd outcome. A regulatory provision leading to the conclusion that an EIS is required, even in situations where it would contribute little or no substantive value to the decision-making process, contradicts the rule of reason; preparing a detailed investigation on a trivial impact just because an environmental resource has sustained a significantly cumulative impact can likewise be viewed as not only unreasonable but a wasteful use of resources. Moreover, the paradox conflicts with direction to prepare EISs that "…concentrate on the issues that are truly significant to the action in question, rather than amassing needless detail" (§1500.1[b]).

Fortunately, there is a systematic and defensible solution for resolving this paradox. The companion book, *Environmental Impact Assessment,* details a procedure referred to as the *Significant Departure Principle*, which provides a systematic, peer-reviewed, and defensible technique for resolving this paradox.[2] This paradox will be revisited again in Section 5.9 in terms of the evaluating greenhouse emissions.

5.9 Performing a greenhouse gas and climate change assessment

Research on GHG emissions and climate change impacts is an emerging and rapidly evolving area of science. While much of the scientific community believes that the increase in atmospheric GHG concentrations are changing the earth's climate, there is still a significant segment that want to see more convincing evidence.

Human activities are producing in the neighborhood of 50 billion tons of GHG annually (measured in carbon dioxide equivalency).[32] Ambient concentrations of GHGs do not cause direct adverse health effects (such as respiratory or toxic effects), but public health risks and impacts as a result of elevated atmospheric concentrations of GHGs might occur via climate change.[33]

Until recently, the issue of potential consequences of GHG emissions in EISs had been all but ignored. However, federal agencies are devoting increased attention to this issue in their EISs. Yet, the issue is fraught with misinformation, controversy, confusion, disputes, and is increasingly the subject of litigation. There is still a great deal of uncertainty, and even conflicting or contradictory evidence. This merely adds to the difficulty of preparing an already complicated GHG assessment. The following section examines the issue of GHG emissions and climate change impacts, and attempts to provide some best professional practices for performing the analysis. The companion book, *Environmental Impact Assessment*, provides a more detailed guide to best professional practices for evaluating GHG and climate change impacts.[34]

5.9.1 General direction for performing the assessment

Some general direction for performing the assessment is provided below. As witnessed earlier, NEPA is governed by the rule of reason. This guidance helps ensure that the EIA is focused on issues that truly merit study, and that those of less importance to the decision-making process are deemphasized.[35] Much of the direction that follows is based on the rule of reason.

5.9.1.1 Dealing with uncertainties

Few other environmental issues are riddled with more confusion and uncertainty. At present, given the state of the art, it may be near impossible for an agency to make definitive statements concerning the consequences of GHG emissions. The Regulations provide for instances where analysis of an impact lies beyond the state of the art or involves *incomplete or unavailable information* (40 CFR §1502.22). This regulatory direction is presented in Table 5.18. The EIS must also discuss any "responsible opposing view(s)" such as scientific opinions that run counter to the IPCC findings (§1502.9).

As difficult as GHG analyses can be, a number of courts have ruled that they must be addressed in an EIS. For instance, one court ordered that an EIS for the proposed Corporate Average Fuel Economy (CAFE) standards on passenger cars and light trucks include a discussion of GHG emissions. Because the proposal involved substantial uncertainty, including incomplete or unavailable information, regarding the potential impacts, one of the statements presented in the EIS read:

Table 5.18 Direction for Dealing with Incomplete or Unavailable Information (§1502.22)

When an agency is evaluating reasonably foreseeable significant adverse effects on the human environment in an environmental impact statement and there is incomplete or unavailable information, the agency shall always make clear that such information is lacking:
(a) If the incomplete information relevant to reasonably foreseeable significant adverse impacts is essential to a reasoned choice among alternatives and the overall costs of obtaining it are not exorbitant, the agency shall include the information in the environmental impact statement.
(b) If the information relevant to reasonably foreseeable significant adverse impacts cannot be obtained because the overall costs of obtaining it are exorbitant or the means to obtain it are not known, the agency shall include within the environmental impact statement:
1. A statement that such information is incomplete or unavailable
2. A statement of the relevance of the incomplete or unavailable information to evaluating reasonably foreseeable significant adverse impacts on the human environment;
3. A summary of existing credible scientific evidence that is relevant to evaluating the reasonably foreseeable significant adverse impacts on the human environment
4. The agency's evaluation of such impacts based on theoretical approaches or research methods generally accepted in the scientific community. For the purposes of this section, "reasonably foreseeable" includes impacts that have catastrophic consequences, even if their probability of occurrence is low, provided that the analysis of the impacts is supported by credible scientific evidence, is not based on pure conjecture, and is within the rule of reason.

> … the magnitudes of the changes in these climate effects that the alternatives produce—a few parts per million (ppm) of CO_2, a hundredth of a degree C [centigrade] difference in temperature, a small percentage-wise change in the rate of precipitation increase, and a 1 or 2 millimeter… sea level change—are too small to meaningfully address quantitatively in terms of their impacts on resources.

The author presents a simplified five-step procedure for assessing GHG emissions in the next section.

5.9.2 Five-step procedure for assessing GHG emissions

As just noted, there is considerable confusion regarding the procedural process that should be used in evaluating GHG emissions and climate change impacts. Table 5.19 illustrates a simplified five-step procedure for assessing potentially significant GHG effects. A more comprehensive and systematic 15-step procedure is presented in the companion book, *Environmental Impact Assessment*.[36]

5.9.3 Investigating alternatives and mitigation measures

To the extent practical, the EIS should evaluate potential measures for mitigating GHG emissions. As appropriate, it should also discuss the permanence, verifiability, and enforceability of such measures. Mitigation measures may include enhanced energy efficiency, lower GHG-emitting technology, renewable energy, planning for carbon

Table 5.19 Simplified Five-Step Procedure for Assessing GHG Emissions and Impacts

1. Identify and quantify the amounts of each GHG emission (and as appropriate provide a total in carbon dioxide equivalents); be conscious of the fact that certain gases such as methane are considerably more potent GHGs than carbon dioxide (CO_2).
2. Investigate potential means for avoiding GHG emissions. As reasonable, include alternatives for reducing emissions; if no reasonable alternatives are available, this fact should be stated.
3. Identify and investigate reasonable mitigation measures that can minimize or compensate for GHG emissions.
4. Document the assumptions and scientific methods used in analyzing the impacts.
5. Analyze the impacts of these GHG emissions (reductions or offsets) based on best existing data (noting incomplete or unavailable data per §1502.22).

capture and sequestration, and capturing or beneficially using fugitive methane emissions.

5.9.3.1 Carbon neutral program

One approach for addressing GHG emissions is to focus alternatives and mitigation measures on a carbon neutral program. In this case, the analysis should take credit for activities that can offset the GHG impacts:

- Environmental awareness programs
- Recycling
- Carbon sequestering (if practical)
- Mulch programs

For example, recycling 1000 pounds of paper as opposed to manufacturing it from virgin materials can save

- 15 trees [The 15 saved trees can absorb between 120 and 220 pounds of carbon dioxide (CO_2) each year. Burning this paper would *create* carbon emissions.]
- 750–1400 pounds of CO_2
- 150 gallons of oil
- 2000 kilowatt-hours of energy
- 4000 gallons of water

This represents a 60% savings in energy, which may translate into less GHGs and other pollutants emitted from fossil-fired plants.

5.9.4 Describing greenhouse emissions and impacts

As explained earlier, the EIS needs to concentrate on potentially significant issues that are truly important (§1502.5, §1502.24). Agencies should ensure that the

- GHG emission and impact descriptions are commensurate with the importance of the issue
- Level of detail is commensurate with the "rule of reason"
- Assessment avoids useless bulk and boilerplate documentation

Analysts may want to research any applicable reporting thresholds in technical documents to help determine the extent to which a GHG analysis is appropriate. To the extent possible

- Potential direct and indirect GHG source emissions should be identified as part of the scoping process.

- Where GHG releases warrant detailed consideration, an effort should be mounted to quantify the emissions.[37]
- In assessing direct emissions, an agency should focus on the emissions over which it has control or authority.[38]

To date, approaches for evaluating GHG emissions vary widely. Some EISs have simply involved reporting a GHG emission such as, "GHG emissions would only contribute an increase of 0.0002% to the total annual global emissions." This may be insufficient to address controversial projects or issues. Instead, some analyses may need to explain, perhaps in a "gross" way, how these emissions might affect environment quality.

Emphasis on investigating potential mitigation measures may be of particular importance. For example, the analysis might focus on best management practices that would conserve energy and reduce GHG emissions.

Where a discussion of cumulative GHG emissions is necessary to support informed decision making, the author recommends that the EIS focus on evaluating the annual and cumulative emissions of the proposed action and the difference in emissions associated with alternative courses of action.

5.9.4.1 Emissions versus impact

As just noted, some EISs merely report quantities of released GHGs. It is important to emphasize that GHG emissions are not actual impacts in themselves. As described in Chapter 2, they are better viewed as "environmental disturbances"—the root cause of a potential climate change impact. In one recent case, a court ruling implied that simply quantifying emissions and comparing them to a baseline was insufficient.[39] Instead, the EIS may need to actually describe how greenhouse emissions affect climate change. For instance, the impacts are the environmental changes that might result from increased GHG concentrations. Examples of GHG impacts are represented in Table 5.20.

Table 5.20 Examples of GHG Impacts

Effects on agricultural production and food supplies
Temperature variations and their effect on species
Spread of diseases
Changes in demographics
Sea-level changes and their effects on coastal zones
More frequent extremes in weather (wetter monsoons or dryer droughts)
Changing precipitation patterns, including droughts or floods
Warmer ocean temperatures affecting weather patterns, coral reefs, fisheries, or tourism

5.9.5 *How to prepare a flawed GHG analysis*

The author recently acted as a consultant in a legal battle involving a proposed wind energy farm. This project is instructive because it shows how even a so-called green energy project can result in substantial environmental impacts including GHG emissions. It also shows how an analysis of GHG emissions can be easily flawed and how to avoid such problems errors.

5.9.5.1 *Just how dirty can a clean energy project be?*

Many people, even scientists and engineers, are surprised to learn that green energy projects might generate substantial amounts of indirect GHG emissions. During construction of the proposed wind farm, GHG emissions could include those from highly visible sources such as construction equipment (such as graders, cranes, and excavators), commuting and personal vehicles, and heavy-haul construction material trips (water, aggregate, and cement for concrete production). It would also result in a small increase in GHG emissions due to the loss of carbon uptake from the removal of vegetation at the plant site. The proposed wind farm would also generate relatively small CO_2 emissions during operation. These emissions would include those produced by equipment and vehicles used for operations, inspection, maintenance, and other activities.

Another important consideration is that the proposed project would likely result in a large reduction in GHG emissions due to the displacement of electricity generated from traditional sources such as fossil fuel power plants. Table 5.21 shows direct GHG estimates presented in the EIS for the proposed wind farm. Most important, the table includes an estimate of the GHG emissions displaced from building a comparable fossil fuel plant. As noted in the table, the project would result in an offset of 289,130 metric tons of carbon dioxide equivalent emissions per year (MTCO2e/year). So far, so good. Let us now consider the flaws that the author identified in this analysis.

5.9.5.2 *How to prepare a flawed greenhouse assessment*

This section describes some of the flaws in the assessment of greenhouse emissions for the proposed wind farm. Its intent is to help the reader learn

Table 5.21 Annual Operation Emissions

Source	MTCO2e/Year
Construction emissions generated	713
Total operation emissions (vehicles and equipment) generated	42
Displaced annual GHG emissions	–289,885
Net project annual GHG emission offset	**289,130**

from and not replicate similar mistakes. The EIS made the following summary statement:

> The proposed project is likely to result in a large reduction in GHG emissions due to the displacement of electricity generated by fossil fuel-fired power plants.

This was a potentially erroneous statement, particularly from the standpoint that the EIS never included a fossil-fueled alternative, so there was little basis for presupposing that the proposed action was acting as a substitute to such an alternative. Moreover, to say that some other "distant" proponent might have proposed a fossil fuel power plant if this wind farm were not constructed amounted to mere speculation. Such questionable statements might mislead a decision maker into reaching a faulty decision. The speculative nature of this statement should have at least been acknowledged in the EIS.

5.9.5.2.1 Failure to adequately consider CO_2 emissions. The EIS neglected to consider the substantial emissions generated in manufacturing wind turbines such as the large amounts of steel, concrete, and aluminum. This manufacturing requires substantial quantities of electricity. Generation of this electricity may release substantial quantities of GHG gases (along with other pollutants). The companion text, *Global Environmental Policy*, has this to say about generation of wind farm GHGs using a nuclear reactor as a comparison:

> A nuclear reactor contains about 500,000 cubic yards of concrete and 120 million pounds of steel. In contrast, a single 45-story wind turbine stands on a base of 500 cubic yards of concrete and contains as much metal as 120 automobiles. As 2000 of these are equal one nuclear reactor that adds up to twice as much concrete and steel; this translates into significant quantities of greenhouse emissions produced over the construction phase (mining, concrete production, forging, transportation, construction).[40]

Richard Donnelly et al. performed an analysis that compared the life-cycle CO_2 emissions for various energy sources (Figure 5.13, modified).[41]

Their study concluded that CO_2 emissions for a typical wind turbine are on the order of 10 to 30 grams (g) of CO_2 equivalent per kilowatt-hour (g CO_2 eq./kWh). They settled on an average value of 19 g CO_2 eq./kWh. As demonstrated in Figure 5.13, the life-cycle emissions for a wind farm

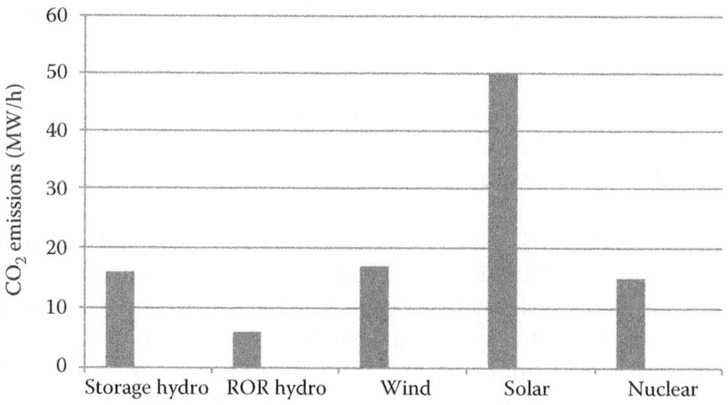

Figure 5.13 Comparison of life-cycle CO_2 emissions for various power sources. (Modified after Donnelly C.R., Carias A., Morgenroth M., Ali M., Bridgeman A., & Wood N., An Assessment of the Life Cycle Costs and GHG Emissions for Alternative Generation Technologies, http://www.worldenergy.org/documents/congresspapers/482.pdf, accessed April 18, 2012.)

are not trivial and in fact are equivalent to sources such as hydro-storage and nuclear power plants.

Most important, the EIS utterly neglected to consider the potentially large CO_2 emissions that result from the *curing of concrete*. Concrete is a large source of worldwide CO_2 emissions. It is so sizeable, in fact, that the concrete industry is one of the largest sources (about 7%) of man-made CO_2 emissions on the planet. While the EIS computed the CO_2 emissions from vehicles and equipment, it totally ignored the emissions from curing concrete. The turbines are secured to massive foundations of concrete. Each individual turbine foundation may require on the order of 500 to 1000 tons of concrete and aggregate. While various figures are taunted around, there is little doubt that the curing of concrete produces large quantities of CO_2. Most of the life-cycle CO_2 emissions from a wind farm may actually result from the massive amounts of concrete used in the foundation of each wind turbine. Neglecting to account for these CO_2 emissions led to a significant underestimation of the CO_2 and a lopsided assessment that made wind energy look unrealistically attractive.

5.9.5.2.2 Incorrect statements regarding benefits of reducing CO_2. The wind farm EIS went on to make unsubstantiated or flawed statements regarding the benefits of the proposed project in terms of displacing fossil fuel GHG emissions. The following statements were made:

1. Benefits of the proposed project in displacing fossil fuel-fired generation and reducing associated GHG emissions from gas-fired generation would occur.

2. There is a net reduction of 288,611 metric tons of CO_2 equivalent per year (MTCO2e/year).
3. This is more than enough [savings], by orders of magnitude, to offset the project's construction and operation GHG emissions, so the proposed project would have negative net GHG emissions.

The first two statements concerning the benefits of displacing fossil-fueled plants were based on an analysis that failed to account for the production of massive amounts of concrete, and a large amount of energy and emissions produced in forging the steel turbines; the estimate of 288,611 in the second statement was incorrect because it failed to account for the life-cycle generation of CO_2 emissions, particularly those from curing concrete. The third statement was likewise unsubstantiated because it was based on a flawed analysis.

5.9.5.2.3 Scientific consensus regarding man-made emissions. The wind farm EIS also made incorrect assertions regarding the scientific consensus on man-made CO_2 emissions:

1. [There] is general scientific consensus that man-made emissions of GHGs are likely to contribute to climate change, if not sufficiently curtailed.
2. It is generally **agreed** within the scientific community that increases in global GHG emission concentration can cause changes to current global climate conditions.

Both of these statements are questionable. The fact is that the scientific jury is "still out" when it comes to the effect that GHG emissions are having on global climate. There is disagreement within the scientific community regarding anthropogenically induced (human) climate change. Even the IPCC has not definitively stated that increased greenhouse emissions are causing climate change.[42]

The author is not arguing that wind farms are a poor idea. The point is that every energy source has impacts, and the analysis needs to be carefully performed so as to avoid the types of flaws just witnessed.

5.9.6 *Other examples of how GHG emissions have been addressed*

Described below are examples of how two recent US Department of Energy EISs addressed GHG emissions. These examples are offered for instructional purposes only, and the author does not necessarily endorse nor reject these approaches. The reader should also note that professional practices and expectations, as well as future litigation, may affect current practices.

5.9.6.1 Gilberton Coal-to-Clean Fuels and Power EIS

The Gilberton Coal-to-Clean Fuels and Power EIS illustrates how one EIS addressed GHG emissions. It included the following statements[43]:

- The CO_2 emissions from the proposed facility would add 2.3 million tons per year to global CO_2 emissions, for an estimated cumulative increase of 29 billion tons.
- "Fossil fuel burning is the primary contributor to increasing concentrations of CO_2.... The increasing CO_2 concentrations likely have contributed to a corresponding increase in temperature in the lower atmosphere."
- "Over the entire fuel lifecycle (from production of the raw material in a coal mine or oil well through utilization of the fuel in a vehicle) and considering all greenhouse gases, production and delivery of liquid transportation fuels from coal has been estimated to result in about 80% more greenhouse-gas emissions than from the production and delivery of conventional petroleum-derived fuels.... Recovery and sequestration of CO_2 at a... production facility... could greatly reduce greenhouse gas emissions from... fuel production, possibly to levels below conventional petroleum-derived fuel production."
- "Although not proposed by the applicant, it may become feasible to reduce the project's contribution to global climate change by sequestering some of the CO_2 captured in the process underground."
- "Using high-range estimates of future oil prices... and assuming the... fuel cycle generates 80% more greenhouse-gas emissions than production and delivery of conventional petroleum-derived fuels, expanded use of [this] technology to produce liquid fuels could cause the U.S. liquid fuel sector to release about 5% more greenhouse gas emissions than if the same quantity of fuel was produced from petroleum."

5.9.6.2 FutureGen project EIS

This EIS addressed climate change impacts using statements and evidence such as[44]

- While "CO_2 is not currently regulated as an air pollutant at the Federal level, it is generally regarded by a large body of scientific experts as contributing to global warming and climate change.[42]"
- The EIS analyzed a coal-fueled electric power and hydrogen production plant integrated with CO_2 capture and geologic sequestration. Such a design would be capable of capturing at least 90% of its CO_2 output.

- The project's individual contribution to global CO_2 emissions and potential climate change is extremely small.

5.9.7 Assessing cumulative GHG emissions

Section 5.8 presented NEPA's standard definition of a cumulative effect. To add more clarity to the assessment of cumulative climate change impacts, the author has revised the standard cumulative effect definition (§1508.7) by injecting references to GHG (in brackets) as follows:

> [A cumulative effect is the]... impact on the environment which results from the incremental impact of the action [GHG emission] when added to [the GHG emissions from all] other past, present, and reasonably foreseeable future actions regardless of what agency (federal or non-federal) or person undertakes such other actions. Cumulative [GHG] impacts can result from individually minor but collectively significant actions taking place over a period of time.

5.9.7.1 GHG emissions: death by a thousand puffs

Any single project's GHG emissions—even that of a very large and long-term project—will likely be small and probably indiscernible from a global perspective. For instance, in the year 2000, the combined worldwide manufacturing and construction industries contributed only about 10% to the total GHG emission inventory for that year.[45] Moreover, the US GHG emissions per year account for just about 20% of the total worldwide contribution; by comparison, if *all* combined US sources make up just a fifth of worldwide emissions, any single US source will undoubtedly be truly minuscule.[46]

Thus, a proposed action might emit 15,800 metric tons of carbon dioxide equivalents per year. When compared to aggregate emissions of 26 gigatonnes of carbon dioxide per year, it would represent a mere 0.006% of worldwide emissions.[47] For such projects and particularly smaller ones, a strong argument can be raised that a detailed investigation of greenhouse emissions is unwarranted if not impractical. On the other hand, this scenario epitomizes the long-recognized NEPA dilemma involving the "tyranny of small decisions" in which[48]

> Thousands of small federal actions, each contributing a trace fraction of global GHG emissions, combine to cumulatively increase atmospheric GHG concentration. Yet by themselves, these individual sources are clearly nonsignificant.

This problem is related to Eccleston's Paradox, described in Section 5.8. The companion book, *Environmental Impact Assessment*, details a procedure referred to as the *Significant Departure Principle*, which provides a systematic, peer-reviewed, and defensible technique for resolving this paradox.[49]

5.10 Performing an accident analyses in an EIS

John Ruskin (1819–1900) once shared this bit of wisdom:

> Quality is never an accident; it is always the result of high intention, sincere effort, intelligent direction and skillful execution; it represents the wise choice of many alternatives, the cumulative experience of many masters of craftsmanship.

Some types of proposals involve potential accidents (including terrorist-related events or natural disasters) with potentially grave consequences on the environment. In some cases, potential accidents represent the greatest single risk to the public and environmental quality. An EIS may not ignore potential accidents simply because the probability is deemed to be low. This section examines the methodology for evaluating such risks. For a more detailed discussion of the issue of an EIS accident analysis, the reader is referred to the companion book, *Environmental Impact Assessment*.[2] As we have seen, the EIS

> … shall provide full and fair discussion of significant environmental impacts and shall inform decision-makers and the public of the reasonable alternatives which would avoid or minimize adverse impacts or enhance the quality of the human environment. (§1502.1)

The EIS must also

> … present the environmental impacts of the proposal and the alternatives in comparative form, thus sharply defining the issues and providing a clear basis for choice among options by the decision-maker and the public. (§1502.14)

Potentially significant accidents necessitate special consideration and enhanced investigation. By their very nature, most potentially catastrophic accidents involve a substantial degree of uncertainty, and unique or unknown risks. Assessing the consequences of events such as a catastrophic accident is of such importance that one of the factors cited in the Regulations for determining significance involves

> The degree to which the possible effects on the human environment are highly uncertain or involve unique or unknown risks. (§1502.[b][5])

Despite this direction, there is substantial disagreement concerning the circumstances that demand a detailed assessment of accidents or natural disasters. A decision tool has been developed for determining when and under what circumstances an accident analysis should be performed. Information on the use of this tool and general topic of addressing accident analyses can be found in the companion text, *NEPA and Environmental Planning*.[50]

The analysis of the probabilities, consequences, and risks can pose particular challenges. The fact that the probability of an event is often uncertain or unknown compounds the problem. A few agencies have been less than candid in disclosing such risks. Recall the case study in Chapter 1 involving the Nuclear Regulatory Commission's program to relicense the nation's antiquated fleet of nuclear reactors. Stakeholders charge that this is a vivid example of how public safety and environmental quality can be compromised when a federal agency shuns its legal responsibility to prepare an open, fair, and objective analysis of potential risks. In violation of the NEPA Regulatory provisions just cited, the Commission's relicensing EISs include a misleading attempt to assess the risk of a severe nuclear accident such as a catastrophic nuclear meltdown. Potential consequences of a catastrophic nuclear accident could include contaminated air and water; human radiation poising, including deaths; genetic mutations, including birth defects; affected species and habitats; contaminated food chains; evacuation of tens or hundreds of thousands of downwinders; property damage in tens or hundreds of billions of dollars; and possible contamination of hundreds or maybe thousands of square miles. To avoid stirring opposition and possibly even jeopardizing its nuclear relicensing initiative, critics charge that the Commission's accident analyses routinely reached the incredulously conclusion that the risk to the environment and public is "small." In approving these EISs, the manager, Mr. Bo Pham, ignored basic legal requirements to prepare open, fair, and objective analyses. The following direction is intended to show practitioners how to prepare accident assessments that openly and objectively disclose potential risks to the decision maker and public.

But accident scenarios are not limited to major facilities like dams or nuclear power plants. Even benign-looking projects and facilities can experience tragic accidents.

5.10.1 Great Molasses Flood disaster

It was a chilly winter day in January 1919. But the temperature around Boston was climbing rapidly from the frigid temperatures of the preceding days. The sudden thaw elevated people's spirits. Locals were out strolling on the street. Little did they realize that a tragedy was brewing 50 ft above street level in an iron tank containing two-and-a-half million gallons of molasses. What was about to happen would become known as the Great Molasses Flood. As shown in Figure 5.14, it would go down as one of the strangest accidents in American history.[51]

The stored molasses was awaiting transfer to the Purity Distilling Company's plant, which fermented the sticky goo into alcohol. The tank was huge, measuring 50 ft high all and 90 ft in diameter. Suddenly, there was a loud rumbling sound, like a machine gun, as rivets blew out of the tank's seams. Witnesses said the ground shook as if a train were passing.

The tank collapsed, unleashing a deadly wave of molasses up to 10 ft high, rolling along at 35 mph. *The Boston Globe* reported that people "were picked up by a rush of air and hurled many feet." Nearby, buildings were crushed like matchsticks. Several blocks were flooded to a depth of 3 ft in sweet goo.

More than 150 were reported injured and 21 people and several horses were killed, some drowned by molasses. Rescuers found it difficult to make their way through the syrupy mess to help the victims. It took 4 days before they stopped searching for victims. Cleanup took weeks (Figure 5.15). Then the lawsuits began flooding in. The court ruled in favor of the plaintiffs, finding that the tank had been overfilled and was not structurally sound. But this did not end the speculation. Rumors circulated that the accident had actually been the act of sabotage or what we might term today as terrorism.

Figure 5.14 The aftermath of the Great Molasses Flood.

Figure 5.15 The flood damaged Boston's elevated railway.

5.10.2 Significance and potentially catastrophic scenarios

The need to consider impacts of natural disasters and accident scenarios (including potential terrorist events) is embedded in at least 4 out of the 10 factors cited in the Regulations for determining significance:

- The degree to which the proposed action affects *public health* or *safety* (§1508.27[b][2])
- The degree to which the effects on the quality of the human environment are likely to be *highly controversial* (§1508.27[b][4])
- The degree to which the possible effects on the human environment are *highly uncertain* or involve *unique or unknown risks* (§1508.27[b][5])
- Whether the action *threatens a violation* of federal, state, or local law, or requirements imposed for the protection of the environment (§1508.27[b][10])

It is clear that potentially significant consequences of such events cannot be ignored and must be evaluated and disclosed in the EIS.

5.10.3 Identifying potential accident scenarios

Experience demonstrates that a well-orchestrated scoping process provides a particularly effective tool for not only focusing on impacts of true concern, but for also dismissing those that are unimportant from further study. As appropriate, the IDT should use the scoping process in conjunction with a safety-assessment specialist to identify potentially significant accident scenarios.

5.10.3.1 Design-basis and beyond-design-basis accidents

The terms *design-basis accident* (DBA) and *maximum credible accident* (MCA)[52] essentially refer to a postulated natural disaster or accident scenario (e.g., nuclear power plant accident) that a facility will be specifically designed and built to withstand.[53] However, this is not always the case, and poor engineering judgment, deception assessments, or mismanagement can lead to faulty or even dangerous conclusions. For instance, assume an agency deems a magnitude 7.8 earthquake to be the design-basis accident that a nuclear power plant will be designed to withstand. But if a magnitude 8.0 earthquake were to occur, it might destroy the plant's safety system, resulting in a major release of radiation into the environment; as described below, the magnitude 8.0 earthquake would constitute a beyond-design-basis accident.

5.10.3.2 Beyond-design-basis accident

Some types of natural disasters or accidents (correctly or incorrectly) are deemed to be so unlikely that proposed facilities are not designed to withstand such an event.[54] These accidents are referred to as *beyond-design-basis accidents* because they exceed the facility's design basis. Such events can include safety system failures, earthquakes, tsunamis, fires, flooding, tornadoes, and even terrorist attacks to name just a few.

Consider the 2011 Japanese earthquake and tsunami, which were deemed to be very unlikely or even impossible. Yet, a powerful 9.0 magnitude earthquake unleashed a mega-tsunami that crashed into the Fukushima Daiichi Nuclear Power Plant complex. This led to meltdowns at three nuclear power plants located at the complex. The reactors were only designed to withstand much smaller earthquakes and tsunamis. Because it was deemed unlikely that such an event would ever occur, the Fukushima accident scenario was categorized as beyond-design-basis accidents, and no attempt had been made to plan for such an incident.

5.10.3.3 Determining a reasonable range of scenarios

An inverse relationship tends to exist between the probability and consequences of an accident or natural disaster. That is, the larger the consequences tend to be, the lower the probability of the event. Conversely, the lower the probability, the larger the potential consequences. For instance, traffic accidents that may harm only a few people are an everyday occurrence. But an accident such as the failure of a dam that could imperil thousands is a rare and infrequent event. Moreover, assessing the probability of a dam failure may be much more difficult than computing the frequency and consequences of an automobile accident.

Partly for this reason, an EIS may need to evaluate a range of potential accident scenarios, representing a "spectrum" of reasonably foreseeable events. As one expert noted:

> NEPA essentially requires analysis of both the lesser risks of greater harm and the greater risks of lesser harm before actions are taken to bring about the risks.

Thus, the spectrum of potential accident scenarios that need to be investigated may include

- Low probability/high-consequence events
- Higher probability/lower-consequence events

In some cases, natural events (large floods, earthquakes, landslides) may need to be evaluated because they can adversely affect the consequences of some types of proposals. For instance, a large flood or earthquake might imperil the safety of a nuclear reactor. Likewise, a landslide or earthquake might lead to the catastrophic failure of a dam. A large flood or hurricane might disrupt a hospital or critical infrastructure, imperiling the lives of those who depend on such services.

5.10.4 Applying the sliding-scale approach in performing an accident analysis

The Regulations state that impacts are to be evaluated in proportion to their significance, i.e., the degree of effort expended on an analysis is commensurate with the level of risk involved (§1502.2[b]). Consistent with direction presented in Chapter 2, Section 2.6, the author recommends that a *sliding-scale approach* be used in identifying and considering potential accident scenarios. Factors shown in Table 5.22 should be considered in determining when and how to apply the sliding-scale approach to the assessment of potential accidents and similar events. These factors dictate the level of analysis appropriate for analyzing potential impacts.

Table 5.23 illustrates how the level of analytical effort varies with a sliding-scale approach in evaluating potential natural disasters and accident scenarios. This table may need to be modified to accommodate specific types of projects performed by an agency. The level of analysis and rigor increases commensurate with increasing consequences and risk. For relatively low levels of risk, a qualitative analysis may be in order, while a detailed and quantitative analysis is warranted for events having larger consequences and risk.

Table 5.22 Factors to Be Considered in Applying a Sliding-Scale Approach to Assessment of Potential Accidents and Natural Disasters

Probability or frequency that an accident or event will occur
Severity of potential consequences
Context of the proposed action and alternatives (e.g., local versus regional or national implications)
Degree of uncertainty of the event
Level of technical controversy involved

5.10.4.1 Remote and speculative accident scenarios

In addition to the rule of reason and sliding-scale approach, the author has identified additional guidance useful in determining if an event accident should be subject to an accident analysis. An analysis of some potentially severe events may require an unnecessary degree of speculation. A review of case law indicates that environmental impacts may not have to be evaluated if they are determined to be very "remote and speculative."

5.10.5 Analytical methodology

This section provides direction for assessing the consequences of potential accidents and natural disasters.

5.10.5.1 Assessing reasonably foreseeable adverse impacts

Many severe accident scenarios involve "catastrophic consequences" that may involve a low probability of occurrence. An analysis of potentially severe accidents and events such as natural disasters frequently involves "incomplete or unavailable" information concerning "reasonably foreseeable significant adverse impacts." As witnessed earlier, the Regulations spell out specific requirements for dealing with situations involving "incomplete or unavailable" information (see Table 5.18). When such information cannot be obtained, the EIS must provide an "… evaluation of such impacts based upon *theoretical approaches or research methods*" that are

1. Generally accepted in the scientific community

The phrase "reasonably foreseeable significant adverse impacts" includes effects that have "catastrophic consequences, even if their probability of occurrence is low," provided that analysis of these effects is (§1502.22[b][4])

2. Supported by credible scientific evidence
3. Not based on pure conjecture
4. Within the rule of reason

It is important to note that in addition to evaluating impacts to humans, an accident analysis must also address impacts on environmental

Table 5.23 Applying Sliding-Scale Approach in Determining Level of Analysis Appropriate for Analyzing Impacts of Potential Accidents or Natural Disasters

Level of analysis (sliding scale) Low risk ◄——► Greater risk		
Qualitative	Semiqualitative	Quantitative
Provide a narrative discussion of potential consequences: • Brief qualitative description and assessment of the accident scenario. • Gross estimate of affected number of workers and public. • Summary evaluation of potential acute and chronic consequences. • Brief explanation of emergency response plan to mitigate potential consequences. • Brief discussion of engineering controls or other potential mitigation.	Provide a semiqualitative discussion and analysis of potential consequences: • Semiqualitative description and assessment of accident scenario(s), including number or workers and public affected. • Semiqualitative estimate of the probability of the event, consequences, and risk. • As applicable, provide gross estimates of chemical, radioactive, or other releases to humans and receptors. As appropriate, use simplistic exposure assumptions. Perform semiqualitative assessment of health-effects analysis (e.g., brief discussion of potential impairment effects, dose-dependent effects, and combined effects). As appropriate, compare effects to exposure limits, including applicable health-related guidelines. • Describe emergency response plans and guidelines to mitigate potential consequences. • Provide semiqualitative assessment of mitigation measures, such as engineering controls, inventory reduction, or design changes.	Provide a detailed quantitative assessment and analysis of potential consequences: • Detailed quantitative description and assessment of accident scenario(s), including quantitative estimates of number of workers and public affected. As applicable, include a range of scenarios. • Detailed quantitative assessment of the probability of the event, consequences, and risk. • As applicable, provide detailed computations of chemical, radioactive, or other releases to humans and receptors (e.g., dispersion modeling, including time/frequency distribution of releases and concentrations). Perform detailed quantitative assessment of health effects (e.g., detailed quantitative discussion of potential impairment effects, dose-dependent effects, and combined effects). Compare effects to exposure limits, including applicable health-related guidelines. • Describe or develop additional emergency response plans and guidelines to mitigate potential consequences. • Provide detailed evaluation of mitigation measures with emphasis of significantly reducing the risk and consequences of the events. Include a detailed evaluation of measures such as engineering controls, inventory reduction, or design changes.

resources. A systematic method for assessing the risk of a natural disaster or severe accident is described below.

5.10.5.2 *Risk–uncertainty significance test*

As we have seen, events involving impacts that are "highly uncertain" or involve "unique or unknown risks" are factors to be considered by a decision maker in reaching a determination regarding potential significance (§1508.27 [b][5]).

5.10.5.2.1 Dealing with uncertainty. Decision makers need to understand the nature and extent of uncertainty in choosing among alternatives and considering potential mitigation measures. Where uncertainties preclude quantitative analysis, the unavailability of relevant information needs to be explicitly acknowledged. The EIS needs to describe the analytical methodology that is used, including the effect that incomplete or unavailable information has on the ability to estimate the frequency/probabilities and consequences of reasonably foreseeable events (§1502.22).

For events where the consequences are relatively low or for which numerical probability estimates are unavailable or difficult to obtain, qualitative descriptions such as "very infrequent" or "highly unlikely" may sometimes need to be used, provided that a basis for such usage is included. A systematic, defensible, and peer-reviewed technique for determining the significance of an impact involving a degree of uncertainty is presented in the following sections.[55,56] The following risk assessment technique was pioneered by Dr. Frederic March of Sandia National Laboratories for evaluating risk in NEPA analyses.

5.10.5.2.2 Risk. Determining the significance of an event involving uncertainty may involve consideration of both the frequency and severity (consequences) of an event. While there is no universally accepted definition, risk is often defined as

1. $R = F \times C$, where
   R = risk
   F = frequency (events expected/year)
   C = consequences

Similarly, the risk associated with a sequence or course of action can be more generally defined as

2. $$R = \sum_{i=1}^{n} F_i \times C_i$$

   where i assumes values from 1 to n, and n is the number of potential events associated with a particular course of action.

5.10.5.2.3 Frequency of an accident or adverse event. Table 5.24 displays a *frequency* scale developed by the US Department of Energy for

Table 5.24 Frequency Scale

Category	Level	Frequency (f)	Description
Frequent	A	$f > 1$	Expected one or more times per year.
Likely	B	$1 > f > 10^{-1}$	Once in 1 to 10 years.
Occasional	C	$10^{-1} > f > 10^{-2}$	Once in 10 to 100 years.
Unlikely	D	$10^{-2} > f > 10^{-3}$	Once in 100 to 1000 years.
Remote	E	$10^{-3} > f > 10^{-6}$	Once in 1000 to 1,000,000 years.
Very remote	F	$10^{-6} > f$	Less than once in 1,000,000 years.

Note: This scale is for assessing the risk–uncertainty significance criterion.

assessing events involving uncertainty. Using a numerical range, Table 5.24 describes the number of times (frequency) a particular event is expected to occur over a given period of time.[57] A category, level, and description (e.g., "Frequent") are included for interpreting and describing the numerical value of the frequency. As appropriate, Table 5.24 might need to be modified to address special problems or circumstances unique to a particular problem or project.

Where possible, the frequency that adverse consequences will occur over the *lifetime* of a proposal should be presented, rather than simply the annual frequency of a single initiating event (e.g., earthquakes, floods).

5.10.5.2.4 Severity of an accident or adverse event. A modified *severity* (consequences) scale developed by the US Department of Defense is presented in Table 5.25.[58] This table has been modified to account for events with extremely catastrophic or "Beyond Catastrophic" consequences. Accordingly, an additional row (Beyond Catastrophic) has been added to the top of this table. This table provides guidance for gauging the severity of potential consequences. Severity is designated using a severity descriptor (i.e., "Negligible" through "Beyond Catastrophic") as well as a numerical scale (i.e., I–IV). The column labeled "Description of consequences" defines the severity in terms of both human and environmental consequences.

5.10.5.2.5 Assessing significance of a potential event. The frequency and severity scales (Tables 5.24 and 5.25) can be combined to produce Table 5.26, which provides a systematic, defensible, and peer-reviewed technique for assessing significance in terms of the frequency and severity of an event. The frequency designation is indicated in the top-most row of Table 5.26, while the severity scale is depicted in the first column. Originally developed by Fred March, Table 5.26 has been modified slightly by the author to account for the revised severity scale. Table 5.26 leads to four possible outcomes with respect to determining the significance of an impact involving a frequency or probability of occurrence. The NEPA

Table 5.25 Severity Scale

Severity	Scale	Description of consequences
Beyond Catastrophic	V	Human: Potential loss of more than 100 lives and/or catastrophic, long-term, large-scale harm, illness or injury to humans Environmental: Potential large-scale, and long-term or permanent damage or losses involving land use, destruction of ecosystems, infrastructure, property, or contamination, and/or major loss of human life
Catastrophic	IV	Human: Potential loss of 10–100 lives and/or large-scale and severe injury or illness Environmental: Potential large-scale damage involving destruction of species, ecosystems, infrastructure or property with long term effects, and/or major loss of human life
Critical	III	Human: Potential loss of less than ten lives and/or small-scale severe human injury or illness Environmental: Potential moderate (medium-scale and short-term) damage to ecosystems, infrastructure, or property
Subcritical	II	Human: Minor human injury or illness Environmental: Minor (small-scale and short-term) damage to ecosystems, infrastructure, or property
Negligible	I	Human: No reportable human injury or illness Environmental: Negligible or no damage to ecosystems, infrastructure, or property

Note: This scale is used for assessing the risk–uncertainty significance criterion.

designations "extremely significant," "significant," "marginally significant" or "insignificant" are defined as follows:

1. **Extremely significant:** If an event falls within this category, the potential consequences are "extremely significant."
2. **Significant:** If an event falls within the category labeled "significant," the threshold of significance is clearly breached. Potentially severe event scenario(s) must be investigated in an EIS.
3. **Marginal:** If an event falls within the category denoted as "marginally significant," the threshold of significance is quantitatively indeterminate. The event might be significant. Professional judgment, combined with conservatism, may need to be exercised in determining if a potentially severe event scenario must be evaluated in an EIS.
4. **Insignificant:** If an event falls within the category labeled "insignificant," the threshold of significance is normally not breached.

Table 5.26 Risk–Uncertainty Significance Test

	A: Frequent ($f > 1$)	B: Likely ($1 > f > 10^{-1}$)	C: Occasional ($10^{-1} > f > 10^{-2}$)	D: Unlikely ($10^{-2} > f > 10^{-3}$)	E: Remote ($10^{-3} > f > 10^{-6}$)	F: Very Remote ($10^{-6} > f$)
(V) Beyond Catastrophic	S_E	S_E	S_E	S_E	S_E	S
(IV) Catastrophic	S	S	S	S	S	M
(III) Critical	S	S	S	S	M	I
(II) Subcritical	S	S	S	M	I	I
(I) Negligible	I	I	I	I	I	I

Note: Determining significance based on the severity and frequency of an event. S_E, Extremely significant, S, Significant to very significant, M, Marginally significant, I, Insignificant.

For example, the significance of a fire with a frequency between 0.1 and 0.01 ($10^{-1} > f > 10^{-2}$), and a severity level of II (subcritical) would be deemed significant.

5.10.5.2.6 Disclosing and describing the consequences. Federal officials must ensure that the risk–uncertainty significance test is not misused to mislead the public about the potential consequences of a potential accident. Recall the case study in Chapter 1. Reputable scientists and engineers charge that the Nuclear Regulatory Commission routinely reaches the deceptive conclusion that the risk (in terms of frequency) of a severe accident such as a full-scale nuclear meltdown is "small" even though the consequences could be catastrophic. This case study vividly illustrates how an agency can misrepresent the concept of risk so as to hide the true nature and consequences of an accident from the public and potentially affected stakeholders.

Moreover, the Regulations clearly require a rigorous analysis of environmental impacts (consequences) of actions including potential accidents. An EIS must

> ... present the environmental impacts [not risk] of the proposal and the alternatives in comparative form, thus sharply defining the issues and providing a clear basis for choice among options by the decision-maker and the public. (§1502.14)

While the EIS can certainly include an analysis in terms of "risk," it is equally clear that it likewise has a duty to describe and disclose the actual consequences (impacts) of a catastrophic accident, even if the risk of such an event is deemed to be small. Information concerning potential consequences (regardless of the assumed risk) is crucial to the aim of public transparency and reaching an informed decision regarding a high-consequence, low-probability accident. Yet, the Commission routinely concludes that the risk of a nuclear accident is "small" and neglects to inform the public and potential stakeholders of the consequences they would face if an accident were to occur: contaminated air and water bodies; human radiation poising, including deaths; health effects such as cancer; genetic mutations, including birth defects; affected species and habitats; contaminated food chains; evacuation of tens or hundreds of thousands of downwinders; property damage, dislocation, and evacuation costs in the tens if not hundreds of billions of dollars; and possible contamination of hundreds or perhaps thousands of square miles.

5.9.5.2.7 Disclosing cumulative risk to the public. As we have seen, an EIS is also required to rigorously investigate cumulative impacts. Also notice that equation 2 above (risk equation) included a summation sign for the number of events that could occur. This equation provides an initial starting point for computing cumulative risk such as the total risk posed by all operating nuclear reactor stations in the United States.

Yet, as we saw in Chapter 1, the Nuclear Regulatory Commission's EISs only consider risk of an accident from a single nuclear station. But there are actually 104 commercial nuclear power reactors located in 31 states. Those schooled in engineering and statistics understand that the actual cumulative risk to the American public from an entire fleet of operating reactors is much greater than that posed by a single lone reactor. Critics charge that the Commission's relicensing EISs have presented the public with a deceptive assessment of the true risk. In neglecting its legal responsibility to evaluate cumulative risk or the cumulative consequences of a severe accident from more than one reactor, the Commission has placed the public at graver risk than is generally realized.

PROBLEMS AND EXERCISES

1. Briefly outline the six-step AIM for analyzing environmental impacts.
2. Outline the five-step procedure for assessing GHG emissions?
3. What are the three types of alternatives recognized in the NEPA regulations?
4. What is the difference between the terms "mitigation" and "monitoring"?
5. What is the region of influence (ROI)?
6. What is the rule of reason?
7. Is there a difference between the terms "effects" and "impacts"?
8. What is the most common definition (mathematical) of "risk"?
9. Describe any two of the ten factors to be considered in evaluating "significance" in terms of intensity (§1508.27[b]).
10. Refer to Table 5.26. Assume that the severity of an accident is considered to be "subcritical" and the frequency falls within the range of $10^{-2} > f > 10^{-3}$. What would be the determination regarding its potential significance?
11. You are managing an EIS and involved in the preliminary scoping phase of the EIS. Your project involves a highway interchange in an undeveloped desert area 20 miles outside a city. Define your own hypothetical project with a sketched map showing major geographic features and environmental resources. You need to prepare a cost estimate for preparing the environmental consequences section of an EIS. You decide to "scope" out the potential impacts using a Leopold Matrix. Prepare a hypothetical Leopold Matrix listing the principal impacts and resources that would be affected. Note: there is no right or wrong answer to this question, so be imaginative.

Notes

1. 40 Code of Federal Regulations [CFR] Parts 1500–1508.
2. Eccleston C. *Environmental Impact Assessment: A Guide to Best Professional Practices*. CRC Press, Boca Raton, FL (2011).
3. Eccleston C. *NEPA and Environmental Planning: Tools, Techniques, and Approaches*. Section 4.5, CRC Press, Boca Raton, FL (2008).
4. Eccleston C. *Preparing NEPA Environmental Assessments: http:// A Guide to Best Professional Practices*. CRC Press, Boca Raton, FL (2012).
5. Eccleston C. *NEPA and Environmental Planning: Tools, Techniques, and Approaches*. CRC Press, Boca Raton, FL (2008).
6. Jediny J. How can GIS support the NEPA process? *NEPA Lessons Learned* (72): (September 5, 2012).
7. Leopold L.B. et al. *A Procedure for Evaluating Environmental Impact*. United States Geological Survey, Geological Survey Circular No. 645, Washington, DC (1971).
8. CEQ. Considering Cumulative Effects Under the National Environmental Policy Act, Page A-11, January 1997. Taken from FERC. Procedures for assessing hydropower projects clustered in river basins (request for comments). *Federal Register* 50: 3385–3403 (1985).
9. CEQ. Considering Cumulative Effects Under the National Environmental Policy Act, Page A-8 (January 1997).
10. Sorensen J.C. *A Framework of Identification and Control of Resource Degradation and Conflict in the Multiple Use of the Coastal Zone*. University of Berkeley, Berkeley, CA (1971).
11. CEQ. Considering Cumulative Effects Under the National Environmental Policy Act, Page A-16 (January 1997).
12. CEQ. Incorporating Biodiversity Considerations into Environmental Impact Analyses under the National Environmental Policy Act. 29 pp. (1993).
13. Interagency Ecosystem Management Task Force. *The Ecosystem Approach: Healthy Ecosystems and Sustainable Economies*. Vol. I Overview, Washington, DC (1995).
14. Eccleston C. *NEPA and Environmental Planning: Tools, Techniques, and Approaches*. CRC Press, Boca Raton, FL (2008).
15. Used in a federal court case involving NEPA.
16. CEQ. Council on Environmental Quality—Forty Most Asked Questions Concerning CEQ's National Environmental Policy Act Regulations (40 CFR 1500–1508), *Federal Register* Vol. 46, No. 55, 18026–18038, Question Number 2a (March, 23, 1981).
17. Schmidt O.L. The Statement of Underlying Need Determines the Range of Alternatives in an Environmental Document, The Scientific Challenges of NEPA: Future Directions Based on 20 Years of Experience, Session 13—The NEPA Process, Knoxville, TN (October 25–27, 1989); Also: The Statement of Underlying Need Defines the Range of Alternatives in Environmental Documents, 18 *Environmental Law* 371-81 (1988).
18. Freeman L.R., March F., & Spensley J.W. *NEPA Compliance Manual*. Government Institutes Inc. (1992).
19. *Fritiofson v. Alexander*, 772 F 2d 1225 (5th Cir. 1985).
20. Fogleman V.M. *Guide to National Environmental Policy Act*. Section 3.5 (1990).

21. Eccleston C. *NEPA and Environmental Planning: Tools, Techniques, and Approaches*. Section 6.7, CRC Press, Boca Raton, FL (2008).
22. A report prepared by the National Research Council of the National Academies, *Improving Health in the United States: The Role of Health Impact Assessment*. (September 2011), http://www.nap.edu/catalog.php?record_id=13229.
23. *Fritiofson v. Alexander*, 772 F.2d 1225, 1243, 1245–6 (5th Cir 1985).
24. Smith M.D. *Cumulative Impact Assessment Under the National Environmental Policy Act: An Analysis of Recent Case Law*, 8 ENVTL. PRAC. 228 (2006).
25. *Sierra Club v. US Department of Agriculture*, 116 F.3d 1482 (7th Cir May 28, 1997).
26. 460 U.S. 766, 774 (1983).
27. 541 U.S. 752 (2004).
28. *Department of Transp. v. Public Citizen*, 541 U.S. at 767.
29. Keeton W.P., Dobbs D.B., Keeton R.E., Owen D.G., & Prosser W.L. *Prosser and Keeton on the Law of Torts*. 5th ed. §41, at 264 (1984).
30. *Kleppe*, 427 U.S. at 414.
31. International Panel on Climate Change (IPCC) Fourth Assessment Report. Synthesis Report at 38 (http://www.ipcc.ch/pdf/assessment-report/ar4/syr/ar4_syr.pdf).
32. 74 *Fed. Reg*. at 66497-98.
33. Eccleston C. *Environmental Impact Assessment: A Guide to Best Professional Practices*. CRC Press, Boca Raton, FL (2011).
34. 40 CFR §1500.4(f), (g), §1501.7, §1508.25.
35. Eccleston C. *Environmental Impact Assessment: A Guide to Best Professional Practices*. CRC Press, Boca Raton, FL (2011).
36. 40 CFR §1508.25.
37. *Public Citizen*, 541 U.S. at 768.
38. *Center for Biological Diversity v. National Highway Traffic Safety Administration, 9th Cir*. (November 15, 2007).
39. Eccleston C.H. *Global Environmental Policy: Concepts, Principles, and Practice*.
40. Figure modified after Donnelly C.R., Carias A., Morgenroth M., Ali M., Bridgeman A., & Wood N. An Assessment of the Life Cycle Costs and GHG Emissions for Alternative Generation Technologies, http://www.worldenergy.org/documents/congresspapers/482.pdf (accessed April 18, 2012).
41. IPCC Fourth Assessment Report: Climate Change (2007, AR4).
42. DOE, DOE/EIS-0357, 2007.
43. DOE, DOE/EIS-0394, 2007.
44. Dow K. and Downing T.E. *The Atlas of Climate Change: Mapping the World's Greatest Challenge 41* (2007).
45. Kass M.J. *A NEPA Climate Paradox: Taking Greenhouse Gases into Account in Threshold Significance Determinations*.
46. 2002 Data drawn from analyses by the Intergovernmental Panel on Climate Change (IPCC) and the Pew Center on Global Climate Change (PCGCC).
47. Odum W.E. *Environmental Degradation and the Tyranny of Small Decisions*, 32 Bioscience 728, 728 (1982). This is not a direct quote.
48. Eccleston C. *Environmental Impact Assessment: A Guide to Best Professional Practices*. Chapter 3, CRC Press, Boca Raton, FL (2011).
49. Eccleston C. *NEPA and Environmental Planning: Tools, Techniques, and Approaches*. Chapter 10, CRC Press, Boca Raton, FL (2008).

50. Puleo S. *Dark Tide: The Great Boston Molasses Flood of 1919*. Beacon Press (2004).
51. Euro Nuclear, http://www.euronuclear.org/info/encyclopedia/m/mca.htm.
52. US Nuclear Regulatory Commission, http://www.nrc.gov/reading-rm/basic-ref/glossary/design-basis-accident.html.
53. Butler D. Reactors, residents and risk. *Nature* (April 21, 2011).
54. March F. Determining the Significance of Proposed Actions. *National Association of Environmental Professionals 21st Annual Conference Proceedings*, NEPA symposium, Session TC3, p. 421 (1996); and March F., *NEPA Effectiveness: Mastering the Process*, Section 3.3.7, Government Institutes, Rockville, MD (1998).
55. Eccleston C.H. *The NEPA Planning Process: A Comprehensive Guide with Emphasis on Efficiency*. Chapter 8, John Wiley & Sons Inc., New York (1999).
56. US Department of Energy, Order 5481.1B.
57. US Department of Defense, MIL-STD-882B.

chapter six

Writing the environmental impact statement

The EIS documentation requirements

> A man who carries a cat by the tail learns something he can learn in no other way.
>
> **Mark Twain**

This wisdom can equally apply to the preparation of a defective environmental impact statement (EIS). Be it regulatory flaws or oversights, project opponents may seek to identify problems that show a lack of adequate planning or adherence to regulatory requirements (Figure 6.1). The National Environmental Policy Act (NEPA) is essentially an environmental planning process. As such, the EIS should capture the results of this planning process. The NEPA implementing regulations (Regulations) spell out strict requirements that the EIS document must meet.[1] Agencies must exercise vigilance in ensuring that all EIS requirements have been identified and adequately addressed.

One problem, however, is that these requirements are strewn throughout the 35 pages of the NEPA regulations. They also transverse many different guidelines, memorandums, and executive orders, as well as lessons learned from case law. This makes for a difficult task in identifying, merging, and complying with all relevant requirements.

This chapter builds on the five previous chapters. While Chapters 4 and 5 focused on the process of preparing the EIS and performing the environmental analysis, this chapter presents a detailed description of the documentation requirements that the EIS must meet. The principal objective is to integrate all documentation requirements into a single coherent and systematic source of information. All pertinent requirements are systematically detailed, including regulatory requirements, guidance issued by the Council on Environmental Quality (CEQ) and Environmental Protection Agency, and presidential executive orders.

Figure 6.1 Mining has important economic benefits as well as environmental issues. (Courtesy images.google.com.)

Lessons from case law as well as best professional practices are likewise described. Some methods for reducing document size and compliance cost are also described. Requirements for preparing other legally mandated EIS documents such as the *notice of intent* and *record of decision* are likewise spelled out. This chapter also draws on lessons learned from the case study presented in Chapter 1; the intent is to help the reader learn from and avoid repeating similar mistakes.

We begin with the *notice of intent* in Section 6.2. Sections 6.3 through 6.5 provide general direction for preparing the EIS document, including guidance on subjects such as recommended page limits, disclosing opposing points of view, and other topics. Section 6.6 provides an in-depth examination of the detailed EIS documentation requirements. We finish with Section 6.7, which presents the documentation requirements for preparing the record of decision (also referred to as the ROD).

A copy of the NEPA implementing regulations is provided in Appendix B.* To assist the reader in preparing or reviewing an EIS, a comprehensive checklist of all important requirements that the EIS must meet is provided in Appendix C. Citations referencing specific regulatory provisions are abbreviated so as to cite the specific "part"

* Specific provisions referenced in the NEPA implementing regulations are abbreviated in this book so as to cite the specific "part" of the NEPA implementing regulations in which it is found. For example, a reference to a provision in "40 Code of Federal Regulations (CFR) 1501.1" is simply cited as "§1501.1."

of the NEPA implementing regulations (Regulations) in which they are found.

6.1 *Learning objectives*

- Requirement for writing the notice of intent (NOI) and notice of availability (NOA)
- General requirements for preparing and writing the environmental impact statement (EIS)
- Hints for writing better EIS
- The content and format of the EIS document
- Describing the range of reasonable alternatives
- Writing the section on the affected environment and sensitive resources
- Describing and writing the section on environmental consequences
- Requirement for writing the record of decision (ROD)

6.2 *Requirement for writing the notice of intent*

Chapter 3 explained the process and procedures for preparing and issuing the notice of intent (NOI) for an EIS. This section details the NOI documentation requirements. We begin with Table 6.1, which summarizes the minimum documentation requirements specified in the Regulations (§1508.22). As indicated in the table, the agency's proposed scoping process and any planned scoping meetings need to be identified. Likewise, a name and address of an individual who may be contacted to answer questions must be identified.

NEPA is a public process. To more effectively promote NEPA's public notification obligation, the author suggests an expanded outline for the NOI (Table 6.2). Professional judgment must be exercised in determining the extent to which these additional items should be included in the NOI.

At this early stage, the agency should generally avoid presenting an overly detailed description of the actual proposal, as this may change based on input from the public; the agency should provide detail sufficient to inform the public about the scope and nature of potential actions.

Table 6.1 Required Contents of NOI

• Description of the proposed action and possible alternatives
• Description of the agency's proposed scoping process, including whether, when, and where any scoping meeting(s) may be held
• Name and address of a person within the agency who can answer questions about the proposal and the EIS

Table 6.2 Suggested NOI Outline Including Additional Items Not Shown in Table 6.1

1. Identify the purpose and need for taking action.
2. Identify any cooperating agency(ies).
3. Provide the agency's website.
4. Brief description of the EIS process for unacquainted members of the public, including the purpose for publishing the NOI and any upcoming public scoping process. Explain that no decision has been made and that the EIS will provide important input in reaching a final decision. Provide pertinent background information, including historical context of the proposal and why action is needed.
5. Brief description of the proposal (proposed action and reasonable alternatives).
6. Proposed schedule of the EIS.
7. Significant environmental issues and impacts that may be involved.
8. Brief description of the agency's proposed scoping process:
    - Dates and locations of any scoping hearings to be held
    - Other means for the public to provide input
    - Location and availability of documents related to the proposal
9. Name, address, e-mail, telephone number, and other contract information for a point of contact within the agency who can answer questions.

Where the scoping process results in a substantial change in the proposal or the scope of the proposal, the NOI needs to be revised and republished in the *Federal Register* (§1501.7[c]).

6.3 General requirements for writing the EIS

Section 6.3 provides general requirements and direction for writing the EIS. Subsequent sections will describe the specific requirements. As outlined in Table 6.3, the EIS must present a rigorous assessment of the alternatives and environmental impacts, which provides the decision maker and public with a rigorous, full, fair, and objective assessment of impacts of the reasonable alternatives (§1502.1 and §1502.14[a]).

The reader should note that the courts tend to grant agencies a degree of latitude in determining the scope of issues and potential significant impacts. For instance, the courts tended to give credence to statements such as "In our best professional opinion…." The agency, of course, is responsible for ensuring that technically competent professionals have rigorously and objectively investigated the environmental issues.

Table 6.3 General Direction and Requirements for Writing EIS

General EIS documentation requirements

- The information must be of high quality (§1500.1[b]).
- The draft EIS is expected to satisfy to the "fullest extent possible," requirements established for final EISs.... Moreover, the draft should be prepared in accordance with the scope determined during the scoping process (§1502.9[a]).
- The EIS must provide "... full and fair discussion of significant environmental impacts and shall inform decisionmakers and the public of the reasonable alternatives..." (§1502.1).
- An EIS is to be "clear," "to the point," and "written in plain language... so that decisionmakers and the public can readily understand them" (§1500.2[b]; §1502.1; §1502.8).
- Every effort should be made to "disclose and discuss... all major points of view on the environmental impacts of the alternatives including the proposed action" (§1502.9[a]).
- "... information must be of *high quality. Accurate scientific analysis...* [is] essential to implementing NEPA" (§1500.1[b]).

Impact assessment requirements

- The EIS must "*Rigorously explore and objectively evaluate* all reasonable alternatives." (§1502.14[a]).
- An EIS is to be "*analytic* rather than encyclopedic..." (§1500.4[b]; §1502.2[a]).
- The EIS provides the "*analytic basis for the comparisons* [of alternatives]..." (§1502.16).
- "Impacts shall be discussed in *proportion to their significance*. There shall be only a brief discussion of other than significant issues." With respect to nonsignificant issues, "there should be *only enough discussion to show why more study is not warranted*" (§1502.2[b]).
- The EIS must provide "the means of assessing the environmental impact of proposed agency actions, *rather than justifying decisions already made*" (§1502.2[g]).
- EISs must include "... *evidence* that agencies have made the necessary environmental analyses" (§1500.2[b]; §1502.1).

Description of alternatives

- The EIS must encompass the "range of alternatives" that will be "considered by the ultimate agency decisionmaker" (§1502.2[e]).
- The EIS must "devote *substantial treatment* to each alternative considered in detail, including the proposed action so that reviewers may evaluate their *comparative merits*" (§1502.14[b]).
- The EIS must identify and evaluate reasonable alternatives to proposed actions "that would *avoid or minimize adverse impacts*" (§1500.2[e] and §1502.1).
- Where alternatives have been eliminated from detailed study, the EIS "... must briefly explain the *reasons* for their having been eliminated" (§1502.14[a]).

6.3.1 Importance of reducing the size of the EIS

The Regulations place emphasis on streamlining and reducing the size of the EIS:

> Agencies shall focus on significant environmental issues and alternatives and shall reduce paperwork and the accumulation of extraneous background data. Statements shall be concise, clear, and to the point, and shall be supported by evidence that the agency has made the necessary environmental analysis. (§1502.1)

The Regulations go on to state that the EIS

> … must concentrate on the issues that are truly significant to the action in question, rather than amassing needless detail. (1500.1[b])

Despite this regulatory direction, a review of recent EISs reveals that many are overly detailed, containing material of nearly useless value in terms of understanding the proposal and its truly significant impacts. This increases the cost of the EIS, lengthens the preparation time, and makes it more difficult for the decision maker and public to focus attention on issues of true merit.

6.3.1.1 A "NEPA miscarriage"

There can be a steep and costly price tag for disregarding this direction. Consider an example involving the US Department of Energy's (DOE) decision to prepare an EIS for a relatively modest proposal to treat radioactive plutonium at its Plutonium Finishing Plant, located at its Hanford site near Richland, Washington.[2] The NEPA compliance officer, Paul Dunigan, decided to prepare an EIS, even though there was substantial reason to believe that a much less rigorous and costly NEPA environmental assessment (EA) would suffice.

The EIS was prepared. When compared with other DOE actions of a nuclear nature, the activity in question was relatively innocuous; yet this *documentation encyclopedium* examined virtually every conceivable environmental and socioeconomic impact in near myopic detail, even those that were clearly nonsignificant.[3] The final EIS exceeded the CEQ's recommended page limit of 150 pages for a "typical" EIS, and was barely within the recommended maximum page limit of 300 pages, which is reserved for projects of "unusual scope or complexity." Excluded from this page count were appendices, comprising nearly 250 additional pages of largely irrelevant material. The font was changed between the draft EIS stage and final EIS to conceal the fact that the EIS had grown so much that if the text size had been left unchanged it would have exceeded CEQ's 300-page limitation. Nonetheless, the NEPA Compliance Officer expressed delight with the quality and detail of the final document.

Under his oversight, the EIS contractor was allowed to prepare a multimillion dollar EIS that spanned a total of 550 pages (including appendices) only to conclude that *every single impact* was *insignificant*. In the end, he approved a 2-year, multimillion dollar EIS that could have been achieved with a relatively simple $50,000 EA that would have reached the same conclusions. A NEPA consultant later confided, "This EIS was a NEPA miscarriage." The lesson from this case study is that poor oversight and management can result in project delays, misallocated resources, and squandered taxpayer money.

6.3.1.2 Incorporation by reference

To promote efficiency, an EIS is required to be concise and to the point. A method known as *incorporation by reference* provides a powerful but underutilized mechanism for reducing the size of an EIS: material that is incorporated by reference is briefly summarized and then referenced (e.g., using a citation such as an endnote) so that the reader can find and review that information. The Regulations provide the following direction:

> The incorporated material shall be cited in the statement and its content briefly described. No material may be incorporated by reference unless it is reasonably available for inspection by potentially interested persons within the time allowed for comment. Material based on proprietary data which is itself not available for review and comment shall not be incorporated by reference. (§1502.21)

Every effort should be made to maximize use of existing material. Table 6.4 provides a list of materials commonly incorporated by reference.

Table 6.4 Material Commonly Incorporated by Reference

- Related NEPA documents
- Other planning and analysis documents
- Environmental permitting documents (e.g., air, water, and waste management permits)
- Facility designs
- Safety analyses and studies
- Biological, soils, geological, hydrological, air quality, meteorological, socioeconomic, and other environmental data, studies, and reports
- Databases, and certain maps, drawings, and charts

6.3.2 Writing in plain language

Mark Twain had this to say about reading: "The man who does not read good books has no advantage over the man who can't read them." This is as true for an EIS as it is with books. The Regulations require that an EIS be "written in plain language" using "clear prose" (§1502.8). This readability requirement is not to be taken lightly. Some courts have applied a principle known as the *reasonable man standard* in their review of EISs. This principle has its roots in old English common law, where a common law was considered to be comprehensible if it could be understood by the common man possessing a reasonable level of intelligence and ability to comprehend such laws. More information on this requirement can be found in the companion book, *NEPA and Environmental Planning.*

6.3.2.1 Clapham Bus Test

Some courts have applied what is known as the *Clapham Bus Test* from English common law. This test uses a hypothetical person to decide if the EIS can be readily understood by the general public. The bus rider is assumed to be reasonably educated and intelligent, but a nonspecialist riding the Clapham bus in south London. With respect to NEPA, the test becomes one of, "Would an average reasonable person riding the Clapham bus in south London be able to read and understand the EIS?" If yes, the EIS passes the readability test.

In the words of one court, an EIS must be

> … readily understandable by government decision-makers and by interested non-professional laymen likely to be affected by actions taken under the EIS.[4]

Of particular importance is the phrase "readily understandable." This has been interpreted to mean that an inordinate amount of time should not be necessary to gain an understanding of the issues or decipher concepts. The term "interested non-professional laymen" was also used in this ruling. The term "interested" has been interpreted to mean a "non-professional layman," interested enough in the issues to have done a minimal amount of background reading or investigation on the proposed project and its potential environmental impacts.

6.3.2.2 Readability direction

Technical and scientific terms should be clearly defined and explained. The EIS should also contain appropriate graphics so that decision makers and the public can readily understand pertinent issues (§1502.8). For instance, common names of biological organisms should be used in addition to the scientific names. To improve readability, a glossary of technical terms and a list of acronyms should be included.

Table 6.5 Example of Active versus Passive Voice

Active voice makes the actor clear: "A qualified contractor will install the water purification system as part of the proposed action."
Passive voice does not identify the actor: "A water purification system will be installed as part of the proposed action."

This readability requirement was reinforced with the issuance of the *Presidential Memorandum on Plain Language*.[5] This memorandum is designed to make the government more responsive and accessible in its public communications. Guidance includes the following directions:

- Lengthy sections of text should be broken into more informative headings and subheadings.
- Short sentences are preferable to long ones.
- Common words should be used whenever practical.
- Paragraphs should be kept concise and focused on one topic.

Graphic aids such as pictures, maps, tables, graphs, and figures can all enhance the reader's comprehension. Writers should also use an *active* versus *passive voice* when feasible. Active voice means that the "subject" performs the action. Table 6.5 shows an example of active voice versus passive voice.

6.3.3 *A full and fair discussion*

The Regulations state that an EIS must provide

> … full and fair discussion of significant environmental impacts and shall inform decisionmakers and the public of the reasonable alternatives which would avoid or minimize adverse impacts or enhance the quality of the human environment. (§1502.1)

Recall the case study in Chapter 1 involving the Nuclear Regulatory Commission's EISs for relicensing the nation's fleet of aging nuclear power reactors. The lesson from Chapter 1 is that federal agencies must strive to prepare a "full," "fair," and "objective" investigation of the proposal, and its impacts and issues. Under no circumstances should an agency ever prepare an EIS that casts doubts on the integrity or objectivity of the analysis.

6.3.4 *A rigorous yet understandable analysis*

Analysts are routinely confronted with two nearly diametrically opposed goals. The EIS must be written in a manner that can be readily understood by the decision maker and the public. Yet, at the same time, it must provide an "accurate," "rigorous," and "scientific" analysis of environmental impacts (§1500.1[b] and §1502.14[a]). Failure to comply with either of these opposing goals may provide a basis for successful litigation.

The EIS should only briefly discuss environmental effects that were considered potentially significant, but upon closer scrutiny are found to be nonsignificant. A brief discussion of nonsignificant issues is frequently necessary to demonstrate that these impacts were indeed considered and found to be nonsignificant, and not simply overlooked or casually dismissed.

6.3.5 A public input, participation, and disclosure process

NEPA is as much a public process as it is a procedural one for making federal decisions. As a *Sunshine Law*, it provides the public with an avenue for shaping federal proposals. The EIS is a key federal mechanism for ensuring that

> … environmental information is available to public officials and citizens before decisions are made and before actions are taken. The information must be of high quality. Accurate scientific analysis, expert agency comments, and public scrutiny are essential to implementing NEPA. (§1500.1[b])

6.3.5.1 Disclosing opposing points of view

Consistent with this direction, the draft EIS must

> … disclose and discuss… all major points of view…. (§1502.9[a])

Added to this, the final EIS must disclose

> … any responsible opposing view which was not adequately discussed in the draft statement…. (§1502.9[b])

Furthermore, an agency is mandated to respond to opposing views and public comments in the final EIS (§1502.9[b]). Returning to the case study in Chapter 1, we witnessed how the Nuclear Regulatory Commission's nuclear power plant EISs reach the incredulous conclusion that the risk from a "serious accident" such as a full-scale, catastrophic nuclear meltdown is "small." A substantial portion of the informed scientific community, including national and international organizations, soundly reject this conclusion. Yet, nowhere in its relicensing EISs does the Commission "disclose and discuss" or respond to these opposing points of view. Beyond promoting the appearance of arrogance, such deficiencies undermine the requirement to provide the public with a full, fair, and objective analysis. The lesson is that an agency has the legal responsibility to "disclose and discuss... *all* major points of view" in an EIS.

6.3.5.2 *How a well-orchestrated public involvement process can lead to a successful project*

Fortunately, most agencies now take this requirement to heart. Consider how the Nuclear Regulatory Commission's sister agency, the Department of Energy (DOE), approaches its public involvement responsibilities. The DOE prepared an EIS for a highly complex and controversial nuclear project involving the safe storage of highly radioactive spent nuclear fuel.[6] The urgent priority given to this project was underscored by the fact that an earthquake or other similar event could have initiated an accident involving the catastrophic release of radiation to the surrounding area and community. The department did not attempt to deceive the public about the consequences. In fact, it publicized the seriousness of such risks. Mr. Eric Gerber, the project manager, used the EIS process to actively embrace the public by bringing opposing parties together in a unified effort that expeditiously determined a safe alternative for securing this nuclear waste.[7] Gerber had this to say about the department's EIS process[8]:

> A decision was made to involve the stakeholders from the beginning. We discussed pending decisions before they were finalized and actually changed our plans based on stakeholder input. After... seeing the impact of their recommendations on our decisions, the Project's credibility became established and stakeholder communications shifted from demands to team participation. An illustration of the success of this effort was the completion of the Project's Environmental Impact Statement in eleven months with few stakeholder comments; previously unheard of for major DOE projects.

6.3.6 *Documenting assumptions*

Uncertainty is an inherent aspect of predicting the future. And so it is with environmental impact assessments. Most EIS analyses involve some degree of uncertainty. Uncertainties are most commonly dealt with by making reasonable assumptions. The balance between success and failure in predicting future outcomes often pivots on one's ability to make rational and defensible assumptions. However, any engineer or scientist will readily testify that assumptions are one of the most common causes for scientific and technical errors. The credibility of an analysis often hinges on the ability to substantiate the assumptions. For this reason, the EIS should clearly

1. Identify and document all uncertainties and the assumptions used to bridge them
2. Provide the basis or rationale for each assumption used

6.3.7 *Incomplete and unavailable information*

As we saw in Chapter 5, agencies are required to make a diligent effort to obtain data and evaluate impacts in the EIS. Nevertheless, there are reasonable limitations on the amount of resources, effort, and cost that an agency can expend. In reality, cost constraints and limitations in the state of the art may prevent analysts from providing a complete picture of the environmental consequences.

The Regulations acknowledge such limitations, by providing for circumstances that involve "incomplete" or "unavailable" information. Table 5.18 provides detailed direction for dealing with incomplete or unavailable information in an EIS (§1502.22). The reader is referred to the companion text, *NEPA and Environmental Planning*, for additional details on this requirement.[9]

6.3.8 *Quantifying the analysis*

As a rule, potentially significant impacts should be quantified wherever practical. Consistent with a sliding-scale approach (see Chapter 2, Section 2.7 and Chapter 5, Section 5.2), the need to quantify a given impact varies with its potential significance (Figure 6.2). Not only does a quantified analysis often yield a shorter, more concise document, it also tends to provide information more useful in discriminating between alternatives and reaching an informed decision. Where it is not possible to quantify important issues, an explanation may be appropriate explaining why it was not done.

6.3.8.1 *Intensity and duration*

To the extent practical, both the intensity and duration of impacts need to be quantified. For instance, rather than stating, "Effluents containing lead would be small and of short duration," the analysis might better indicate,

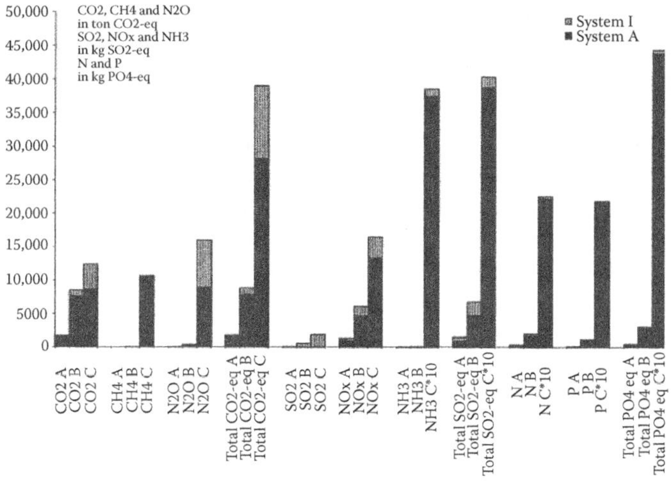

Figure 6.2 Example of quantifying and comparing impacts. (Courtesy images.google.com.)

"Effluents containing lead would be less than 0.1 gram per week, lasting for a period of five weeks."

As practical, analysts should avoid relying solely on relative measurements to describe impacts. For example, statements such as "Sulfur dioxide emissions would increase by 7%." should not be used in lieu of absolute metrics; this statement does not provide the decision maker and public with an absolute measure of the environmental impact, including gauging the actual environmental and health effects that would be expected. If possible, analysts may want to include a statement that includes both a relative and absolute measure of the impact. For instance, an EIS might indicate something to the effect, "The action would produce effluents containing 3 grams of lead per week for a period of nine weeks. This represents a 21% increase in lead released to the outflow stream."

6.3.8.2 Comparison to regulatory standards

It is not uncommon to encounter evidentiary declarations stating that a particular action "would be conducted in accordance with all applicable regulatory requirements." Such statements are sometimes the sole or primary source of evidence that a particular effect would have no significant impact. Statements such as this might provide supporting verification, but they should never be relied on as the sole or primary evidence of nonsignificance.

It is essential to note that an action may (and frequently does) comply with all applicable laws and regulations, yet can still result in a significant environmental impact. More to the point, compliance with applicable statutes, regulations, and environmental standards does not necessarily ensure that an action would not significantly affect or degrade an environmental resource. Consider a proposal to construct a commercial airport. Assume that this action would be permitted, constructed, and operated in accordance with all applicable federal and state laws, regulations, and requirements. But consider the impacts in local air quality degradation, noise, housing values, traffic congestion to and from the airport, and the risk of an aircraft crash. Such a project would almost certainly pose a significant impact.

6.3.9 Economic and cost–benefit considerations

In the author's opinion, an EIS should identify considerations, including factors not related to environmental quality, that are likely to be relevant and important to a decision. Direction concerning incorporation of economic analysis and considerations, including preparation of a cost–benefit analysis, are noted in Table 6.6.

Caution must be exercised to ensure that costs versus derived benefits are objectively compared and evaluated. Consider a case where the Natural Resources Conservation Service was challenged for preparing an inadequate EIS that was used in issuing a permit to construct a dam that could affect species, two of which were endangered. When challenged, the court concluded that exaggerated estimates of economic benefits had been used in the analysis.[12] In particular, the EIS cited gross rather than net economic benefits that would be derived from constructing the dam. The court further concluded that inflated economic benefits, considered

Table 6.6 Selected References Pertaining to Analysis and Consideration of Economic Factors

- The EIS must "identify environmental effects and values in adequate detail so they can be compared to *economic* and technical analyses" (§1501.2[b]).
- The analysis may incorporate use of a "cost–benefit analysis" (§1502.23).
- An agency may discuss preferences among alternatives based on relevant factors including *economic* and technical considerations and agency statutory missions (§1505.2[b]).
- The analysis of impacts includes "economic" effects (§1508.8[b]).
- Reasonable alternatives "include those that are practical or feasible from the technical and *economic* standpoint and using common sense, rather than simply desirable from the standpoint of the applicant."[10]
- The "agency's preferred alternative" is the alterative that the agency believes would fulfill its statutory mission and responsibilities, giving consideration to *economic*, environmental, technical and other factors.[11]

"crucial" in reaching the decision to approve the dam, had impaired the ability to balance benefits against environmental damage.

6.3.9.1 Cost–benefit analysis

As depicted in Figure 6.3, environmental considerations must often be weighed against other pertinent socioeconomic considerations. The NEPA Regulations provide direction for addressing such considerations. As noted in Table 6.6, the Regulations encourage incorporation of economic and other decision-making factors into the EIS. The Regulations provide the following direction regarding preparation or incorporating of a cost–benefit analysis (§1502.23):

> If a cost–benefit analysis relevant to the choice among environmentally different alternatives is being considered for the proposed action, it shall be incorporated by reference or appended to the statement as an aid in evaluating the environmental consequences… when a cost–benefit analysis is prepared, discuss the relationship between that analysis and any analyses of unquantified environmental impacts, values, and amenities… the weighing of the merits and drawbacks of the various alternatives need not be displayed in a monetary cost–benefit analysis and should not be when there are important qualitative considerations. In any event, an environmental impact statement should at least indicate those considerations, including factors not related to environmental quality, which are likely to be relevant and important to a decision.

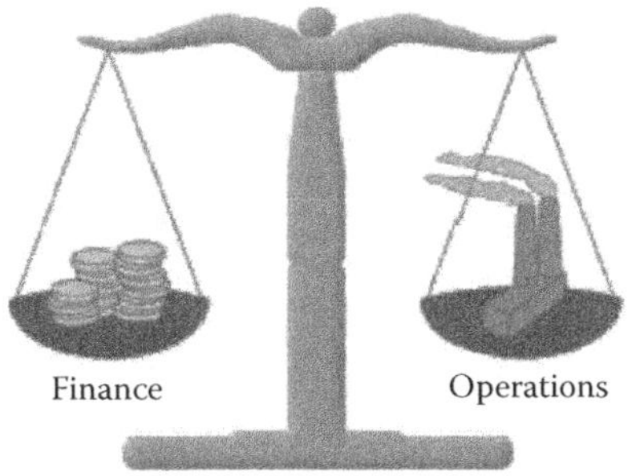

Figure 6.3 Environmental considerations must often be weighed against other pertinent socioeconomic considerations. (Courtesy images.google.com.)

6.4 *Techniques and hints for writing the EIS*

Some techniques and recommendations for improving the quality of the EIS are presented below.

6.4.1 *Citation methods*

Important data and information should be properly cited. The two most popular citation systems are endnotes and the scientific reference system. Under the scientific reference system or "parenthetical referencing," citations are enclosed within parentheses along with the date of publication, and embedded in the appropriate paragraph. An example of a parenthetical reference is "(Smith 2010)." Then under the references section, the specific citation is listed against the author's name and date of publication.

Table 6.7 Comparison of Benefits and Disadvantages of Using Scientific Citations versus Endnote Referencing System

Considerations	Scientific reference system	Endnote reference system
Cost/time	Consumes additional cost, time, and effort.	Reduces the cost, time, and effort.
Changes/modifications to the text	More difficult to update and maintain. When new text is inserted, deleted, moved, or changed, it can necessitate resequencing the scientific citations.	Changes in the text do not normally affect endnote references (i.e., the software automatically tracks, inserts, deletes, and updates endnotes cited in the text).
Accuracy and potential referencing errors in the EIS	More prone to potential errors/mismatches.	Reduces the potential for errors in text citations and the reference list.
Scientific usage	Commonly used in scientific literature, particularly scientific or scholarly studies.	Less commonly used in scientific literature.
Length of reference list	List of references tends to be shorter.	List of references tends to be longer.
Regulatory guidance	No requirement in the NEPA regulations to use a scientific citation system. While the regulations vaguely refer to referencing, they do not specify the system to be used.	Same.

Each system has its own set of advantages and disadvantages. Endnotes are generally easier and faster to incorporate, and maintain, because they are tracked and updated by software such as Microsoft Word. If a new endnote is added or deleted, the software automatically tracks and renumbers all endnotes accordingly. Table 6.7 compares the advantages and drawbacks of both referencing systems.

6.4.2 *Use of the word "would" versus "will"*

As witnessed in the case study of Chapter 1, agency managers must make every reasonable effort to avoid even the appearance of impartiality. For this reason, discussions of potential actions should be written using the conditional tense, as if the action *might* take place. This is done to clearly delineate the fact that no decision has been made. Thus, it would be inappropriate to make a statement such as, "The facility *will* affect an area of 24,320 square meters." Instead, this should more correctly be stated as, "The *proposed* action *would* affect an area of 24,320 square meters." In the second case, the words "proposed" and "would" indicate to the reader that a final decision has not been made.

6.4.3 *Units of measurement*

Consistent units of measurement need to be incorporated throughout the document. For instance, if a cubic meter is chosen as the unit for expressing waste streams in one section of the EIS, waste streams should not be presented in terms of kilograms in other sections. Metric units are frequently used, followed by English units in parentheses:

"The pipe would be 300-meter (984-foot) long."

A table of conversion factors should be included. Any use of scientific notation should be applied consistently throughout the text. Where scientific notation is used, an explanation of this system should likewise be provided. Small numbers should not be expressed differently from larger numbers. For instance, it is more difficult to compare a number against another that is expressed in scientific notation, i.e., 0.006 compared with 4.4×10^{-4}.

6.4.4 *Definitions, abbreviations, and acronyms*

All technical terms need to be defined. One technique has involved highlighting definitions in "text boxes." Common practice dictates that acronyms and abbreviations be written out the first time they are used in a chapter. However, it can be burdensome to search back for the first occurrence of an unfamiliar acronym or abbreviation. For this reason, a list of acronyms and abbreviations should be provided in the EIS. It is also recommended that an acronym or abbreviation be redefined each time it is used in a chapter of the EIS.

6.4.4.1 The magical number seven

Use of acronyms and abbreviations should be minimized. Princeton psychologist George Miller wrote one of the most highly cited papers in psychology. His paper, titled *The Magical Number Seven, Plus or Minus Two*, is often interpreted to argue that the number of objects an average human can hold in working memory is 7 ± 2.[13] This is now known as *Miller's law*. On the basis of Miller's law, the author recommends that the EIS manager try to limit the number of acronyms used in each chapter of the EIS to between seven and nine.

6.5 Page limits and size of the EIS

The Regulations place emphasis on preparing concise EISs. Consistent with this direction, Table 6.8 summarizes CEQ's recommended page limits for various sections of an EIS. As indicated in this table, no page length direction has been provided for more than half of the sections in an EIS. To the extent feasible, NEPA practitioners should strive to comply with this direction, deviating only in special circumstances. In reality, EISs commonly exceed CEQ's suggested page limits.

6.5.1 Page limits and the "main body" of the EIS

As shown by Figure 6.4, the assessment of some issues can go on indefinitely. But this is not the case with NEPA. The NEPA Regulations provide clear direction for limiting the length of the EIS analysis. As noted in Table 6.8, a typical EIS should normally be less than 150 pages in length, with unusually complex EISs limited to less than 300 pages. These page limits refer to the "main body" of the EIS, which includes the following sections (§1502.7):

- Purpose of and need
- Alternatives (including proposed action)
- Affected environment
- Environmental consequences

Table 6.8 Summary of CEQ Direction on EIS Page Limits

Section of EIS	Page limit	Reference
Cover sheet	Not to exceed 1 page	§1502.11
Summary	Normally not to exceed 15 pages	§1502.12
Table of contents	No direction	Not applicable
Purpose and need	Agencies are directed to "briefly specify" the purpose and need for taking action.	§1502.13
Alternatives section	No direction	Not applicable
Affected environment section	Agencies are directed to "succinctly describe the environment of the area(s)." The CEQ has also stated that this section is to be considerably smaller than the section on the environmental consequences.	§1502.15; "Talking points on CEQ's Oversight of Agency compliance with the NEPA Regulations." (CEQ, 1980)
Environmental consequences section	No direction	Not applicable
List of preparers	This section is "… not to exceed two pages." The CEQ also advices that this section should contain "A line or two for each person's qualification…"	§1502.17; CEQ's 40 Questions, Number 27c
List of agencies consulted	No direction	Not applicable
Index	No direction	Not applicable
Appendices	No direction	Not applicable
Total "text" (main body) of the final EIS	The CEQ also recommends that an EIS should: 1. Normally be less than 150 pages 2. Less than 300 pages for proposals of unusual scope or complexity	§1502.7

Figure 6.4 The NEPA analysis must be properly managed and limited in its scope. (Courtesy images.google.com.)

As noted in the last item of Table 6.9, the Regulations use the term "should" and "normally." These terms were purposely chosen because the CEQ understood that agencies must be given a degree of latitude in responding to unusual circumstances.[14] In reality, NEPA documents are commonly plagued by what the author refers to as *documentum infinitum*, in which unimportant aspects of the proposal are documented in excruciating detail.

Table 6.9 Direction on Reducing Size and Streamlining Preparation of EIS

- To reduce paperwork, agencies may, if changes are minor, attach and "… circulate only the changes to a draft EIS rather than rewriting and circulating the entire statement…" (§1500.4[m]).
- The draft EIS is to be "*analytic* rather that encyclopedic," emphasizing material "useful to decisionmakers and the public" while "reducing emphasis on background material" (§1500.4[f]). "Most important," it "must *concentrate on the issues that are truly significant* to the action in question, rather than amassing needless detail" (§1500.1[b]).
- An EIS "shall be kept *concise* and shall be no longer than absolutely necessary to comply with NEPA and with these regulations. Length should *vary first with potential environmental problems* and then with project size" (§1502.2[c]).
- Emphasis should be placed on describing "*significant* environmental issues and alternatives" while reducing "accumulation of extraneous background data" and "needless detail" (§1500.1[b]; §1500.2[b]; §1502.1).
- With respect to nonsignificant issues, only enough discussion should be presented to "demonstrate why more study is not warranted" (§1502.2[b]).
- The text of a final EIS should normally be less than 150 pages in length. Proposals of "unusual scope or complexity" should "normally be less than 300 pages" (§1502.7).

In the author's experience, most EISs not only exceed the 150-page limit but also commonly surpass the 300-page limit. In fact, there are many examples where the total length of an EIS has exceeded several thousand pages.

In the author's experience, the chapter on the affected environment is often excessively long. It is not uncommon to encounter a description of the affected environment that describes an environmental resource in unwarranted detail even though it would not even be affected by the proposal. This is one of the reasons that EISs often cost so much and why they take so long to prepare.

As described earlier, it is recommended that a *sliding-scale approach* (see Chapter 2, Section 2.7 and Chapter 5, Section 5.2) be used in determining the amount of attention devoted to a particular issue. This approach recognizes that the degree of attention and detail devoted to a given impact/issue varies with the circumstances and potential for significance.[15]

6.5.2 Reducing document size

The Regulations place particular emphasis on streamlining preparation of the EIS. Experience shows that much of this regulatory direction is either overlooked or, in some cases, blatantly disregarded. Table 6.9 lists specific direction for reducing the size, duration, and level of effort expended on preparing the EIS.

As shown in Table 6.9, agencies are directed to place emphasis on preparing "analytic" analyses over "encyclopedic" ones. This direction can substantially reduce the size of the EIS, since quantitative data can normally be presented more concisely, while at the same time providing decision makers with a more solid foundation on which to base decisions.

As noted in the last item of Table 6.9, the term "text" refers to the main body of the EIS, which starts with *purpose and need* and continues through the chapter on *environmental consequences* (i.e., see items 4 through 7, Table 6.10).

Table 6.10 CEQ's Recommended Outline and Format for an EIS

1. Cover sheet
2. Summary
3. Table of contents
4. Purpose of and need for the proposed action
5. Alternatives including the proposed action
6. Affected environment
7. Environmental consequences
8. List of preparers
9. List of agencies, organizations, and persons to whom copies of the EIS are sent
10. Index
11. Appendices (if any)

One technique for reducing the size of the main body (i.e., text) of an EIS is to move material of less importance to the appendices. Although this technique will not actually reduce the total length of the EIS, it will at least allow the reader to focus on material that is truly important.

6.5.3 *How much detail is enough? The sufficiency question*

No two experts or reviewers are likely to completely agree on the amount of discussion that is necessary to provide coverage sufficient to allow the decision maker or public to reach an informed decision. This observation is born out, perhaps humorously, by what has become known as *Cohn's law,* which states

> The more time you spend documenting what you do, the less time you have to do what you do. Equilibrium is reached when you do nothing but it's fully documented.

6.5.3.1 *The sufficiency question*

Project proponents will typically argue that the description and analysis provides sufficient coverage to pursue an action, while adversaries may argue that the analysis is insufficient. Even where two reviewers agree that a particular discussion does not provide adequate coverage, they may still disagree on the degree of additional analysis and discussion needed.

In a NEPA lawsuit, the plaintiff may have one opinion, while the agency has a different opinion, and the judge yet another. The plaintiff will almost always argue that the analysis is insufficient while the agency will argue the opposite. In the end, a judge may decide which party is correct.

The author refers to this problem as the *sufficiency question*—"how much information is enough?" Since NEPA's inception, no definitive direction has been established for determining the amount of detail, discussion, and analysis that is sufficient to adequately cover an action. Yet, agency decision makers are routinely called upon to do just that. Inevitably, such determinations tend to be subjective. For instance, when describing a proposal to construct a federal facility, it might be considered sufficient to merely mention that the project would include construction of a short walkway from the office building to a maintenance workshop. But if the same project were built in an area where sensitive habitat or species could be harmed, an extensive description of the walkway might be necessary, including alternative pathways or even suspending the walkway above groundcover. How much consideration, description, information, and analysis is justified?

Consider a second example involving construction of a proposed hazardous waste treatment plant. Is it sufficient to simply provide a five-page description of the proposed plant, equipment, and processes that would be used? Or is a detailed 50-page description warranted, complete with a floor plan including entrance and emergency exit doors? Some project opponents may not be satisfied with even a detailed 150-page description.

6.5.3.1.1 The Sufficiency-Test Tool. Lacking definitive guidance, the decision maker, analysts, and public may all point to a host of different factors and considerations in defending their contention that a particular topic is or is not adequately described. Assertions are often based on ambiguous opinions that are difficult to definitively prove or disprove. Although common sense and the rule of reason are an integral part of the EIS preparation process, definitive direction would greatly reduce the degree of ambiguity and subjectivity.

The author has developed a systematic tool, referred to as the *Sufficiency-Test Tool*, for resolving this problem. This tool consists of four simple criteria (or tests) that can be used in determining if the discussion of a particular topic or issue is sufficient for the purposes of NEPA. The Sufficiency-Test Tool has been successfully used in NEPA court cases. A description of this tool and how it is used is detailed in the companion book, *NEPA and Environmental Planning*.[9]

6.6 EIS content and format

Mark Twain once counseled:

> Substitute "damn" every time you're inclined to write "very." Your editor will delete it and the writing will be just as it should be.

This chapter is about preparing an EIS "just as it should be." The EIS must clearly demonstrate that all requirements have been met and all potentially significant environmental issues have been investigated. Table 6.10 depicts CEQ's recommended format for an EIS (§1502.10). As shown in the table, the standard EIS outline contains 11 sections or chapters, beginning with a cover sheet and ending with appendices. This outline should be followed unless there is a compelling reason to deviate.

While other formats may be used, the EIS must at a minimum include Sections 1, 2, 3, 8, 9, and 10 and the essence of the requirements depicted by Sections 4, 5, 6, 7, and 11 (§1502.10). Although it is generally best to minimize deviations from the CEQ's recommended format, the

Table 6.11 Expanded EIS Outline Based on CEQ's Recommended Format

Abstract
Table of contents
List of figures
List of tables
Abbreviations and acronyms
Executive summary
1.0 Purpose and need for the proposed action
2.0 Alternatives including the proposed action
3.0 Affected environment
4.0 Environmental consequences
5.0 List of preparers
6.0 List of agencies, organizations, and persons to whom copies of the EIS are sent
Index
Appendices (optional)

most important consideration is that the EIS satisfy the substantive content depicted in CEQ's format. Requirements governing the content and preparation of each of the sections shown in Table 6.10, beginning with the "cover sheet," are described shortly. The author offers the enhanced outline shown in Table 6.11. The actual outline, of course, must be tailored to meet the agency's specific needs.

The following sections address some general requirements that the EIS must meet.

6.6.1 *Addressing public scoping and draft EIS review comments*

Comments received on the public scoping process must be included in the EIS. These comments and the agency's responses are typically placed in an appendix to the EIS.

6.6.1.1 *Comments on review of the draft EIS*

There are some differences, albeit minor ones, between the draft and final statements. The most important difference is that the final EIS includes comments received from public circulation of the draft statement; it also incorporates the agency's responses to those comments. As noted in Chapter 4, Section 4.9, all substantive comments received on the draft EIS (or summaries thereof where the comments are exceptionally voluminous) are attached to the final EIS regardless of whether or not the comment is thought to merit individual discussion by the agency in the text of the statement (§1503.4 [b]). The comments and the agency's responses to these draft EIS comments are typically placed in an appendix to the EIS. Often the comment responses require making changes to the content of the EIS.

Potential agency responses are indicated in Table 4.15. Other documentation differences are described in appropriate sections of this chapter.

6.6.2 Preparing the "draft" versus "final" EIS

As we saw in Chapter 4, preparation of an EIS is a two-stage process—a draft followed by the final EIS. The Regulations require that

> To the fullest extent possible, the draft [EIS] must meet requirements established for final EISs. (§1502.9[a])

The draft EIS is expected to conform to the scope agreed upon during the scoping process. While the Regulations use the term "draft EIS," this term is actually a misnomer. The statement is not a "draft" in the traditional sense of the word. As indicated above and in other related regulatory requirements, the draft EIS must be complete, as accurate as possible, meet the EIS regulatory requirements, and be capable of standing on its own merits. The EIS is referred to as a draft because it does not yet incorporate comments received from public review of the statement. The principal purpose of the final EIS is to respond to comments on the draft, *not* to address deficiencies in a draft that was not ready to be issued to the public.

6.6.2.1 When schedule trumps accuracy and quality

Unfortunately, to meet a scheduled deadline, a few agencies have sideswiped the EIS requirements noted above. In the process, deficient and sometimes misleading draft EISs have been issued to the public. Recall the case study in Chapter 1 involving the Nuclear Regulatory Commission's program to relicense the nation's aging fleet of nuclear reactors. Individual relicensing EISs were prepared under Mr. Pham's direction for each relicensing project. Multiple sources, including the EIS contractor, replied that an EIS could not be adequately completed within NRC's established 18-month schedule. These concerns were ignored. However, just as predicted, the project schedule began to slide. To reduce the slippage, Pham ordered the EIS contractor to stop further work so that the EIS could be issued to the public on the scheduled deadline. This order was given although the EIS contained inaccuracies, missing information, and was not of sufficient quality to publicly release. In response, the EIS contractor replied that more time was needed to prepare a thorough and accurate analysis. Pham responded, "We'll fix it up later during the final EIS stage."[16] This direction violated any number of NEPA regulatory provisions and case law. Even more troubling is the fact that Mr. Brian Holian, the NRC division manager, reviewed and approved this EIS for public release; if Holian could not even catch blatant errors, one is left to wonder what larger but harder to catch flaws may be passing by. Perhaps more important, it demonstrates how schedule commitments can trump potentially catastrophic environmental and safety concerns.

6.6.3 EIS cover sheet

Table 6.12 lists the specific items that the cover sheet must contain (§1502.11). The cover sheet is not to exceed one page in length. Beware of the fact that a wrong or misleading title can misrepresent a proposal, confuse the public, and possibly fuel controversy; the title should be tailored so that it accurately conveys the nature and scope of the proposal.

The instructions presented in Table 6.12 are simple and straightforward. Yet, a few agencies cannot even prepare a cover sheet that meets NEPA's regulatory requirements. Consider the Nuclear Regulatory Commission's EIS process led by Mr. Brian Holian. The Commission's EISs for relicensing nuclear reactors fail to comply with at least two, if not three, of the requirements spelled out in Table 6.12. While these errors certainly do not rise to the level of a fatal flaw, they are telling in more ways than one. If an agency cannot even comply with five trivial requirements for preparing a one-page cover sheet, it should come as little surprise that their assessment of alternatives and impacts are riddled with errors and inaccuracies; as we will see, the Commission's faulty process begins on the first page (cover sheet) and continues throughout the hundreds of pages of each relicensing EIS. This flaw could have been easily prevented if the Commission's management had simply taken time to read the requirements spelled out in the Regulations. The lesson here is that an agency needs to strive to meet NEPA requirements, and this effort begins on the first page of the EIS and continues to the last page.

Table 6.12 Requirements for Preparing EIS Cover Sheet

1. List of the responsible agencies including the lead agency and any cooperating agencies
2. Title of the proposal or proposed action (and if appropriate the titles of related cooperating agency actions), together with the state(s) and county(ies) (or other jurisdiction if applicable) where the action is located
3. Name, address, and telephone number of the person at the agency who can supply further information
4. Designation of the statement as a draft, final, or draft or final supplement
5. One-paragraph abstract of the statement
6. Date by which comments must be received (consistent with the minimum comment review period under §1506.10)

6.6.4 EIS summary

The *summary* provides a succinct mechanism for informing agency officials and the public about potential actions and their resulting environmental impacts. It can be viewed as a concise overview of the EIS, where each topic summarized is written in proportion to its importance. For many readers, the summary forms their first and lasting impression of the proposal. As the summary is often the only section read by many people, its importance is that much more. With respect to decision making, the summary allows the reader to quickly assess and balance environmental implications of the decision against technical, economic, and other factors. Thus, the summary bears a greater than normal obligation to clearly communicate the essence of the EIS to the reader. Table 6.13 outlines requirements that the summary must address.

As indicated in Table 6.13, the summary stresses "major conclusions, areas of controversy (including issues raised by agencies and the public), and the issues to be resolved (including the choice among alternatives)" (§1502.12). Consider how the Nuclear Regulatory Commission's EIS relicensing process complied with this requirement. Recall that the Commission routinely reaches the near-whimsical conclusion that the risk of a "severe [nuclear] accident" such as a full-scale nuclear meltdown is "small." Needless to say, a sizeable portion of the scientific community and American public sharply disagree with this statement. Yet the Commission's relicensing EISs do not even acknowledge, let alone respond, to such controversy.

Now consider the requirement that the summary also address "issues to be resolved." Chapter 1 described how NRC has steadfastly refused to consider the impacts of highly radioactive nuclear power plant waste piling up around the nation. This certainly falls into the category of a major and national "issue to be resolved." Yet, the Commission's relicensing EISs, directed by Mr. Brian Holian, have neglected to evaluate one of the most prominent and controversial issues of our time; these EISs simply state that the issue of nuclear waste disposition will be "addressed in the future." Consistent with this dismissal, the Summary Sections for these relicensing EISs have likewise ignored this troubling issue. That was until the Commission was sued in 2012 by 24 organizations for failing to address this issue in its relicensing EISs. The Commission lost in a resounding defeat.[17] The Commission was forced to suspend all relicensing until this nuclear waste issue has been adequately investigated in its relicensing EISs.

Table 6.13 Items That Must Be Discussed in the Summary

The EIS summary must stress (§1502.12):

- Major conclusions
- Areas of controversy (including issues raised by agencies and the public
- Issues to be resolved (including the choice among alternatives)

Critics complain that such omissions are intentional because the Commission does not want to spotlight controversial issues that could alarm the public and jeopardize its nuclear relicensing initiative. But here is the real problem. If Pham's project branch cannot even comply with basic requirements for preparing an EIS summary, why should the public trust the Commission's other analyses and conclusions? The lesson here is that to establish trust and confidence, officials need to disclose all pertinent decision-making information, particularly regarding areas of public and scientific controversy. The next section provides guidance for preparing the summary and addressing issues of concern and controversy.

6.6.4.1 *Preparing the summary*

In reality, few members of the public have the time or interest in reading a large, complex, and detailed study of a technical proposal. Consistent with this observation, the Regulations identify cases where the summary may sometimes provide an appropriate vehicle for saving resources by reducing paperwork. Specifically, agencies are allowed to reduce excessive paperwork by "Summarizing the environmental impact statement and circulating the summary instead of the entire environmental impact statement if the latter is unusually long" (§1500.4[h]).

Because the summary allows the reader to quickly focus on issues of greatest concern, a skillfully composed summary can provide a vehicle essential to the success of a planning process. This effort requires the EIS manager and staff to sift through a large volume of material in an effort to identify the succinct information that will be summarized as important topics of interest. In some cases, the EIS manager may find it advantageous to assign responsibility for preparing each section of the summary to the corresponding specialists involved in preparing that section of the EIS. A professional writer or editor can also play a pivotal role in preparing a summary that communicates essential information to the decision maker and public. Staff unfamiliar with the subject matter can also be instrumental in critiquing the summary in terms of how well information has been summarized and explained.

As information is extracted from the body of the EIS, it should be packaged into a coherent narrative. This implies more than a simple copying and pasting of paragraphs from the body of the EIS. To maximize its utility, the summary needs to be prepared as a "stand-alone" document; that is, technical terms and analyses need to be defined and described so that they can be readily understood by the average reader without having to refer back to the body of the EIS.

6.6.4.1.1 Guidance on the contents of a summary. A well-crafted summary is informative, concise, and can be readily understood by nontechnical members of the general public. An effective summary should sharply define differences between the environmental consequences of the analyzed alternatives. It should also indicate any decision(s) that need to be considered and eventually made. The summary should accurately emphasize issues and impacts of greatest concern to the public and decision maker. The author provides the following synopsis of guidance for preparing the summary:

- **Major conclusions:** As the analysis and comparison of alternatives is the "heart" of an EIS, it must also be the focus of the summary. The summary needs to emphasize principal conclusions regarding (1) significant impacts, (2) key differences among the alternatives, and (3) environmental implications associated with the choice of alternative.
- **Areas of controversy:** As just witnessed, the summary needs to plainly delineate areas of concern and controversy. Clearly announcing important and controversial issues is important in the event of later EIS litigation; acknowledging areas of controversy provides the agency with a means of demonstrating that it considered all relevant information, *including views contrary to its position;* failure to acknowledge such controversy may be construed as evidence that such issues were neither considered nor afforded the attention they deserved.
- **Issues to be resolved:** The summary identifies unresolved issues such as scientific and technical *uncertainties,* particularly those that may need to be resolved in lower-tier or supplemental EISs.

The summary should normally not exceed 15 pages in length (§1502.12). While a summary for a complicated or programmatic EIS may sometimes need to exceed this 15-page target, the EIS manager must nevertheless strive for conciseness. Note that brevity is not tantamount to conciseness. Brevity simply implies shortness, while conciseness implies that this section contains information deemed essential to the decision maker and public but does not include superfluous or needless detail.

6.6.4.1.2 Enhancing the usefulness of the summary. The summary should briefly explain aspects of the proposal and how it would be constructed and/or implemented (i.e., "who," "what," "where," "when," "why," and "how"). A question-and-answer format has sometimes been used as a means for engaging the public. Proven methods for enhancing effectiveness of the summary include

- Briefly describing the purpose of the EIS in the decision-making process so that an unfamiliar reader understands why the statement has been prepared, what it will be used for, and how to participate in the process.

- Emphasizing impacts that are truly significant (e.g., changes in health and environmental resources); intermediate steps in the causal chain of events should not be described.
- Exercising discretion in discussing nonsignificant impacts, since such discussion often tends to obscure the pivotal decision-making issues. Focus attention on the significant impacts and comparing the key findings and differences between the alternatives.
- Appropriate graphics and tables may provide a means of concisely summarizing complicated or voluminous data. The summary might also contain a map illustrating the location of the proposal, including nearby facilities, population centers, and pertinent geographic features.

6.6.5 *Table of contents*

While every EIS must include a *table of contents,* the Regulations provide no specific guidance or requirements for preparing the table. The format and level of detail is left to the discretion of the EIS manager. Various nomenclature systems for organizing the EIS are in common usage. Many agencies use the scientific system of headings (i.e., 1.0, 1.1, 1.1.1, ...). In recent years, the trend has been away from a scientific nomenclature system, toward use of different fonts and type styles for organizing headings and subheadings. Still other systems employ a hybrid, combining a scientific system with that of different font and type styles.

6.6.6 *Statement of purpose and need*

The Regulations state that the EIS *briefly* (§1502.13)

> ... specify the underlying purpose and need to which the agency is responding in proposing the alternatives including the proposed action.

The importance of the statement of purpose and need (SPN) is often not afforded the attention it deserves. Chapter 5 explains how a properly crafted SPN can provide an invaluable tool for determining a reasonable range of alternatives for investigation in the EIS. Defined correctly, the SPN provides a rationale for distinguishing reasonable alternatives from those that are not, providing the agency with a defensible basis for dismissing courses of action that do not meet the agency's underlying need.

An inaccurate or improperly defined description of the *underlying* need for taking action has sometimes led to serious problems. A need that is vaguely or inaccurately defined may be difficult to publicly defend. In the past, a recurring problem has involved mistakenly discussing the SPN

Figure 6.5 There can be many different ways of crafting the SPN. (Courtesy images.google.com.)

for preparing the EIS document rather than the need for the proposal.[18] For example, the author recently reviewed an EIS for a client that stated the purpose and need of the EIS was to "comply with the requirements of NEPA." This gaffe immediately suggested that the EIS manager did not understand the EIS process or its requirements. Instead, the EIS should have explained the need for *taking action*. As described in Chapter 5, another commonly encountered problem is failure to clearly distinguish between the terms "purpose" and "need."

As noted in Figure 6.5, there are many different ways in which the SPN can be crafted. While determining the *underlying* SPN may appear to be deceptively simple, it can sometimes be a challenging exercise. The author has been party to more than one project in which it took several weeks or more to reach a complete consensus on the SPN. One technique for properly defining the SPN is to simply ask the following simple questions: Why are we considering the proposed action? Why is it necessary to take an action?

6.6.6.1 How to prepare a flawed statement of purpose and need

As just noted, the author recently consulted on a lawsuit involving a party that was opposed to a proposal to consider an application to construct a wind farm on land managed by the agency. The agency was responsible for preparing the EIS and making a decision to approve the applicant's proposal to construct the wind farm. The EIS defined its SPN as follows:

> In accordance with Federal Land Policy and Management Act (FLPMA), public lands are to be managed for multiple use that takes into account the long-term needs of future generations for renewable and non-renewable resources. The Secretary… is authorized to grant ROWs on public lands for

> systems of generation, transmission, and distribution of electric energy. Taking into account the agency's multiple use mandate, the purpose and need for the Proposed Action is to respond to a FLPMA ROW [right-of-way] application submitted by the Applicant to construct, operate, maintain, and decommission a wind energy-generating facility and associated infrastructure on public lands administered by the agency….

Thus, the EIS incorrectly defined the SPN as the need to respond to the applicant's proposal for a right-of-way so that it could build a wind farm. True, one of the agency's responsibilities was to review and authorize right-of-ways, but responding to or approving a right-of-way was definitely *not* the *underlying* need for the project. The underlying need was the applicant's desire to construct a wind farm to *supply energy*. The EIS mixed the underlying need for taking action with the agency's responsibility to review and enable the applicant's proposal. Because the SPN was flawed, the EIS failed to adequately evaluate a reasonable range of alternatives for meeting the underlying need.

6.6.7 The proposed action and alternatives chapter

The Regulations place emphasis on identifying and investigating reasonable courses of action for satisfying the agency's SPN. For this reason, the centerpiece of the EIS is the chapter describing the proposed action and reasonable alternatives (alternatives chapter in an EIS, Table 6.10); so much so, that this chapter is described as the "heart" of an EIS (§1502.14).

As explained in Chapter 4, an agency is not bound to choose an alternative based solely on environmental considerations. Nor is there a substantive legal requirement to mitigate environmental impacts. However, the EIS must thoroughly investigate reasonable alternatives and mitigation measures, and present this information in a form that will assist the decision maker in making an informed decision. Chapter 3, Section 3.3 and Section 6.6.6 of this book described how the SPN can be used to define a range of reasonable alternatives for detailed analysis. Chapter 5, Section 5.5 provided additional direction on assessing and describing alternatives.

6.6.7.1 Terminology

As used herein, the term "proposed action" denotes the action that an agency is specifically proposing to satisfy the SPN. The terms "proposal" and "alternatives" are used in referring to the agency's proposed action (if one is defined) and the range of reasonable alternatives. Not every reasonable alternative needs to be investigated in detail; only a "reasonable range" of alternatives need be investigated in detail. Alternatives that are both reasonable and are also examined in detail are referred to as "analyzed alternatives."

6.6.7.2 Alternatives versus environmental consequences

Confusion sometimes arises over the difference between the chapter on alternatives (see Table 6.10) and the chapter on environmental consequences. The chapter on alternatives of the EIS describes each of the alternatives, including those that are dismissed from detailed examination. In contrast, Chapter 7 (environmental consequences) examines the impacts in detail for each of the alternatives described in Chapter 5. In other words, Chapter 5 of the EIS describes the alternatives and provides information used for evaluating the impacts in Chapter 7.

An EIS must compare the impacts of alternatives. While this comparison can be presented in Chapter 7, it is most commonly provided in Chapter 5. Thus, Chapter 5 *summarizes* and compares the impacts, but should *not replicate* the analysis presented in the environmental consequences chapter.[19]

6.6.7.2.1 Component actions. A course of action is often composed of a number of *component actions*. For example, a proposal to construct a natural gas pipeline may actually involve a number of discrete component actions, including grading and devegetation, construction of the trench, construction of the actual pipeline, construction of an access road and right-of-way, construction of one or more pumping stations, construction of a maintenance and field office, and of course a maintenance program. Each of these activities can be viewed as an individual component action of the proposed action or one of its alternatives. For the proposal and each of its analyzed alternatives, these individual component actions must be identified and described in Chapter 5 of the EIS. The impacts of these component actions are then evaluated in Chapter 7.

The alternatives chapter also explains how the proposal may be related to any other action undergoing a NEPA review. In addition to explaining project interrelationships, this discussion can assist the agency in demonstrating that any interim actions (i.e., actions that need to proceed before the EIS process has been completed) comply with the interim action requirements spelled out in the Regulations (§1506.1).

6.6.7.3 Examining a range of reasonable alternatives

As noted in Chapter 4, Section 4.4, alternatives include (§1508.25[b])

1. No-action alternative
2. Other reasonable courses of actions (including the proposed action)
3. Mitigation measures (not in the proposed action)

To the extent practical, the EIS must identify and examine all three of these alternatives. The EIS need not discuss every conceivable alternative when an unmanageably large number of reasonable options exist. An alternative is deemed to be "reasonable" if it is considered to be "practical" or "feasible" from the standpoint of[20]

- Common sense
- Technical feasibility
- Economic viability

The author has supplemented CEQ's direction, defining a reasonable alternative to be an option that meets the criteria shown in Table 6.14.

6.6.7.3.1 "Magical number seven." As depicted in Figure 6.6 and in Section 6.4 of this book, Princeton psychologist George Miller formulated what became known as the Miller's law or the Magical Number Seven. Miller's law, which essentially argues that the number of objects an average human can hold in working memory is 7 ± 2.[13] Miller found that people's maximum performance corresponds to the ability to distinguish between 4 and 8 alternatives. On the basis of Miller's law, the author recommends that where practical, the number of analyzed alternatives described in the EIS be limited to the low end of Miller's range or between 5 and 7. Of course, in rare instances, the number of alternatives may need to exceed this number particularly where it is deemed necessary to provide the decision maker with an extended range of potential courses of action. As always, professional judgment needs to be exercised in determining the wisdom of limiting alternatives.

6.6.7.3.2 A large or infinite number of alternatives. Where there are a very large number of potential options, only a reasonable number of cases, representing the full spectrum of alternatives, must be analyzed.[20] For instance, some proposals may involve a very large or even an infinite number of reasonable alternatives. Consider a proposal for designating a wilderness area in a national forest. This proposal might involve

Table 6.14 Suggested Criteria for Determining if an Alternative Should Be Deemed a "Reasonable Alternative" Subject to Examination in an EIS

An alternative is deemed to be reasonable if it:

1. Satisfies the underlying need for taking action (§1502.13)
2. Is "practical" or "feasible" from the standpoint of
    - Common sense
    - Technical feasibility
    - Economic viability

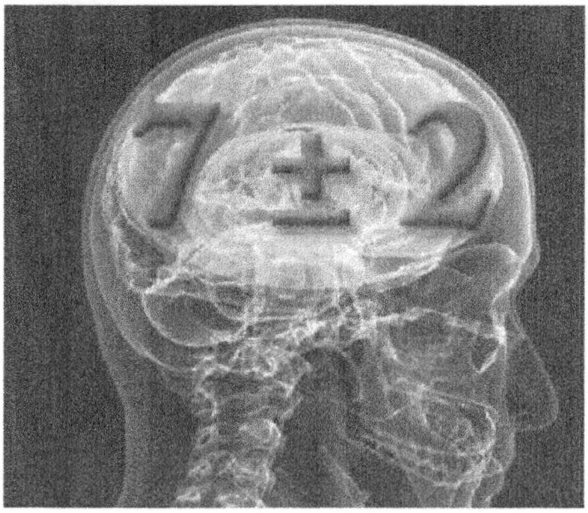

Figure 6.6 Magical Rule of Seven. (Courtesy images.google.com.)

an infinite number of alternatives ranging from 0% to 100% of the forest. When there are a very large or infinite number of alternatives, only a reasonable number of options, covering the full spectrum of alternatives, must be analyzed. In this example, an appropriate series of alternatives might include the alternatives of dedicating 0%, 20%, 40%, 60%, 80%, or 100% of the forest to wilderness[21]; the EIS should explain why the agency believes the range of analyzed alternatives covers a full spectrum of reasonable alternatives.

6.6.7.3.3 Types of alternatives. Alternatives can involve options as diverse as alternative siting, transportation methods, different modes of transportation, or alternative technologies. Analysts should not neglect options such as leasing a service or facility from a private party since these are often not only "reasonable," but also economical courses of action; moreover, if an agency is able to avoid having to construct a new facility, the environmental footprint and impacts might be reduced.

Be sure to search for approaches or measures that can reduce environmental impacts. It is important to note that a reasonable alternative must be considered and where appropriate, evaluated, even if it lies outside the legal jurisdiction of the agency (§1502.14).[22] If such an alternative is deemed reasonable, the agency should clearly explain why the alternative cannot be chosen; this can provide a basis for changing the law so that the agency can pursue that alternative. Failure to adequately analyze alternatives outside the agency's jurisdiction has been a problem in some EISs.[18]

6.6.7.3.4 Dismissing alternatives. Briefly discuss alternatives considered but dismissed from detailed evaluation, particularly any raised during the public scoping process. When dismissing an alternative from detailed analysis, thoroughly explain the rationale for dismissing the alternative (e.g., the cost is unreasonable or that it is technical impractical) (§1502.14[a]). Alternatives that have been considered and dismissed are often placed under a section labeled something to the effect of "Alternatives Considered but not Carried Forward." Discussion of such alternatives should generally be minimized, no more than is necessary to give the reader an adequate understanding of what the alternative involves and *why* it was dismissed. Evidence should, clearly but briefly, demonstrate why the alternative is unreasonable, from the standpoint of economic, technical, or other considerations.[23]

6.6.7.4 The no-action alternative

As described below, the EIS must include the alternative of taking no action (§1502.14[d]).

6.6.7.4.1 The no-action alternative versus the affected environment. The *no-action alternative* and the description of the *affected environment* (Chapter 6, in an EIS, Table 6.10) are frequently confused. A common mistake is that the impacts of taking no action are equivalent to that of the baseline or "affected environment." The description of the affected environment constitutes a snapshot of *present* conditions of resources and the geographic area that could potentially be affected by the proposal. Thus, the affected environment defines the current environmental baseline for assessing potential impacts of a proposal.

In contrast, the no-action alternative provides a different environmental baseline, allowing the decision maker to compare future impacts of the proposed action or its alternatives with the long-term effects of taking no action. The potential impacts of taking no action are based on a projection of current and any evolving conditions into the future. The no-action alternative is not necessarily a static condition. This acknowledges the fact that the environment can be affected and can evolve even though the agency takes no action. For instance, suppose an agency proposes to take measures to reduce beach erosion. The no-action alternative would describe the degradation of the beach area over a time period if no action were taken to mitigate the erosion.

6.6.7.4.2 Describing the no-action alternative. Consider a scenario in which an existing landfill is nearly filled to capacity. A federal agency needs to issue a permit (triggering NEPA) for a proposal to construct a replacement landfill. An EIS is prepared. In evaluating the alternatives, the agency must consider the environmental and socioeconomic impacts of taking no action to replace the existing landfill. The analysis of taking no action involves projecting the impacts of doing nothing into the future. For example, without a replacement landfill, people within the community would lack a location or means of disposing their garbage. Without a replacement landfill, some people might dig holes in their backyards to dispose of garbage; this could lead to contaminants leaching into the groundwater. Others might dump garbage along roadsides. Still others might allow the garbage to accumulate in their backyards, which could attract rodents and spread disease. Others might dump their garbage in remote areas. The resulting impacts of taking no action on the present environmental baseline could be quite significant. The no-action alternative describes these reasonably foreseeable effects.

6.6.7.5 Describing the analyzed alternatives

As we have seen, the EIS must investigate a range of reasonable alternatives for achieving the agency's SPN. As shown in Figure 6.7, an interdisciplinary team is often assembled to identify, describe, and investigate alternatives. Some alternatives may be identified and briefly described, and then dismissed from detailed study because they are deemed to be either unreasonable or because they do not meet the need for taking action. Alternatives that are investigated in detail are often referred to as the "analyzed alternatives."

As noted earlier, the alternatives chapter draws on and summarizes the scientific discussion presented in the environmental consequences chapter of the EIS. It needs to be written in a way that allows the reader to compare and draw a sharp distinction between the impacts of each alternative (§1502.14, §1502.16).

Figure 6.7 An interdisciplinary team is often assembled to identify, describe, and investigate alternatives.

As described shortly, all analyzed alternatives described in the EIS must be given "substantial" treatment. Descriptions must contain sufficient detail to provide analysts with information needed to assess the potential impacts. The alternatives must also be clearly described so that the decision maker and public have a thorough understanding of what would take place if that course of action were chosen. Key regulatory requirements for describing the alternatives are summarized in Table 6.15. Some of these requirements are explained in more detail in subsequent sections.

Table 6.15 Key Regulatory Requirements for Describing Alternatives (§1502.14)

a. Rigorously explore and objectively evaluate all reasonable alternatives, and for alternatives that were eliminated from detailed study, briefly discuss the reasons for their having been eliminated.
b. Devote substantial treatment to each alternative considered in detail, including the proposed action, so that reviewers may evaluate their comparative merits.
c. Include reasonable alternatives not within the jurisdiction of the lead agency.
d. Include the alternative of no action.
e. Identify the agency's preferred alternative or alternatives, if one or more exists, in the draft statement and identify such alternative in the final statement unless another law prohibits the expression of such preference.
f. Include appropriate mitigation measures not already included in the proposed action or alternatives.

6.6.7.5.1 Presenting a rigorous and objective alternatives analysis. As indicated in Table 6.15, a solid scientific assessment requires that the analysis "rigorously explore and objectively evaluate all reasonable alternatives" (§1502.14[a]). The requirement to present the decision maker and public with an objective analysis is no less important. To this end, the Regulations require that the alternatives analysis provide a "full and fair discussion of significant environmental impacts…" (§1502.1).

6.6.7.5.2 Devoting "substantial treatment" to each of the reasonable alternatives. Each analyzed alternative must be described in sufficient detail such that its scope is clear and its potential impacts can be understood. To meet the requirements cited in Table 6.15, some descriptions may require substantially more detail and explanation that others. With respect to the rigor of analysis, the EIS must devote "substantial treatment" to each of the reasonable alternatives (§1502.14[b]). The term "substantial treatment" is used in the Regulations in lieu of the phrase "equal treatment" because the degree of attention devoted to the analysis tends to vary with the complexity of the alternative under consideration and extent of its impacts. This direction applies equally to the investigation of the proposed action; unless justified by its complexity or other related assessment issues, the level of consideration given to the proposed action should *not* differ substantially from that devoted to the other reasonable alternatives.

6.6.7.5.3 Life cycle description. To the extent practical, alternatives need to be investigated over their entire life cycle (site preparation, construction, operation, and post-closure). To the extent feasible, the following phases should be described:

- *Construction activities*—including preconstruction activities such as site surveys, site clearing and preparation, road access construction, and other activities that support construction
- *Operational activities*—including operational activities, maintenance and transportation activities, waste streams, and other activities that can effect environmental quality
- *Post-operational requirements*—description of reasonably foreseeable future requirements for closing a project such as site closeout, decontamination, and site restoration

Also examine any reasonably foreseeable modifications and circumstances that may change over time. Aim to include rather than exclude activities.

6.6.7.6 Comparing alternatives

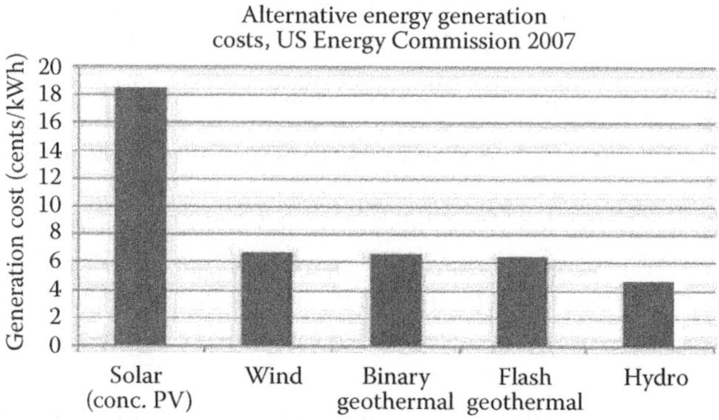

One of the most important functions of an EIS is to provide the decision maker and public with a crisp comparison of the impacts, thus sharply contrasting the differences between alternatives. As specified in the Regulations, the alternatives chapter must

> ... present the environmental impacts of the proposal and the alternatives in comparative form, thus sharply defining the issues and providing a clear basis for choice among options by the decisionmaker and the public. (§1502.14)
>
> Devote substantial treatment to each alternative considered in detail... so that reviewers may evaluate their comparative merits. (§1502.14[b])

Identify environmental effects and values in adequate detail so they can be compared to economic and technical analyses (§1501.2[b]).

Describe each analyzed alternative so that it is clear what the distinctions and differences are. Each alternative must be described in detail sufficient to allow a reasonable comparison. Alternatives are to be compared with each other, not just with the proposed action or with the no-action alternative (see the example of the alternative energy figure above).

6.6.7.6.1 Techniques for comparing alternatives. Figures, tables, and graphs can provide particularly effective techniques for comparing and summarizing important information about alternatives. All figures and tables should have instructive titles and informative text descriptions. Many techniques are in common usage for comparing alternatives. Construction of an *alternatives-versus-impacts* matrix can provide a particularly useful technique

Table 6.16 Example of How a Matrix Can Be Used to Compare Important Impacts and Decision-Making Characteristics among Alternatives

Environmental consequences	No action	Proposed action	Alternative 1	Alternative 2
Potentially affected population	11,500	11,500	9250	13,400
Total land disturbed (acres)	245	680	290	980
Habitat disturbed (acres)	55	490	485	640
Water consumption (gal/day)	12,800	195,000	125,000	88,000
Peak noise level (dBA)	22	39	41	28
Number of deer killed	45	350	220	390
Miles of road built	0	13	12	32
Annual NO_2 ($\mu g/m^3$)	2.7	3.1	1.95	3.1
Annual CO ($\mu g/m^3$)	23.7	25.8	29.7	23.7.9
Increase in health-related illnesses over 20 years	5	3	2	3

for illuminating differences among alternatives. Such a matrix need not be limited to comparing environmental impacts. Other important decision-making considerations may also be included. Table 6.16 provides a simplified example of how a matrix may be used to compare important impacts and decision-making characteristics among alternatives. Table 6.17 provides

Table 6.17 Example of How a Matrix Can Be Used to Compare Important Impacts and Decision-Making Characteristics among Alternatives

	Impact area							
Alternatives	Land use	Air quality	Groundwater	Surface water	Aquatic and terrestrial resources	Human health	Socioeconomics	Waste management
Proposed wind farm	S	S	S	S	S	S	S	S
Coal-fired alternative	M	M	S	S to M	S to M	S	S to M	M
Natural-gas-fired alternative	S to M	S to M	S	S	S	S	S to M	S
No-action alternative	S	S	S	S	S	S	S to M	S

Note: S, small impact; M, moderate impact; L, large impact.

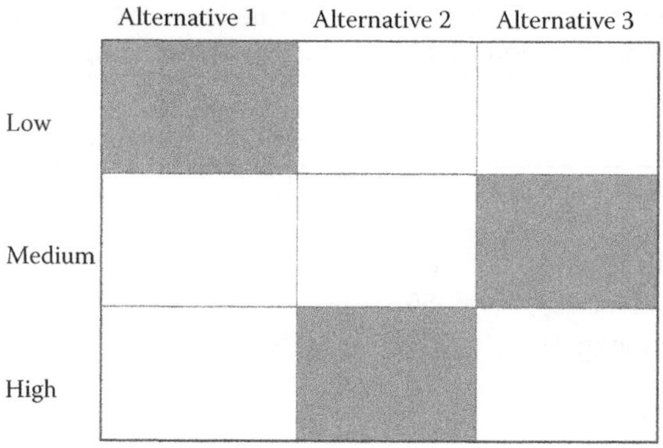

Figure 6.8 Example using a visual aid. This simplified example illustrates the use of a visual aid to quickly convey information to the reader. In this example, an EIS analyzes three alternatives and compares an environmental impact such as air degradation in terms of values: "low, "medium," and "high." A specific cell is shaded to indicate the size of the impact.

a second example of how a matrix may be used to compare environmental consequences between three power plant options.

Figure 6.8 provides a simplified example of how a figure can be used in comparing the impacts of various alternatives. In this case, the figure illustrates how each alternative would affect an environmental resource such as air quality.

6.6.7.7 The "preferred" versus "environmentally preferable" alternative

From a decision-making standpoint, the agency must identify two different types of alternatives:

- Agency's preferred alternative
- Environmentally preferable alterative

6.6.7.7.1 Agency's preferred alternative. The draft EIS is expected to identify the course of action that the lead agency favors. Specifically, the EIS must (§1502.14[e])

> Identify the agency's preferred alternative or alternatives, if one or more exists, in the draft statement and identify such alternative in the final statement unless another law prohibits the expression of such a preference.

The agency is expected to identify its preferred course of action in the draft EIS. Identifying its "preferred alternative" during the draft stage

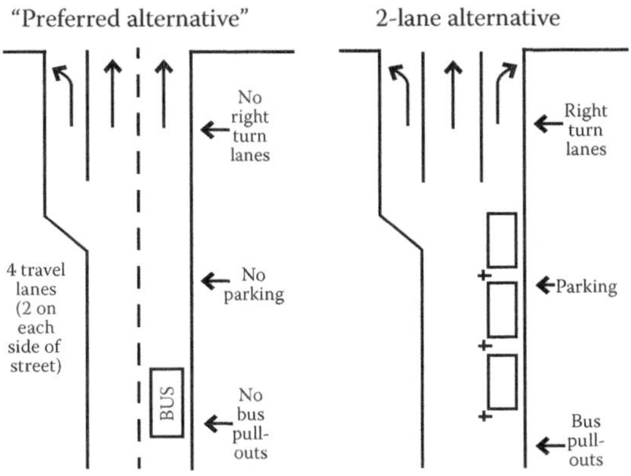

Figure 6.9 Example of preferred alternatives for a road that facilitates bicycling and lacks a right-turn lane and bus pullout. (Courtesy images.google.com.)

of the EIS is advantageous to the public and stakeholders because it provides an early indication of the direction in which the agency is leaning (Figure 6.9). This allows reviewers and the public to focus attention on the alternative that, in all likelihood, will be chosen in the final EIS. As new information comes to light, views, opinions, and preferences may change. Sometimes, an agency changes its preference from the "proposed action" to one of analyzed alternatives.

If the agency does not have a preferred alternative upon completing the draft, it must disclose this fact and explain that the preferred alternative will be identified in the final EIS. The only exception to this requirement occurs when another law prohibits identification of such an alternative (§1502.14[e]).[24]

As shown in the following regulatory requirement, the draft EIS must explain factors and considerations that are likely to shape the agency's final decision. Surprisingly, many EISs neglect to identify such considerations. Specifically, the draft must indicate

> … those considerations including, factors not related to environmental quality, which are likely to be relevant and important to a decision. (§1502.23)

6.6.7.7.2 "Preferred" versus "environmentally preferable" alterative. The terms *preferred alternative* and *environmentally preferable alterative* are sometimes confused As just noted, the agency's preferred alternative is the course of action that the agency favors; where feasible, this alternative must be identified in the draft EIS.

In contrast, the environmentally preferable alterative is the one that, on balance, is most desirable from the standpoint of environmental

Table 6.18 A Partial Checklist for Reviewing the Adequacy of Alternatives Analysis

- Does the EIS devote "substantial treatment" to each of the reasonable alternatives (§1502.14[b)])[26]?
- Does the EIS rigorously explore and objectively evaluate all reasonable alternatives (§1502.1, §1502.14[a])?
- Have any alternatives been excluded simply because they were not consistent with an action that the agency favors?
- Do the analyzed alternatives encompass the range of alternatives that will be considered by the ultimate agency decisionmaker (§1502.2 [e])?
- Have reasonable alternatives been considered even if they lie outside the legal jurisdiction of the agency (§1502.14[c])[27]?
- Does the EIS explain how each alternative as well as any decision based on it, would or would not achieve the goals of NEPA and other environmental laws (§1502.2[d])?
- Does the EIS indicate considerations, including factors not related to environmental quality, that are likely to be relevant and important in reaching a final decision (§1502.23)?
- Do the analyzed alternatives include appropriate mitigation measures (§1502.14 [f])?

quality. The agency's preferred alternative may or may not be the same as the environmentally preferable alternative. The environmentally preferable alterative is not required to be identified in the final EIS but must be identified in the ROD, along with any other alternatives that were considered in the EIS (§1505.2[b]).[25] The environmentally preferable alternative will be discussed in more detail in Section 6.7.

A checklist, useful in assuring that the alternatives chapter meets the principal regulatory requirements, is provided in Table 6.18.

6.6.7.8 Mitigation measures

Under the Regulations, mitigation measures are categorized as alternatives that need to be investigated in the EIS. While an agency does not have a substantive mandate to mitigate impacts, it is required to identify and evaluate mitigation measures in the EIS; it must also at least consider the adoption of such measures during the decision-making process. Specifically, the Regulations state that an EIS must discuss

> ... appropriate mitigation measures not already included in the proposed action or alternatives. (§1502.14[f])

Consistent with the methodology employed for describing alternatives, mitigation measures should be identified and described in the alternatives chapter of an EIS, while analysis of their effectiveness is evaluated in the environmental consequences chapter. The investigation of mitigation options should include measures indicated in Table 6.19 (§1508.20).

6.6.7.8.1 Scope of mitigation. The scope of mitigation measures must address the *range* of potential impacts that can be expected to occur. With the exception of the no-action alternative, mitigation measures are described for all analyzed alternatives (including the proposed action if one is defined).

In the CEQ's opinion, the term "mitigation measures" does not normally include methods or technology that is considered standard engineering practice or is required by law or regulations. Where applicable, mitigation measures *must* include methods such as land use controls and alternative designs for decreasing emissions, construction impacts, and aesthetic intrusion.[28]

All reasonable mitigation measures must be considered, even if they lie outside the jurisdiction of the agency (§1502.16[h], §1505.2[c]). This requirement extends even in cases where the measures are unlikely to be adopted or enforced by the responsible agency. While such a requirement might at first appear to be unjustified, it serves to alert officials and other agencies that possess the capability to implement these measures or to institute changes in existing rules or regulations.[29]

6.6.7.8.2 Evaluating mitigation measures. Once a proposal is determined to result in potentially significant impacts, mitigation measures must be considered, developed, and analyzed for all impacts, whether such impacts are significant or not (§1502.14[f], §1502.16[h], §1508.14).[28] Thus, the requirement to investigate mitigation applies even to impacts that are not, by themselves, considered significant.

It is not uncommon to find that an EIS has merely identified mitigation measures without actually evaluating their potential effectiveness. The EIS must evaluate the *effectiveness* of mitigation measures in reducing potential impacts.[30] The description of such measures must therefore be sufficiently detailed to allow analysts to evaluate their effectiveness. For instance, instead of simply stating that an environmentally sensitive area would be revegetated to mitigate construction disturbances, the discussion should specifically describe how and which areas would be revegetated, including the types of vegetation that would be used. Analysts would then evaluate how effective this revegetation effort would be in rectifying the construction impacts.

To ensure that environmental effects are given fair assessment, the agency's EIS and the ROD should indicate the likelihood that such measures will be adopted or enforced by the responsible agencies. Where

Table 6.19 Potential Measures for Mitigating Impacts (§1508.20)

1. Avoiding the impact altogether by not taking a certain action or parts of an action
2. Minimizing impacts by limiting the degree or magnitude of the action and its implementation
3. Rectifying the impact by repairing, rehabilitating, or restoring the affected environment
4. Reducing or eliminating the impact over time by preservation and maintenance operations during the life of the action
5. Compensating for the impact by replacing or providing substitute resources or environments

there is a history of nonenforcement or opposition to such measures, the EIS/ROD should acknowledge and discuss such opposition/nonenforcement. If the necessary mitigation measures will not be ready for a long time, this fact should likewise be acknowledged.[29]

6.6.7.8.3 Investigating and documenting mitigation measures. The author provides the following guidance for performing, investigating, and documenting the assessment of mitigation measures:

- Indicate whether the implementation of a mitigation measure is within the agency's jurisdiction.
- Investigate mitigation for all impact areas, emphasizing steps for addressing those impacts with the greatest potential for significance.
- Identify any external parties (e.g., state, local, or tribal government agencies; land owners) who must be involved in establishing or implementing the mitigation.
- Evaluate pollution prevention strategies and technologies beyond those inherent in the alternative.

6.6.8 Affected environment chapter

As we have seen, the significance of potential impacts is a function of both *intensity* and *context* (§1508.27). The affected environment chapter provides the baseline description of the existing environment against which both the intensity and context of potential impacts are compared (§1508.15). This allows the decision maker and public to gauge how and to what extent environment resources will be affected. Chapter 5, Section 5.5 provides additional direction on assessing the affected environment.

6.6.8.1 Describing the affected environment

The affected environment chapter of the EIS must (§1502.15)

> … succinctly describe the environment of the area(s) to be affected or created by the alternatives under consideration.

As illustrated in Figure 6.10, this description depicts the baseline environment as it presently exists—*not* as it would be affected if the proposal were implemented. The introduction to this chapter should indicate that the purpose of this chapter is to provide a baseline against which the potential impacts are measured.

This chapter should be organized and presented on a resource-by-resource basis. Resources should be organized in the same order as they are discussed in the environmental consequences chapter.

6.6.8.1.1 Limiting range of resources and level of detail. It is not uncommon to find that the affected environment chapter is unduly detailed and excessively long. Some EISs provide an extensive discussion of environmental resources, even those that clearly have no potential to be affected by the proposal. As an example, a discussion on local air quality may be dismissed if the proposal does not result in any releases that could

Figure 6.10 The affected environment section depicts the baseline environment as it presently exists—*not* as it would be affected if the proposal were implemented.

conceivably affect air quality. Alternatively, baseline air contaminant concentrations may need to be described if an action could potentially affect air quality. A brief rationale should be provided explaining why any such resources have been dismissed from a more detailed examination.

The affected environment chapter should normally be considerably smaller than the environmental consequences chapter.[18] As a rule, each resource should be described in only enough detail to allow the reader to understand how it could be affected. Descriptions of the affected environment should be commensurate with the importance of the potential impact (e.g., sliding-scale approach, described in Chapter 2, Section 2.7). The EIS manager must exercise vigilance in ensuring that the scope of environmental resources is limited to those which are potentially threatened. One means of reducing excessive verbiage is to simply incorporate unessential data by reference (§1502.15).

6.6.8.1.2 Generic outline for the affected environment chapter. A suggested generic outline for the affected environment chapter is provided in Table 6.20. This outline should be tailored as necessary to meet specific circumstances.

The following specific direction is offered for describing the affected environment:

- Remember that the extent of the "affected environment" may not be the same for all potentially affected resource areas or for all the analyzed alternatives. The spatial domain frequently varies with the resource and impact under consideration.
- Take heed to provide information that is necessary to assess or understand the impacts.
- Limit the description of the existing environment to information that directly relates to the scope of the analyzed alternatives. For example, do not provide detailed information on air quality if none of the analyzed alternatives would generate any substantial air emissions.
- As appropriate, summarize and incorporate by reference more detailed descriptions of the affected environment. Limit the description of environment resources to information that directly pertains to the scope of the analyzed alternatives; only provide information that is necessary to assess or understand the impacts. For example, do not provide information on water quality if none of the analyzed alternatives would produce any discernible effluents.

Table 6.20 Suggested Generic Outline for Affected Environment Chapter

3.0 Affected environment (as applicable, discuss the following disciplines)
3.1 Purpose for describing the affected environment
3.2 Local and regional environment
- 3.2.1 Location of the proposal
- 3.2.2 Geography and socioeconomic context
- 3.2.3 Nearby Native American reservations
- 3.2.4 Other introductory information

3.3 Air resources
- 3.3.1 Meteorology
- 3.3.2 Air quality
- 3.3.3 Air quality issues (e.g., acid rain, global warming)

3.4 Hydrology
- 3.4.1 Surface water sources and quality
- 3.4.2 Groundwater, and hydrogeology sources and quality

3.5 Geologic resources
- 3.5.1 Soils
- 3.5.2 Geomorphology
- 3.5.3 Structure
- 3.5.4 Geologic hazards

3.6 Ecology
- 3.6.1 Plants
- 3.6.2 Animals
- 3.6.3 Aquatic life
- 3.6.4 Habitats and fish/wildlife reserves
- 3.6.5 Sensitive, threatened/endangered species and habitats
- 3.6.6 Biodiversity

3.7 Socioeconomic resources
- 3.7.1 Population, local communities, employment and income
- 3.7.2 Local and regional, industries and agriculture
- 3.7.3 Traffic, transportation, public services and utilities
- 3.7.4 Subsistence (e.g., hunting and gathering)

3.8 Land use
- 3.8.1 General land use plans, policies, and restrictions
- 3.8.2 Transportation and utility restrictions
- 3.8.3 Residential and recreational restrictions
- 3.8.4 Industrial and agricultural restrictions
- 3.8.5 Infrastructure (e.g., roads, sewage systems) restrictions

(*continued*)

Table 6.20 (Continued) Suggested Generic Outline for Affected Environment Chapter

3.9 Other disciplines
3.9.1 Ambient noise levels
3.9.2 Ambient electromagnetic forces
3.9.3 Background chemical contamination levels
3.9.4 Background radiation levels
3.9.5 Aesthetics

6.6.8.1.3 Describing sensitive resources. Particular emphasis needs to be placed on describing any environmentally sensitive resources. Table 6.21 lists categories of environmentally sensitive resources that may need to be addressed.

6.6.8.1.3.1 Prime and unique farmland. As depicted in Table 6.21, the terms *prime* and *unique* farmlands deserve special mention. Prime and unique farmlands are designations assigned by the US Department of Agriculture (USDA). The term *prime farmland* refers to land having the best combination of characteristics for producing food, forage, fiber, oilseed crops, and which is available for use. The land is also used as cropland, pastureland, and rangeland.

Unique farmland is land, other than prime farmland, that has a combination of characteristics that can support production of high-value food or fiber crops. Such land has a special combination of soil quality, location, growing season, and moisture needed to produce sustained specific high-quality crops. Examples include citrus, tree nuts, olives, cranberries, fruit, and vegetables. The USDA and the Soil Conservation Service can provide additional information on addressing these resources.

6.6.8.1.3.2 Restriction on releasing sensitive resource information. In some instances, it is illegal to reveal locations of threatened and endangered species, archaeological sites, fossil beds, and other environmentally sensitive resources. References to such subjects in the EIS, especially in regard to specific locations, may need to be restricted from

Table 6.21 Some Environmentally Sensitive Resources

Prime and unique agricultural lands
Threatened/endangered, federal or state, listed or potentially listed species or critical habitats
Properties of historic, archaeological, or architectural significance
Protected natural resources such as national or state forests and parks
Special water resources such as a sole source aquifer, floodplains, or wetlands

public dissemination. Legal counsel may need to be consulted before such material is publicly disseminated. The reader is directed to §1507.3(c) for procedures that may be helpful in addressing sensitive environmental resources.

6.6.8.1.4 Commonly encountered problems. One occasionally encountered problem involves mixing discussions of potential impacts with the description of the affected environment. It is important to reemphasize that the description of the impacts is presented in the environmental consequences chapter.

Another commonly encountered problem has involved insufficient or inadequate field data. In lieu of conducting a necessary monitoring or field survey, some EISs have simply dismissed potentially significant environmental resource issues with simple statements such as "... is not known to exist in the location of the proposed site." This may not be sufficient, particularly if the subject involves a potentially significant impact. Monitoring or a site-specific survey may be required when information regarding potentially affected environmental resources is unavailable, insufficient, or of questionable accuracy.

6.6.9 Environmental consequences chapter

As we have seen, the detailed investigation of the environmental impacts is addressed in the environmental consequences chapter (see Table 6.10, item 7, and §1502.16). The introduction to this chapter should begin by explaining that its purpose is to evaluate potential impacts and that it provides the scientific and technical basis for comparing alternatives in the alternatives chapter (see Table 6.10, item 5). The Regulations define three types of impacts that must be investigated in the EIS (§1508.25[c]):

- Direct
- Indirect
- Cumulative

While an action results in both direct and indirect effects, their discussion should not be segregated into separate sections. Discussion of these impacts is simply integrated into a common analysis. Because of their complexity and special analytical methodologies that may need to be employed, the cumulative impact analysis is frequently relegated to a separate section of this chapter.

The reader should note that the nouns "effects" and "impacts" are synonyms, but the verb synonymous with "impacted" is "affected." Some fastidious NEPA writers avoid using the word "impacted," using "affect" as the verb instead; but using "effect" as a verb is clearly incorrect.

6.6.9.1 Required environmental issues and impacts

In addition to ecological effects, environmental impacts also include but are not limited to aesthetics, historic, cultural, economic, social, and health effects (§1508.8[b]). Table 6.22 details what this chapter must include (§1502.16). Each of these items is detailed in this chapter.

Table 6.22 Items Addressed in Chapter on Environmental Consequences

Standard requirements that must be addressed (§1502.16):

a. Direct effects and their significance (§1508.8).
b. Indirect effects and their significance (§1508.8).
c. Possible conflicts between the proposed action and the objectives of federal, regional, state, and local (and in the case of a reservation, Indian tribe) land use plans, policies and controls for the area concerned (§1506.2[d]).
d. The environmental effects of alternatives including the proposed action.
e. Energy requirements and conservation potential of various alternatives and mitigation measures.
f. Natural or depletable resource requirements and conservation potential of various alternatives and mitigation measures.
g. Urban quality, historic and cultural resources, and the design of the built environment, including the reuse and conservation potential of various alternatives and mitigation measures.
h. Means to mitigate adverse environmental impacts (if not fully covered under §1502.14[f]).

Additional requirements

1. As appropriate, include a cost–benefit analysis (§1502.23).
2. Relationship between short-term use of man's environment and the maintenance and enhancement of long-term productivity (Section 102(2) of the NEPA statute, §1502.16).
3. Adverse impacts that cannot be avoided should the proposal be implemented (Section 102(2) of the NEPA statute, §1502.16).
4. Any irreversible or irretrievable commitment of resources should the proposal be implemented (Section 102(2) of the NEPA statute, §1502.16).
5. How alternatives considered in the EIS and decisions based on it, will or will not achieve the requirements of Sections 101 and 102(1) of NEPA and other environmental laws and policies (§1502.2[d]). Note: This requirement is sometimes met in the alternatives section.
6. List of all federal permits, licenses, and other entitlements that must be obtained in implementing the proposal. If it is uncertain whether a federal permit, license, or other entitlement is necessary, the EIS must so indicate (§1502.25[b]). Note: This requirement is sometimes placed in the alternatives chapter or in an individual chapter.

6.6.9.2 *Suggested general purpose outline*

The description of consequences should be organized and presented in an order corresponding to the way the alternatives are presented in the alternatives chapter. On the basis of Table 6.22, the author suggests the general purpose outline for this section of the EIS (see Table 6.23). The actual format, of course, must be tailored to meet the agency's specific needs. Specific direction for preparing this chapter is presented in the following sections.

6.6.9.3 *Commonly encountered problems*

Recently, the author was retained to review the adequacy of a draft EIS. The EIS described emissions and effects on air quality, yet had no corresponding discussion of the air quality baseline in the affected environment chapter. This is an all too common oversight. Lacking a corresponding baseline, it can be difficult (if not impossible) to accurately gauge how a specific impact would change or affect an existing resource. A cross-check should always be performed to verify that a corresponding discussion has been included in the affected environment section.

Another common problem involves identification of environmental disturbances without analyzing their environmental consequences (Chapter 5, Section 5.3.2). In other words, the analysis simply identifies the disturbances without explaining how they would affect or alter environmental quality.

Some EISs also neglect to assess and assign a significance value to the impact. Yet another problem involves not investigating mitigation measures and their effectiveness in reducing potential impacts.

6.6.9.4 *Identifying scientific methodologies*

Agencies are responsible for ensuring the professional and scientific integrity of the EIS analyses. Scientific methodologies including any models used in the analysis need to be described (§1502.24). Details of these methodologies and models are often presented in an appendix.

Prudence must be exercised in choosing particular models or methodologies. For instance, critics may claim that a particular methodology was arbitrarily disregarded because it would lead to results unfavorable to the agency's objectives. The rationale used in determining why a particular methodology was chosen should be detailed. Similarly, this discussion should also briefly describe other methodologies considered and why they were rejected.

6.6.9.5 *Direction for describing the environmental consequences*

It is recommended that the six-step technique *action–impact model* described in Figure 5.1 should be used in assessing impacts. Using a

Table 6.23 Suggested General Purpose Outline for Environmental Consequences Chapter

4.0 Environmental consequences (by resources and disciplines)
- Explanation of the purpose of this chapter

4.1 Environmental consequences of taking no action (direct, indirect, and cumulative impacts) (as applicable, discuss the impacts described below).
- 4.1.1 Air
- 4.1.2 Surface water and groundwater
- 4.1.3 Geology, soils, and geological hazards
- 4.1.4 Ecology
  - Plants, animals, and habitats
  - Sensitive and endangered species, and habitats
- 4.1.5 Visual resources, land use, and land features
- 4.1.6 Health and safety
- 4.1.7 Facilities and infrastructure
  - Transportation and telecommunication systems
  - Electric, water, and sewage systems
- 4.1.8 Socioeconomics
- 4.1.9 Historical, archaeological, and cultural
- 4.1.10 Amenities
  - Local, state, and national parks/recreational areas
  - Recreation
  - Hunting and fishing
- 4.1.11 Other environmental disciplines and issues
  - Radio or television interference
  - Electromagnetic field levels
  - Radiation levels
  - Noise and odors
- 4.1.12 Mitigation measures

4.2 Environmental consequences of the proposed action (similar to outline shown in Section 4.1)

4.3 Environmental consequences of alternative A (similar to outline shown in Section 4.1)

4.4 Environmental consequences of alternative B (similar to outline shown in Section 4.1)

⋮

⋮

⋮

4.5 Comparison of alternatives and their consequences
- Environmental consequences and other factors such as schedule and cost (This section is optional depending on how it is covered in the alternatives chapter of the EIS.)

(continued)

Table 6.23 (Continued) Suggested General Purpose Outline for Environmental Consequences Chapter

4.6 Discussion of agency's preferred alternative
- Discussion of factors leading to the selection of this alternative (This discussion can either be included in this section or in the alternatives chapter of the EIS.)

4.7 Adverse impacts that cannot be avoided (as applicable, discuss the following disciplines)
- 4.7.1 Air
- 4.7.2 Surface water and groundwater
- 4.7.3 Geology and soils
- 4.7.4 Ecology
- 4.7.5 Visual resources, land use, and land features
- 4.7.6 Facilities and infrastructure
- 4.7.7 Health and safety
- 4.7.8 Socioeconomics
- 4.7.9 Historical, archaeological, and cultural
- 4.7.10 Amenities
- 4.7.11 Other environmental disciplines

4.8 Irreversible and irretrievable commitment of resources (as applicable, discuss the following disciplines)
- 4.8.1 Air
- 4.8.2 Surface water and groundwater
- 4.8.3 Geology and soils
- 4.8.4 Ecology
- 4.8.5 Visual resources, land use, and land features
- 4.8.6 Facilities and infrastructure
- 4.8.7 Health and safety
- 4.8.8 Socioeconomics
- 4.8.9 Historical, archaeological, and cultural
- 4.8.10 Amenities
- 4.8.11 Other environmental disciplines

4.9 Energy requirements and conservation potential

4.10 Natural or depletable resource requirements and conservation potential

4.11 Relationship between short-term use of the environment enhancement of long-term productivity

4.12 Relationship between the proposal and use plans

sliding-scale approach, describe impacts in proportion to their potential significance (see Chapter 2, Section 2.4); consistent with the sliding-scale approach, quantify impacts to the extent practicable, taking into account existing data. Do not evaluate impacts in detail that are clearly nonsignificant. Describe impacts for as long of a period as is reasonably foreseeable. Describe the likelihood of potential impacts whenever possible; if possible, provide both the frequency and magnitude of the impacts. Do not labor on impacts that are deemed "remote or speculative."

Provide sufficient data and references to allow the reader to review and verify the analysis and results. Identify environmental standards or requirements that are pertinent in limiting environmental impacts; briefly describe any of these conditions (e.g., emission or effluent limits). Indicate if the proposal threatens a violation of any applicable environmental protection standards or requirement.

Environment impacts are investigated within the individual geographic boundaries established in the affected environment chapter. Thus, the spatial bounds for an air quality investigation may be different from those of water resources or biota. As indicated earlier, the investigation of environmental consequences is performed on a resource-by-resource basis. The order used for describing impacts on environmental resources should follow the same sequence used in the affected environment chapter.

6.6.9.5.1 Disclosing uncertainty and any missing data. As we have witnessed, it is essential that the EIS acknowledge uncertainty and any important missing data. Where some degree of uncertainty, such as incomplete or unavailable information, is involved, explain how the uncertainty affects the analysis (Section 6.3 and §1502.22). In some instances, the degree of uncertainty can be explained qualitatively. In other cases, a more detailed analysis might be necessary.

6.6.9.5.2 Assessing significance, mitigation measures, and monitoring. In addition to describing environmental consequences, the EIS must assess and assign a significance value to the impact. The analysis may conclude that an impact originally considered to be significant was, on closer examination, found to be nonsignificant. In such cases, the analysis should clearly explain why the impact is now considered nonsignificant. As we have seen, the assessment of significance depends on

1. The context in which the impact occurs (§1508.27[a])
2. Ten intensity factors (§1508.27[b])

The EIS must identify and evaluate means for mitigating potential impacts. This discussion must describe the effectiveness of these measures in reducing potential impacts. A monitoring program may also need to be designed and described, particularly where there is a chance that impact projections could be exceeded.

6.6.9.5.3 Disclosing scientific controversy and opposing points of view. As we have seen, the draft EIS must disclose and discuss (§1502.9[a])

- All major points of view on the environmental impacts of the alternatives, including the proposed action

The final EIS must (§1502.9[b])

- Respond to public comments on the draft EIS
- Discuss… any responsible opposing view that was not adequately discussed in the draft statement and shall indicate the agency's response to the issues raised

Now consider these requirements in light of the Nuclear Regulatory Commission's EISs for relicensing the nation's fleet of nuclear power reactors. Experienced and reputable scientists and engineers have raised serious doubts about the safety and potentially calamitous effects of this program. As one example, critics within the scientific community find the Commission's conclusion that the risk of a severe nuclear accident is "small" to be incredulous. Yet the Commission's management has ignored such criticism and concerns in their relicensing EISs—a clear violation of the regulatory requirements just cited. The lesson to be learned is that all major points of view, not simply those of the agency, need to be disclosed to the public.

Direction for addressing some special impacts and issues of concern are described in the following sections.

6.6.9.6 Impacts on human health and safety

Some proposals may affect the health and safety of workers and the public. While the Regulations state that impacts to human health must be considered, they are silent about any corresponding requirement to investigate safety issues. However, one of the factors established for determining significance specifies that the assessment of significance depends on

> The degree to which the proposed action affects public <u>health or safety</u>. (§1508.27[b]]2])

This significance factor implies that, as appropriate, an EIS must investigate potentially significant health and safety issues (Figure 6.11). Chapter 5, Section 5.7 provided direction for performing a health assessment. This

Figure 6.11 Landslide destroyed part of a town near San Salvador. (Courtesy kids.britannica.com.)

section focuses on how to present the results of this analysis to the decision maker and public. To the extent feasible, human health impacts should be compared against recognized standards such as regulatory limits established by the Safe Drinking Water Act or the Occupational Safety and Health Act.

6.5.9.6.1 Describing human health effects. Some specific guidance for describing human health effects includes

- Applying the sliding-scale approach when characterizing human health effects.
- Considering all potential routes of exposure, not just the most obvious route.
- Determining the estimated period of exposure. In general, impacts should be analyzed for as long as they are reasonably expected to occur.

6.6.9.7 Natural disasters and accident scenarios

As we saw in Chapter 5, Section 5.10, some types of proposals involve the risk of a potentially significant accident and natural disaster. Where this is the case, the EIS must investigate potentially significant accident or natural disaster scenarios. Chapter 5, Section 5.10 described procedures for analyzing accident scenarios. This section focuses on how to present the results to the decision maker and public.

The investigation of potentially significant accident or natural disaster scenarios is particularly important where an accident scenario may pose grave consequences. Such an analysis may need to consider potential consequences of low-probability, high-consequence incidents, as well as high-probability, low-consequence events. For instance, the US Corps of Engineers and the Bureau of Reclamation evaluate accident scenarios such as the overtopping of dams and dam failures. The author has developed a decision tool for determining when an accident analysis should be assessed in an EIS. Information on the use of this tool and the general topic of addressing accident analyses can be found in the companion text, *NEPA and Environmental Planning*.[31]

6.6.9.7.1 Natural disasters. Natural phenomena may profoundly influence the impacts, including accidents, of certain types of actions (e.g., dams, nuclear facilities, waste facilities, hazardous or bioengineering laboratories, and chemical processing plants). For instance, a natural disaster can

1. Adversely disrupt or destroy critical infrastructure, facilities, or essential services, which may harm citizens or society at large. For instance, a major flood could disrupt a critical highway or destroy a bridge isolating a community from the outside; a tornado might destroy an urgently needed medical facility that could lead to loss of life.
2. Trigger an accident at a potentially hazardous facility, structure, or system that could result in grave repercussions. For instance, a large earthquake might destroy the cooling system of a nuclear reactor, resulting in a meltdown and release of large quantities of highly radioactive constituents into the biosphere. Severe flooding or an earthquake might jeopardize the integrity of a dam that sits upstream of a community.

The EIS may need to assess natural phenomena, such as how geological or atmospheric hazards could affect human safety and environmental quality. An analysis of geological hazards, for instance, may involve evaluating how phenomena such as earthquakes, floods, tornadoes, volcanic eruptions, or landslides could affect a proposal. If the impacts on the proposal are considered potentially significant, the analysis may need to investigate alternatives such as a different design or location; a critical highway, for instance, might need to examine an alternative route, such as rerouting the road around the floodplain. Mitigation measures such as backup safety systems, emergency response plans and procedures, or emergency shelters may likewise need to be investigated.

6.6.9.7.2 Describing accidents scenarios and natural disasters. As applicable, the EIS needs to investigate realistic scenarios that represent a full spectrum of reasonably foreseeable accidents or natural disasters. This analysis should consider maximum reasonably foreseeable accidents. A maximum *reasonably foreseeable* accident is an event with the most severe consequences that can be reasonably expected to occur.

A *worst-case accident*, on the other hand, is the worst conceivable accident imaginable. Such scenarios are often unreasonable. For example, a worst-case accident might assume an asteroid striking a train carrying nerve gas or perhaps hitting a large dam; the probability of such an event is so remote and speculative as to render it unreasonable and, therefore, not helpful to the decision-making process.

The EIS accident assessment should consider

- Impacts on the public, involved workers, and non-involved workers. For each group, the EIS may need to examine impacts on the maximally exposed individuals, as well as the collective impact to the group as a whole.
- Factors contributing to uncertainties in the accident analysis.
- Both the probability of occurrence and the consequences of the accident.
- It is general practice to consider scenarios with frequencies of 10^{-6} to 10^{-7} per year if the consequences may be very large; scenarios with frequencies less than 10^{-7} per year rarely need to be examined.

Where chemical or radioactive releases are involved, the analysis should not simply report dose (chemical or radioactive) estimates. Instead, dose-to-risk conversion factors should be used to determine the potential consequences on a given population. These data are used to identify and quantify potential health effects (e.g., latent cancer fatalities).

6.6.9.8 Socioeconomic impacts

The issue of socioeconomics is not an environmental attribute. While such impacts may need to be considered, they do not necessarily merit the same level of attention as do environmental impacts. The author recommends that a sliding-scale approach be used in determining the level of socioeconomic analysis that is appropriate.

As applicable, the EIS explains how social and economic resources from both a local and regional perspective could be affected, especially with respect to induced population growth. This analysis may need to assess impacts on employment patterns and income levels, as well as impacts on public and private institutions, such as housing, schools, hospitals, public utilities, and recreational resources. For instance, impacts resulting from increased use of sewage and waste disposal facilities, water, and other services may need to be evaluated.

As appropriate, the analysis may need to consider impacts on the housing and rental market as a result of primary and secondary population growth. This may also necessitate an assessment on the local school system and the existing transportation system (i.e., roads, railroads, airports, and port facilities) and if there would be a need to expand such systems. For more information on addressing socioeconomic impacts, the reader is referred to the companion text, *Environmental Impact Assessment*.[32]

6.6.9.8.1 Environmental justice. Presidential Executive Order 12898 promotes the goal of *environmental justice* (EJ), focusing federal attention on human health and environmental conditions within low-income and minority neighborhoods.[33] As appropriate, this order directs each federal agency to identify and address

> … disproportionately high and adverse human health or environmental effects of its programs, policies, and activities on minority populations and low-income populations…

In a nutshell, this guidance establishes a two-step procedure for determining disproportionately high and adverse human health or environmental effects:

1. First determine if the effects are *high* and *adverse*.
2. Next determine if the effects disproportionately affect minority and low-income populations.

A presidential memorandum accompanying this executive order instructs federal agencies to[34]

> … analyze the environmental effects, including human health, economic and social effects, of Federal actions, including effects on minority communities and low-income communities, when such analysis is required by the National Environmental Policy Act.

The presidential memorandum calls on federal agencies to address significant adverse environmental effects and mitigation measures on affected communities in

- Environmental assessments (EAs)
- Findings of no significant impact (FONSIs)
- Environmental impact statements (EISs)
- Records of decision (RODs)

6.6.9.8.1.1 Performing an EJ assessment. Consistent with the aforementioned direction, an EIS may need to evaluate factors that could unfairly burden or place disproportionate adverse health or environmental impacts on minority and low-income segments of a community. The author recommends a sliding-scale approach be used in determining the level of effort expended on an EJ analysis. The CEQ has provided direction for analyzing EJ impacts.[35] This guidance includes six principles for performing EJ analyses (Table 6.24).

As appropriate, mitigation measures for reducing or eliminating disproportionately significant adverse impacts to minority/low-income population may need to be evaluated. For example, a proposal involving transportation of explosives might need to consider if the proposed transportation route would unfairly jeopardize the health or safety of a minority or low-income segment of the population. Potential alternatives or mitigation measures might involve spreading potential risk across a number of port cities, docks, and railroad/highway corridors.

Consider a case where a preliminary screening analysis indicates that the effects of a proposal could be significant in terms of both environmental and human health impacts. A review of this issue determines that the closest identified minority or low-income community is located 40 miles from the proposed site, and the environmental and human health effects on this community would not be adversely high and would not disproportionately affect this community when compared with the general population. On the basis of such a finding, it would be reasonable to conclude that there would be no disproportionately high or adverse impacts on minority or low-income populations. No additional analysis would be warranted.

As detailed below, the analysis of disproportionately high and adverse impacts should be considered in terms of both *health* and *environmental* effects.

Table 6.24 Six Principles for Performing Environmental Justice Analyses

1. Consider the composition of the affected area to determine whether low-income, minority, or tribal populations are present and whether there may be disproportionately high and adverse human health or environmental effects on these populations.
2. Consider relevant public health and industry data concerning the potential for multiple exposures or cumulative exposure to human health or environmental hazards in the affected population, as well as historical patterns of exposure to environmental hazards.
3. Recognize the interrelated cultural, social, occupational, historical, or economic factors that may amplify the natural and physical environmental effects of the proposed action.
4. Develop effective public participation strategies.
5. Assure meaningful community representation in the process, beginning at the earliest possible time.
6. Seek tribal representation in the process.

6.6.9.8.1.2 Assessing health effects. Factors considered in determining the appropriateness of evaluating health impacts on minority and/or low-income populations include

- Could the risk of impact on a minority/low-income population be (1) significant and (2) appreciably exceed those to the general population or other appropriate comparison groups?
- Could the health impact be significant or above generally accepted norms? Adverse health effects may include death, illness, or bodily impairment.
- Could the effect on health result from multiple or cumulative adverse exposures from other environmental hazards?

6.6.9.8.1.3 Assessing environmental effects. Factors considered in determining the appropriateness of evaluating disproportionately adverse environmental impacts on minority and/or low-income populations include

- Could a minority/low-income population involve a (1) significant adverse impact that is (2) likely to appreciably exceed those to the general population or other appropriate comparison groups?
- Could an impact on the natural or physical environment occur that significantly and adversely affects a minority/low-income population? Potential impacts include ecological, cultural, economic, and/or social impacts on minority/low-income communities when those impacts are interrelated to impacts on the natural or physical environment.
- Could the environmental impact involve multiple or cumulative adverse exposures from other environmental hazards?

6.6.9.8.2 Protection of children. Executive Order 13045 stipulates that where an agency's action may pose a disproportionate effect on children, the agency is responsible for ensuring that its policies, activities, and standards address environmental, health, and safety risks.[36] Consistent with this requirement, an EIS needs to consider disproportionate impacts on children. The analysis should pay special attention to products or substances that a child is likely to come in contact with or ingest.

6.6.9.9 Urban, historic, and cultural resource impacts

As we have seen, the EIS must address

> Urban quality, historic and cultural resources, and the design of the built environment, including the reuse and conservation potential of various alternatives and mitigation measures. (§1502.16[g])

Requirements for investigating historic impacts including the Section 106 review were described in Chapter 4, Section 4.5. This chapter examines requirements for documenting the findings in the EIS. In complying with this requirement, analysts should evaluate impacts to both the prehistoric and historic settings, and how these impacts can be avoided or mitigated. Information should be presented with the goal of allowing the reader to judge whether the merits of the project outweigh the potential adverse impacts to cultural resources.

As described in Chapter 4, Section 4.5, emphasis is placed on evaluating facilities and sites listed on the *National Register of Historic Places* (NRHP), as well as potential candidates for such listings.[37] Assessing the significance of impacts on historic places is typically based on evaluation criteria established by the NRHP and guidelines established by the Secretary of the Interior.[38] Effects on Native American resources are often evaluated against the factors established by the NRHP and the American Indian Religious Freedom Act.[39]

As practical, this investigation addresses how potential actions would be coordinated with the State Historic Preservation Officer (SHPO). Geographic areas that have yet to be surveyed for historic resources but which are deemed to have such potential should be clearly delineated. If a Programmatic Memorandum of Agreement is prepared, it should describe how it will be coordinated with the EIS planning process.

6.6.9.10 Air emissions and air conformity determinations

The EIS needs to examine air emissions from many different sources. For instance, construction activities normally include combustion of fuel and exhaust from construction equipment and vehicle traffic, grading, and use of volatile organic compounds (VOCs) (e.g., paints and lubricants). Fugitive dust is dust that is not emitted from definable point sources such as smokestacks; sources include unpaved roads and barren fields. Fugitive emissions may be exacerbated by earth-moving activities such as dozing, grading and material loading/handling, and vehicle trips on unpaved roads. Remember that operational emissions sometimes exceed construction emissions.

Each state is required to establish a State Implementation Plan (SIP) identifying how it will attain or maintain the National Ambient Air Quality Standards (NAAQS) defined under the Clean Air Act. The SIP provides a mechanism for enforcing criteria pollutant standards for carbon monoxide, lead, nitrogen dioxide, ozone, particulate matter, and sulfur dioxide; some greenhouse gases, including carbon dioxide, may be added to this list.

Federal actions may not contribute emissions that could cause any new violation of criteria standards within a nonattainment or maintenance area.[40] Federal agencies are required to assure that their actions comply with the NAAQS and SIPs. Where possible, *conformity determinations* must be integrated with the NEPA process. An EIS must consider conformity with State Implementation Plan criteria where the proposal could affect a nonattainment or maintenance area. As necessary, mitigation measures must be evaluated.

However, a conformity determination is not required for any major stationary source already covered by a Prevention of Significant Deterioration (PSD) program; moreover, some types of activities are exempted from a conformity determination.

6.6.9.10.1 Approach for performing an air conformity assessment. An air conformity analysis involves a two-phased approach.

1. A *conformity review* is performed to determine whether the conformity requirement would apply to the proposal (i.e., whether a conformity determination is needed).

   The conformity review is performed on all analyzed alternatives to facilitate comparison of air quality issues among the alternatives. A conformity review normally is not needed for the no-action alternative.
2. A *conformity determination* process is performed to demonstrate how an alternative would conform to the applicable SIPs.

   The conformity determination process (if required) is frequently only performed for the preferred alternative. The rationale for this guidance is due to the extent of analysis required to perform a conformity determination, which, coupled with the potential need to negotiate binding mitigation measures or offsets, can make it impractical to complete the determination on all alternatives. Moreover, the conformity regulations do not require that conformity determinations be performed on all alternatives.

6.6.9.10.2 Describing air conformity. The air-conformity analysis commonly involves computing estimates of construction and operational air emissions, and comparing those emissions against established threshold standards. For example, Table 6.25 compares the maximum daily mitigated construction air emissions for a proposed wind farm against established threshold standards.[41]

As indicated in Table 6.25, the mitigated maximum daily emissions are estimated to be below the thresholds for VOCs, carbon monoxide (CO), particulate matter 2.5 μm in diameter (PM2.5), and sulfur oxides (SO_x). However, daily construction emissions of nitrogen oxides (NO_x) and PM10 (10 μm in diameter) emissions are estimated to exceed applicable thresholds. One method for mitigating PM10 emissions involves using soil binders or paving dirt roads to the construction site.

If the NO_x and PM10 emissions cannot be mitigated below significance threshold values, it means that the proposed project would result in temporary (construction period) significant and unavoidable NO_x and PM10 impact.

Table 6.25 Example of a Simplified Air Conformity Determination Using Data for a Project in California

	NO_x	CO	PM10	PM2.5	SO_x
Emissions	451.7	140.3	521.4	104.1	1.2
ICAPCD threshold	100	150	150	150	N/A

Note: Emissions from the proposed action are compared against established air quality threshold standards. In this example, emissions are compared against significance criteria established for the Imperial County Air Pollution Control District (ICAPCD) in California.

6.6.9.11 *Describing biological impacts*

In most circumstances, the principal biological issue of concern involves the population level of a species, with a focus on the potential decline in species population and/or a habitat.

6.6.9.11.1 Section 7 consultation. As described in Chapter 4, Section 4.5, Section 7 of the Endangered Species Act (ESA) applies to management of federal lands, as well as other federal actions that may affect listed species. It directs all federal agencies to protect threatened and endangered species. When a listed species or designated critical habitat may be affected by a federal action, Section 7 requires the federal agency to initiate early informal consultation with Fish and Wildlife Service (FWS) or National Marine Fisheries Service (NMFS). The EIS describes any informal or formal Section 7 consultations. Any consultation correspondence between the FWS or NMFS is included, usually in an appendix. Likewise, any biological assessment or biological opinion is also included in an appendix.

6.5.9.11.2 Describing a floodplain and wetland review. Special requirements are triggered when a federal agency proposes to take an action that may affect floodplains and wetlands. Federal agencies are required to take various actions to protect floodplains and wetlands, including preparing a floodplain or wetlands assessment for any action proposed in a floodplain and new construction proposed in a wetland.[42] In many instances, a floodplain or wetland assessment must be coordinated with the NEPA process and included in the EIS.

6.6.10 *Four special NEPA requirements*

As indicated in Table 6.22, the NEPA statute spells out three special requirements that must be specifically addressed in an EIS (§1502.16)[43]:

> The relationship between local short-term uses of man's environment and the maintenance and enhancement of long-term productivity
>
> Any adverse environmental effects which cannot be avoided should the proposal be implemented
>
> Any irreversible and irretrievable commitments of resources which would be involved in the proposed action should it be implemented

These three requirements are frequently addressed in a standalone section at the end of the environmental consequences chapter. However, they can also be included in other sections of the EIS. They are of such importance that they deserve special consideration. Once these three requirements have been described, we will examine other special requirements that must also be addressed in the EIS. A fourth requirement identified in the Regulations is also described below.

1. *Short-term uses versus long-term productivity*

As just indicated, the section on environmental consequences must address

> ... the relationship between short-term uses of man's environment and the maintenance and enhancement of long-term productivity.... (§1502.16)

The precise intent of this requirement is interpretive, and has been the subject of considerable confusion and debate. Using perhaps more lucid text, the author recommends that this requirement be interpreted to mean

> Evaluate tradeoffs between a short-term benefit (economic or otherwise) which would be derived from pursuing an action, versus the long-term benefit or productivity that would be derived from not exploiting the environmental resources.

Unfortunately, many if not most EISs give little more than vaguely crafted, ineffective lip service to this requirement. This is regrettable as a carefully thought-out response can provide the decision maker and public with information *essential* in making an informed decision regarding the proposal's benefits and trade-offs. In fact, properly thought out and presented, it can be one of the most important sections for disclosing the impacts to the decision maker and public.

The response to this requirement should focus on presenting a fair, open, and balanced assessment of trade-offs. For example, the EIS might consider the short-term benefit derived from constructing a hydroelectric dam to generate electricity, versus the long-term degradation of habitat and species, and the scenic and recreational use of the river that would be lost. Shown below is a partial summary of how the author responded to this requirement in an EIS for the proposed power plant:

> As used in this section, the term "short term" refers to the period of time during which power generating activities would occur. The principal short-term benefit derived from this action would be the generation of a relatively clean and economical supply of energy.
>
> Construction of the plant site and the utility corridor would result in a short-term impact (until the plant is shut down and decommissioned) on the surrounding biological habitat and resources, and would limit other land use options.
>
> Use of cooling water could result in a small short-term decrease in groundwater productivity. However, once the plant was shutdown and withdrawal of water from the river ceased, the groundwater aquifer could be recharged; that water would then be available for other purposes.
>
> Hazardous air emissions would have some adverse affect on public health and long-term productivity of the ecosystem. The plant would also produce GHG emissions with potentially more serious and longer-term health concerns.
>
> Construction of the power plant would result in a long-term or permanent consumption of materials and resources such as steel, concrete, diesel and gasoline fuels, electricity, water, land, and potential loss of biological habitat. In addition to construction resource usage, this plant would burn significant quantities of natural coal.
>
> Hazardous waste generated from this action could result in a long-term detrimental impact on the biosphere and environmental productivity.
>
> Construction and operational staff would be an overall benefit to the surrounding community. Tax revenues generated as a result of this proposal

> would directly benefit local, regional, and State economies over the short term. Local agencies investing tax generated revenues into local infrastructure and other public services could enhance socioeconomic productivity over a longer-term.
>
> When compared with the no-action alternative, the short-term benefit would be the production of electricity. Conversely, there would be no short-term electrical generation benefit derived from pursuing the no-action alternative.

Now consider how the Nuclear Regulatory Commission responds to this requirement in their EISs for relicensing the nation's aging fleet of nuclear reactors (see the case study discussed in Chapter 1). In responding to this requirement, management does not even acknowledge, let alone investigate, the impacts that a severe nuclear accident would pose on the maintenance and enhancement of long-term productivity; the potentially most catastrophic impacts of a severe accident on long-term productivity are not even conveyed to the public or decision maker. Critics rightly charge this is the equivalent of concealing critical information from the public.

There are many other similar flaws as well. Consider the fact that highly radioactive waste from nuclear power plants is accumulating around the country. Because it remains highly radioactive for tens of thousands of years, it certainly affects the "maintenance and enhancement of long-term productivity." Yet again, the Commission's relicensing EISs have not even acknowledged let alone disclose such impacts to the decision maker, let alone the public. This faulty practice is coming to an end. In 2012, the Commission was sued and lost in a resounding defeat because it had failed to adequately address nuclear waste issues in its relicensing EISs.

2. *Adverse effects that cannot be avoided*

As just indicated, the section on environmental consequences must address

> … any adverse environmental effects which cannot be avoided should the proposal be implemented.

Essentially, this requirement constitutes a disclosure statement regarding adverse impacts. The reader should note that this requirement is not limited to those impacts deemed to be significant. Is also applies to adverse impacts that have not breached the threshold of significance, including cumulative effects. A summary of how an

EIS prepared by Pham's project branch for relicensing a nuclear power plant responded to this requirement is cited below:

> Unavoidable adverse environmental impacts are those effects that would occur after implementation of all feasible mitigation measures. Under the license renewal alternative, the existing plant and transmission corridors would continue to be used for their current mission. This alternative would continue to limit other land use options. However, no additional land would be required to support this alternative.
>
> Withdrawing surface water from the river could result in a drawdown in the underlying groundwater system, which could limit water use for other purposes. This impact would be small, even during periods of low flow. For both the river and underlying groundwater system, current practices for managing the impact of plant water usage are considered to be adequate.
>
> Under the alternative of license renewal, the existing plant and transmission corridors would continue to be used for their current mission. This land would continue to pose a small impact on biological resources. However, no additional biological disturbances would occur under this alternative.
>
> Workers and members of the public would face exposure to small amounts of radioactive emissions. Workers would be exposed to small levels of radiation. Workers would have higher levels of exposure than members of the public. Workers would also face unavoidable exposure to small amounts of radiation from radioactive spent nuclear fuel and waste operations. Management and disposal of this waste would require long-term funding and monitoring, and would consume space at treatment, storage, or disposal facilities to prevent release to the biosphere.
>
> Potential disturbance to historic and archaeological artifacts could result in a moderate impact to these cultural resources.

Again, it is noteworthy that the Commission's discussion failed to even consider the potential adverse impacts that could not be avoided as a result of a severe accident such as a full-scale nuclear meltdown; the single most important adverse impact was not even acknowledged or disclosed to the public or decision maker. While it at least made a reference to radioactive exposures, it completely neglected the very controversial issue of adverse impacts from the generation, storage, and disposal of highly radioactive nuclear waste. Thus, two of the most significant issues of concern to the public and much of the scientific community (i.e., nuclear accident and disposition of nuclear waste) were purposely ignored. This is a clear violation of the Commission's statutory duties.

3. *Irreversible and irretrievable resources*

Often referred to as the *I&I requirement*, an EIS must address

> ... any <u>irreversible</u> or <u>irretrievable</u> commitments of resources which would be involved in the proposal should it be implemented. (§1502.16)

Neither of the above-mentioned terms are defined in the Regulations. Not surprising, the difference between the terms "irreversible" and "irretrievable" has led to confusion. Webster's dictionary defines these terms as[44]

Irreversible: "... incapable of being reversed..."
Irretrievable: "... that cannot be retrieved, restored, recovered, or repaired; irrecoverable"

With respect to NEPA, a resource commitment may be considered:

Irreversible: when the effects cannot be reversed; such effects typically limit future use options.
Irretrievable: when use or consumption is neither renewable nor recoverable for use by future generations. Irretrievable commitment can be assumed to involve loss of production, harvest, or natural resources.

An irreversible commitment of land use might involve construction of a permanent structure over buried archaeological resources. An irretrievable commitment of natural resources might involve burning oil or mining minerals.

In complying with the I&I requirement, analysts should discuss the commitment of resources expended during the construction and operational phases of a proposal. In the past, the I&I requirement has often been given only cursory treatment. As described in the following sections, careful adherence to this requirement may have important implications in terms of saving untold sums in natural resource damage assessments. Private parties and applicants, in particular, should pay close attention to ensuring that the I&I requirement has been adequately addressed. A summary of how an EIS for a manufacturing facility responded to this requirement is presented below:

> This section describes the irreversible and irretrievable commitments of resources described in this EIS. With respect to the proposed action, irreversible actions include the short-term commitment of land for the plant and utility corridors, which would limit other land use options. Also related to this issue is the irreversible loss of biological habitat and species, at least until the plant is decommissioned and the land is released.
>
> The proposal would result in an irretrievable commitment of cooling water which is diverted from other potential uses, including support of natural and biological resources. While surface water consumption represents a short-term loss of a renewable resource, lack of adequate groundwater recharge could constitute a relatively small longer-term irretrievable loss to the underlying aquifer.
>
> An irretrievable commitment of material resources includes materials that cannot be recovered or recycled, materials that cannot be economically decontaminated, and materials consumed or reduced to unrecoverable forms of waste.
>
> One of the principle irreversible impacts is the generation of hazardous waste such as heavy metals. This could result in long-term adverse effects on human health and biological resources. The treatment, storage, and disposal of large quantities of hazardous waste, and nonhazardous waste would require long-term or permanent irretrievable commitment of land, as well as capital and personnel to manage and monitor the waste at storage, treatment, and disposal facilities. As an irreversible action, such waste might also have the potential to adversely affect the biosphere and other natural resources. In general, the commitment of capital and labor to provide long-term monitoring of this waste is an irretrievable commitment of socioeconomic resources.

> Another irreversible impact involves the production and release of hazardous air emissions that could result in long-term adverse effects on human health and biological resources. The plant would also release substantial amounts of CO_2 and other GHGs. These GHGs might contribute to a global irretrievable degradation or loss of ecological and natural resources.

6.6.10.1 *Natural resource damage assessments*

Superfund is the common name for the Comprehensive Environmental Response, Compensation, and Liability Act of 1980 (CERCLA). The Superfund Act authorized the EPA to identify parties responsible for contamination of sites and compel those parties to clean up the sites.

Under Section 107 of the Superfund Act, responsible parties may be held liable for damages to publicly owned or managed natural resources.[45] Better known as the Natural Resource Damage Assessment (NRDA), this provision allows natural resource trustees to recover damages (i.e., money) for injury incurred to natural resources.[46] NRDAs can be assessed for damages resulting from a discharge of oil or release of a hazardous substance that damages an environmental resource.

Claims including those filed by state and tribal trustees may be brought even though a site is not on the Superfund National Priorities List. Claims can also be filed against contractors managing federal facilities. NRDAs only apply to natural resource damages that occurred after enactment of CERCLA.

As used in this context, the term *natural resources* is far-reaching and includes resources as diverse as land, fish, wildlife, biota, air, surface water, groundwater, and other such resources controlled, managed, held in trust, or belonging to the United States, or any Indian tribe. Natural resources can also include lakes, rivers, streams, and coastal waters bounded or unbounded by public lands. Liability most clearly extends to lands that are owned or managed by federal or state agencies, or Indian tribes.

Damages to both use and non-use values are recoverable. Natural resource trustees may assess damages as high as $50 million per release of a hazardous substance plus, standard cleanup costs assessed under CERCLA, the cost of assessing such injury, and any prejudgment interest.

6.6.10.1.1 *Protection from natural resource damage claims.* NRDAs can easily run into the tens of millions of dollars. With respect to NEPA's I&I requirement, Section 107 of the Superfund Act provides an important exemption from natural resource damage assessment claims if the following two conditions are met[47]:

- A decision to grant a permit or license authorizes the commitment of resources; and the action must be operated within the terms of any permit or license that is granted.
- Potential damages to natural resources are specifically identified as an irreversible and irretrievable commitment of natural resources in an EIS or other comparable environmental analysis.

Thus, an action that generates a hazardous substance subject to CERCLA is exempt from a future NRDA if

1. It is conducted under a permit or license
2. Its potential impacts have been adequately identified and evaluated as an I&I commitment of natural resources in an EIS

This allows potentially risky actions to be taken so long as the long-term consequences have been identified and considered by the decision maker (i.e., I&I section of an EIS). The rationale for this exemption can be found in a US Senate report[48]:

> … Federal officials make decisions in which resource tradeoffs must necessarily be made, and in such cases liability for resource damage… should be limited… In such a case, where the specific trade-offs are understood and anticipated in issuing the permit for such releases and the agency takes into account this knowledge and allows the trade-off, then no liability will accrue for resources damaged pursuant to those permitted releases.

Consistent with this exemption, prudence should be exercised in identifying and describing any potential loss or damage to a natural resource that may be viewed as an I&I commitment of resources. For instance, if an environmental remediation project on a contaminated waste site has the potential to wash contaminants into an adjacent stream, the potential degradation to fish and human use of that water supply should be carefully described as a potential I&I commitment of resources in the EIS. Such an assessment could protect both the agency and the environmental contractor from future NRDA claims.

3. Identifying likely decision-making factors

   The Regulations provide the following direction regarding the identification and assessment of factors likely to be considered in reaching a final decision regarding a course of action. Specifically, the EIS needs to (§1502.23)

> … indicate those considerations, including factors not related to environmental quality, which are likely to be relevant and important to a decision.

This requirement allows the public to understand the considerations and factors likely to shape the agency's final decision. This is an important public disclosure requirement, as it allows the public to focus attention on influencing those considerations and factors. It is not uncommon to find that an EIS has not addressed this regulatory requirement.

6.6.11 *Land use conflicts, and energy and natural resource consumption*

As detailed in the following sections, the EIS must address five special environmental issues:

- How alternatives and decisions will or will not achieve NEPA's goals (§1502.2[d])
- Energy consumption and conservation potential of various alternatives (§1502.16[e])
- Natural resource consumption and conservation potential of alternatives (§1502.16[f])
- Conflicts between land use plans, policies, and controls (§1502.16[c])
- Identifying inconsistencies with other plans and laws

These five requirements are frequently addressed in a standalone section at the end of the chapter on environmental consequences. However, they can also be addressed in other sections or chapters of the EIS as well. Each of these five requirements is described below.

6.6.11.1 *How alternatives achieve NEPA's goals*

The purpose of NEPA is to establish and promote a national policy to protect and preserve environmental quality. Section 2 of the statute states NEPA's purpose[49]:

> To declare a national policy which will encourage productive and enjoyable harmony between man and his environment; to promote efforts which will prevent or eliminate damage to the environment and biosphere and stimulate the health and welfare of man….

Section 101 of NEPA goes on to state that the federal government is to use all practicable means to improve and coordinate federal plans, programs, and actions in a way designed to[50]

1. Fulfill the responsibilities of each generation as trustee of the environment for succeeding generations
2. Assure for all Americans safe, healthful, productive, and aesthetically and culturally pleasing surroundings
3. Attain the widest range of beneficial uses of the environment without degradation, risk to health or safety, or other undesirable and unintended consequences
4. Preserve important historic, cultural, and natural aspects of our national heritage, and maintain, wherever possible, an environment that supports diversity, and variety of individual choice
5. Achieve a balance between population and resource use, which will permit high standards of living and a wide sharing of life's amenities
6. Enhance the quality of renewable resources and approach the maximum attainable recycling of depletable resources

Consistent with NEPA's purpose and the aforementioned goals, an EIS must

> … state how alternatives considered in it and decisions based on it will or will not achieve the requirements of Sections 101 and 102(1) of the Act and other environmental laws and policies. (§1502.2[d])

Section 102(1) of NEPA states that

> … the polices, regulations and public laws of the United States shall be interpreted and administered in accordance with the policies set forth in this act….

Thus, the EIS must explain how alternatives and decisions will or will not achieve NEPA's goals as spelled out in Sections 101 and 102(1) of the statute. It is not uncommon to find that an EIS has totally ignored this requirement.

6.6.11.2 Energy consumption

Many energy sources are finite and nonrenewable (e.g., coal, natural gas, uranium). Federal actions can consume large amounts of energy. Moreover, energy generation can extract a significant toll on environmental quality. To this end, the EIS must evaluate

> Energy requirements and conservation potential of various alternatives and mitigation measures. (§1502.16[e])

This provision is interpreted to mean that agencies must analyze energy requirements and consumption, including costs and benefits, as well as hidden and indirect costs associated with implementing a proposal.[51] The conservation potential of various alternatives and mitigation measures must likewise be evaluated. As applicable, mitigation measures for reducing and conserving energy should be considered. It is not uncommon to find that an EIS has all but disregarded this requirement.

6.6.11.3 Natural resources consumption

Federal actions can consume large amounts of natural or nonrenewable resources (e.g., steel and mineral resources, land, timber, and water). Even where they can be reused, such materials can represent a long-term commitment of such resources. To this end, the EIS must evaluate

> Natural or depletable resource requirements and conservation potential of various alternatives and mitigation measures. (§1502.16[f])

As applicable, analysts should identify, compute, and evaluate any substantial consumption of important natural or depletable resources. The conservation potential of various alternatives and mitigation measures must likewise be evaluated. As applicable, mitigation measures for reducing and conserving natural or depletable resources must be considered. Again, it is not uncommon to find that an EIS has largely disregarded this requirement.

6.6.11.4 Land use conflicts

As we have seen, NEPA is an environmental planning process. As such, an EIS must address conflicts with established land use plans as well as any inconsistency with such plans and laws. An EIS provides an ideal tool for integrating a federal proposal with other planning and land use constraints. Consistent with this objective, an EIS must address

> Possible conflicts between the proposed action and the objectives of federal, regional, state, and local (and in the case of a reservation, Indian tribe) land use plans, policies and controls for the area concerned. (§1502.16[c])

The EIS evaluates how the proposal might impact land use plans and laws. Where an inconsistency is identified, the EIS must discuss the inconsistency or discrepancy, and describe the extent to which the agency would reconcile its proposal with the land use plan or law (§1506.2 [d]). The EIS investigates possible methods for resolving any land use conflicts. Comments from officials with responsibility over the affected area should be solicited early and should be adequately addressed in the EIS.[52]

The phrase *land use plans* includes any formally adopted zoning or land use plans, including any plans that have been formally proposed by a government body and are under active consideration. Similarly, the phrase *land use policies* includes formally adopted statements of land use policy, including those embodied in laws or regulations; it also includes land use policies that have been formally proposed but have not yet been adopted.[53]

6.6.11.5 Identifying inconsistencies with other plans and laws

An EIS provides an ideal tool for planning and developing consistent and integrated plans, coordinated with other laws and regulations. To this end, the EIS must also discuss

> Any inconsistency of a proposed action with any approved State or local plan and laws (whether or not federally sanctioned). Where an inconsistency exists, the statement should describe the extent to which the agency would reconcile its proposed action with the plan or law. (§1506.2[d])

Thus, the EIS identifies any inconsistencies with other state or local plan and laws. Where an inconsistency is identified, the EIS must discuss the inconsistency, and describe the extent to which the agency would reconcile its proposal with that plan or law. The purpose of this requirement is to force federal agencies to investigate methods for resolving any conflicts.

6.6.12 Listing permits, licenses, and other entitlements

Consistent with the requirements just described, the EIS must list

> … all Federal permits, licenses, and other entitlements which must be obtained in implementing the proposal. If it is uncertain whether a Federal permit, license, or other entitlement is necessary, the draft environmental impact statement shall so indicate. (§1502.25[b])

As a planning process, an EIS provides an ideal tool for reducing project "surprise" by alerting officials of other unidentified requirements that may also need to be satisfied. Permit and other requirements are often identified through agency consultation. Work on preparing this list of requirements should begin early during the EIS consultation process, as it often provides a basis for identifying other federal agencies that may need to be consulted or brought into the scoping process. This requirement has several advantages:

1. Identifying other related requirements early in the planning process, reducing the risks of later "surprises" during the project implementation phase
2. Assisting the agency in integrating various requirements so as to enhance efficiency, reducing overall compliance costs and subsequent delays
3. Facilitating a more comprehensive and robust environmental planning and investigation process

Some EISs address this permitting and licensing requirement in the alternatives chapter, while others present it in the chapter on environmental consequences. Often this requirement is dealt with in an appendix or standalone chapter devoted solely to this regulatory requirement, particularly in circumstances involving many permits and licenses. The EIS should specify the entity responsible for granting and obtaining each permit or license.

6.6.12.1 Regulatory compliance matrix

It is recommended that the agency consider developing an environmental regulatory compliance plan for ensuring that all applicable plans, laws, permits, and licenses are identified. A regulatory compliance plan provides an ideal tool for coordinating and scheduling such requirements, and for identifying potential problems early in the planning process. Such a plan may be incorporated by reference into the EIS. For instance, if a Programmatic Memorandum of Agreement is prepared, the EIS may summarize this agreement and describe how it would be coordinated with the proposal.

Each alternative may require a unique set of permits, licenses, and other entitlements. If there is considerable variation in these requirements, a regulatory compliance matrix may prove invaluable in allowing the decision maker and public to compare the laws, and permitting and licensing requirements of various alternatives. Table 6.26 provides a simplified example of such a matrix.

Table 6.26 Example of a Regulatory Compliance Matrix

Permits, licenses, or entitlements	No action	Proposed action	Alternative 1
Dangerous waste management permitting	N/A	A State Hazardous Waste Management Act License	A State Air Pollution Control License
EPA permitting	N/A	EPA approval for PCB treatment under TSCA	EPA approval for PCB treatment under TSCA
Transportation approvals	State transportation permit	US Department of Transportation certificate	US Department of Transportation certificate
RCRA permitting	N/A	RCRA Part A and B Permit	RCRA Part B Permit
State hazardous materials permitting	State Materials License No. 1217-1	N/A	State Materials License
State hazardous waste generation permitting	State Generator's User Permit, No. C-1122/B	New Effluent Source Construction Permit	State Health and Welfare Construction and Operation Permit

Note: The matrix compares permits, licenses, and entitlements against three alternatives.

6.6.13 *List of preparers and entities to whom the EIS is sent*

Some EISs fail to provide adequate information on the preparer's education and experience, making it difficult to identify individual areas of responsibility. This information is important as it can demonstrate that an interdisciplinary approach was followed in preparing the EIS.[18] To this end, the EIS must include a list of persons who prepared the EIS (§1502.17):

> … list the names, together with their qualifications (expertise, experience, professional disciplines), of the persons who were primarily responsible for preparing the environmental impact statement or significant background papers, including basic components of the statement.…

There are three reasons for including this list[51]:

1. It introduces accountability, which tends to enhance professional competency and integrity of the EIS (§1502.24).

2. It provides a basis for determining if an interdisciplinary approach has truly been used (§1501.2).
3. It promotes professional standing of EIS staff by recognizing their contribution to the analysis.

The list must indicate both the names and qualifications of the individuals who were "primarily" responsible for preparing the EIS. An individual's qualifications should *briefly* describe their expertise, experience, and professional discipline (§1502.17).[54] The author also recommends that it include, at a minimum, an individual's academic degrees and years of experience. As a rule, individuals who have had only minor input or responsibilities need not be included.

If the EIS is prepared by a consulting firm, those members who made substantial contributions to the EIS should be listed.[55] Individuals responsible for preparing important background papers should likewise be included (§1502.17). If the EIS uses information submitted by an applicant, names of the persons responsible for independently evaluating and verifying the accuracy of these data must be included (§1506.5[a]). Individuals responsible for reviewing or editing the EIS should also be included.[56]

The CEQ has suggested that one or two lines of text is sufficient to cover an individual's qualifications. In general, the list should not exceed two pages in length.

6.6.14 *List of entities to whom the EIS is sent*

As indicated in Table 6.10, the EIS must list agencies, organizations, and persons to whom copies of the statement are sent. The Regulations provide no direction regarding the content or preparation of this list. Although the Regulations do not provide direction for preparing this section, Chapter 4 of this book describes parties to whom a copy of the EIS must be furnished (§1502.19).

6.6.15 *Index, glossary, and bibliography*

As indicated in Table 6.10, an EIS must contain an index (§1502.10). While not specifically required by the Regulations, it is recommended that the EIS also contain a glossary and bibliography.

6.6.15.1 *Index*

The Regulations are silent regarding the content of the index. The CEQ recommends that the index reference more than simply key topics and issues; however, it does not need to reference every conceivable term or phrase.[57] A *keyword index* uses descriptive terms to identify important concepts, issues, and topics; it enables readers to quickly locate particular areas of interest such as types of alternatives, affected resources, impacts and issues, and so forth. As a rule, the index should reference a particular topic, if it is believed to be of reasonable interest to a reader.[58]

6.6.15.2 *Glossary and list of references*

As just noted, the Regulations do not require an EIS to include either a glossary or list of references. It is considered good professional practice, however, to include these sections. At a minimum, a glossary should contain standard NEPA, and scientific and technical terms used in the EIS. A bibliography listing references allows the reader to quickly locate sources referenced in the EIS.

6.6.15.3 *Table of acronyms and measurements*

It is further recommended that the EIS include

- List of symbols, acronyms, and abbreviations
- Table of measurement conversions

6.6.16 *Appendices*

Although incorporation of appendices in an EIS is optional, it can provide an excellent mechanism for preparing a *succinct* description of the proposal and its environmental consequences. The main body of the EIS should be designed to present the decision maker with information necessary to focus on aspects of the proposal and significant impacts truly important in reaching an informed decision. This can markedly improve both the readability and usefulness of the EIS.

Public comments on the EIS and the agency's response to those comments are perhaps the most important and common use of appendices. Material that is highly technical in nature, such as analytical methodologies

Table 6.27 Direction on Appropriate Use and Content of Appendices

- Public comments and the agency's responses on the draft EIS.[60]
- The appendices should normally contain material that substantiates any analysis fundamental to the EIS (§1502.18[b], §1502.18[c]).
- The appendices should contain material prepared in connection with the EIS. Material not prepared directly in connection with the EIS should be incorporated by reference (§1502.18[a]).
- Lengthy or detailed descriptions of the scientific methodology used in the analysis should be placed in the appendix (§1502.24).[60]
- Appendices should be circulated with the EIS. If the appendix is not circulated with the EIS, it must be made readily available upon request (§1502.18[d]).

and models used in the analysis, should normally be presented in the appendices. Other technical material frequently placed in the appendices includes lists of affected species and related studies. When such material is placed in the appendices, the main body of the EIS need only provide a summary discussion of the material; the reader can then be directed to the appropriate appendix for details.[59] Table 6.27 summarizes direction regarding appropriate use and content of the appendices (§1502.18).

6.6.16.1 Incorporation by reference versus appendices

Material prepared directly in support of the EIS should normally be placed in the appendix. In contrast, material prepared only in connection with the EIS should be incorporated by reference (§1502.18[a]). Material incorporated by reference must be accessible to the general public, but does not need to accompany the EIS. This can be done either by

- Citing publicly available information.
- Providing copies of incorporated information to central locations such as public reading rooms.
- Sending copies to commenters upon request. Such material must be publicly available for the full length of the minimum public comment period.[61]

In contrast, all appendices should normally be circulated with the EIS for public review. If the appendices are not circulated, they must be placed in locations that are accessible by the general public or furnished to commenters upon request.[61]

6.7 The record of decision

Mark Twain once offered this bit of solace:

> I didn't attend the funeral, but I sent a nice letter saying I approved of it.

As illustrated in Figure 6.12, the agency's EIS process ends with the approval and publication of the ROD. The decision maker is legally mandated to consider alternatives and balance the environmental consequences against other decision-making factors in reaching a final decision to pursue a course of action. It is important to note that a decision maker may not choose a course of action unless it has been adequately described and analyzed (§1505.1[e]). The agency prepares a ROD to publicly record its final decision. The ROD is a concise statement describing the agency's final choice among the alternatives considered. It may be integrated into any other record prepared by the agency.

Figure 6.12 The decision maker is legally mandated to consider alternatives and balance the environmental consequences against other decision-making factors in reaching a final decision.

In addition to the environmental considerations, a final decision regarding the agency's course of action can be based on many factors, including *economic* and *technical considerations* as well as the agency's *statutory mission*. Accordingly, the EIS must identify and discuss all factors considered and how they were weighed by the agency in reaching its final decision (§1505.2[b]).

6.7.1 Contents

The ROD may be integrated into any other record prepared by the agency. The Regulations require that the ROD (§1505.2)

1. State what the decision was.
2. Identify all alternatives considered by the agency in reaching its decision, specifying the alternative or alternatives that were considered to be environmentally preferable. An agency may discuss preferences among alternatives based on relevant factors, including economic and technical considerations and agency statutory missions. An agency shall identify and discuss all such factors, including any essential considerations of national policy that were balanced by the agency in making its decision, and state how those considerations entered into its decision.
3. State whether all practicable means to avoid or minimize environmental harm from the alternative selected have been adopted, and if not, why they were not. A monitoring and enforcement program shall be adopted and summarized where applicable for any mitigation.

As indicated in item 3, the ROD must indicate whether all practicable mitigation measures have been adopted, and if not, why they were not.

6.7.1.1 Compilation of all principal guidance and regulatory requirements

In addition to this regulatory provision, Table 6.28 provides a compilation of all key guidance and regulatory requirements that the ROD must address.

Table 6.28 Compilation of Pertinent Guidance and Regulatory Requirements That the Record of Decision Must Address

- Identify alternatives considered by the agency in reaching its decision. Preferences among alternatives based on relevant factors including *economic* and *technical considerations* as well as the agency's *statutory missions* may be discussed (§1505.2[b]).
- Identify and discuss all factors considered and how they were weighed by the agency in reaching its final decision (§1505.2[b]).
- Provide a statement of the agency's decision (§1505.2[a]). This statement explains the decision, how it was made, and any mitigation measures that are being imposed to lessen adverse environmental impacts.[62]
- Specify the alternative(s) considered to be environmentally preferable; this is referred to as the "environmentally preferable" (§1505.2[b]).
- State whether all practicable means have been adopted to avoid or minimize environmental consequences associated with the selected alternative, and if not, why they were not (§1505.2[c]).
- A monitoring and enforcement program must be adopted and summarized where applicable for any mitigation (§1505.2[c]).
- Provide a concise summary of any mitigation measures that the agency has committed to adopt. The ROD must identify any mitigation measures, and monitoring and enforcement programs that will implemented, and clearly indicate that they are being adopted as part of the agency's decision. Discussion of mitigation and monitoring must be more detailed than a general statement that mitigation will be adopted, but not so detailed as to duplicate discussion of mitigation in the EIS.[63]
- The ROD should indicate the likelihood that mitigation measures will be adopted or enforced by the responsible agency(ies). If there is a history of nonenforcement or opposition to such measures, the ROD should acknowledge such opposition or nonenforcement. If any necessary mitigation measures will not be ready for a long period of time, this fact should also be acknowledged.[64]
- If the proposal involves issuance of a permit or other approval, specific details of any adopted mitigation measures must be included as appropriate conditions in whatever grants, permits, funding, or other approvals are being made by the agency. If the proposal is to be carried out by the agency itself, the ROD should delineate any adopted mitigation and monitoring measures in sufficient detail to constitute an enforceable commitment, or they should be incorporated-by-reference into the sections of the EIS that do so.[63]

Table 6.29 A General Purpose Outline for Record of Decision

1. Heading
   a. Agency name
   b. Type of decision document (i.e., record of decision)
   c. Title of the proposed action
   d. Location of the proposed action, including administrative unit, county, and state
2. *Decision and rationale for the decision*: Describe the decision, including permits, licenses, grants, or authorizations needed to implement the decision. Identify the specific location of the action. This section should also address the following rationale:
   a. How the selected alternative best meets the purpose and need
   b. How environmental issues were considered and addressed
   c. Factors other than environmental consequences considered in making the decision
   d. Applicable laws, regulations, and policies
   e. Identification of environmental document(s) considered in making the decision
3. *Public involvement*: Identify public issues considered in determining the scope of the analysis. Briefly summarize the public participation process. Important agencies, organizations, or persons who raised issues and those who offered opposing viewpoints.
4. *Alternatives considered*: All alternatives considered in reaching the final decision are briefly discussed with specific references to the EIS. Management and mitigation measures, and monitoring provisions pertinent to environmental concerns are discussed.
5. *Environmentally preferable alternative*: Identify the environmentally preferable alternative(s).
6. *Avoid or minimize environmental consequences*: State whether all practicable means have been adopted to avoid or minimize environmental consequences associated with the selected alternative, and if not, why they were not.
7. *Summarize mitigation measures*: Provide a concise summary of any mitigation measures which the agency has committed itself to adopt.
8. *Findings required by other laws*: Include any findings required by other laws. Include findings such as consistency with other plan, land use plans, coastal zone consistency determination, 404 permitting, Section 7 consultation or permitting, Section 106 review etc.
9. *Implementation date*: Identify date when the responsible official implements the decision.

(continued)

Table 6.29 (Continued) A General Purpose Outline for Record of Decision

10. *Any applicable administrative review or appeal opportunities*: State whether the decision is subject to review or appeal (citing the applicable regulations); identify when and where to file a request for review or appeal.
11. *Contact person*: Identify the name, address, and phone number of a contact person who can supply further information.
12. *Signature and date*: The responsible official signs and dates the ROD on the date the decision is made.

6.7.1.2 Suggested general purpose outline of the ROD

A general purpose outline for preparing an ROD is presented in Table 6.29.

6.7.1.3 Preparing the ROD

Plaintiffs sometimes focus their legal challenge against the ROD rather than the EIS itself, and as a result, prudence should be exercised in preparing the ROD. A well-structured ROD clearly demonstrates that the responsible official(s) understand the potential actions and alternatives (i.e., whom, what, where, when, why, and how) and the resulting environmental consequences. Some RODs have been found deficient because they neglected to

- Identify the environmentally preferable alternative
- Fully describe considerations that led to a decision not to adopt the environmentally preferable alternative
- Describe whether all practical means were employed in mitigating environmental impacts

Surprisingly, some RODs have reached conclusions based on environmental issues or information that was not even clearly spelled out in the EIS. All conclusions regarding environmental impacts must be specifically tied to the analysis contained in the EIS. Under no circumstances should the ROD choose a course of action that has not been evaluated or at least bounded by the alternatives analysis in the EIS.

The rationale used in choosing a final course of action should be carefully documented; if the agency is later challenged, a well-documented rationale can strengthen its defense, particularly in cases where it is accused of making an arbitrary or capricious decision. For these reasons, the trend in recent years has been toward lengthier RODs.

In rare occasions, an alternative that meets the goals of the lead agency may not meet those of a cooperating agency. One example includes EISs prepared for multiple land use. Where this may be the case, the CEQ recommends that each agency identify its own preferred alternative within the EIS. Each agency may then prepare and issue its own separate ROD, identifying the course of action it will pursue. In such a case, the environmentally preferable alternative (see the next section) identified in the ROD by one agency may not necessarily be the same as that identified by the second agency.[65]

6.7.1.4 Environmentally preferable alternative

While the *environmentally preferable alternative* must be identified in the ROD, there is no corresponding requirement to identify it in the EIS (§1505.2[b]). The environmentally preferable alternative is the one that, on balance, is considered to best promote the goals expressed in Section 101 of NEPA. The CEQ has interpreted this to mean the alternative that "... causes the least damage to the biological and physical environment" and "... which best protects, preserves, and enhances historic, cultural, and natural resources." Thus, at least three environmental and cultural attributes are to be considered in identifying the environmentally preferable alternative[25]:

- Biological resources
- Physical environment
- Historic, cultural, and natural resources

The reader should note that failure to choose an environmentally preferable alternative deemed to be practical is one factor that can be used in determining if an action should be referred to the CEQ (§1504.2[f]).

6.7.1.5 Mitigation and monitoring plans

Under federal administrative law, commitments made in the ROD are considered to be legally binding, and agencies may be held accountable for their implementation. Such commitments are enforceable by other agencies and private entities alike.[66] Partly for this reason, the ROD should discuss the likelihood that mitigation measures will actually be enforced by the responsible agency(ies). Where there is a history of nonenforcement or opposition to such measures, this fact should be acknowledged in the EIS/ROD.[29]

As witnessed earlier, a monitoring and enforcement plan should be adopted and summarized where applicable for any mitigation measures chosen. Any adopted mitigation measures must be adequately evaluated in the EIS. A monitoring program for ensuring decisions are appropriately implemented should be adopted and summarized in the ROD, especially in important cases (§1505.2[c], §1505.3). To reduce paperwork, the ROD may incorporate discussion of mitigation measures by reference from the EIS.

PROBLEMS AND EXERCISES

1. An agency is currently housing its regional staff in an aging office building designated for closure in three years because it is deemed unsafe to occupy. An EIS is being prepared to build a large office building to house the agency's staff. The EIS must be completed rapidly in order to support the agency's plan to replace the old and unfit office building. To speed up the EIS, the agency administrator directs the EIS manager to not include a no-action alternative because the agency has no other recourse but to build a new office building. Do you think this direction is correct or does it violate EIS regulatory requirements? Explain your answer.
2. An agency is preparing an EIS for a gold mine lease on property that it controls. The mining operation will involve use of hazardous materials and generation of hazardous waste including mercury. Why can an EIS help protect an agency or applicant against potential natural resource damage assessments (NRDA)?
3. An agency is preparing a proposal to construct a coal-fired power plant. The EIS manager plans to evaluate a range of alternatives including a gas-fired plant, solar energy farm, nuclear reactor, and even a conservation program that might eliminate the need for the plant all together. A member of the public submits a comment demanding that a fusion-power alternative be included. The agency's engineering department responds with numerous studies showing that fusion power is a futuristic and technically unproven technology. The agency argues that a fusion-energy alternative should therefore be dismissed from consideration. The commenter will not budge in their opinion that a fusion-power alternative be included. Who is right? Should the EIS include a fusion-energy alternative or dismiss it? Explain your rationale.
4. What is the difference between the agency's "preferred alternative" and the "environmentally preferable alternative"?
5. An EIS is completed and circulated for internal review. The EIS involves a highly technical assessment of technologies for remediating air pollution. The public relations department uses a commercial text assessment software program to review the EIS and determine the educational level it has been written to. Its readability index indicates it has a technical complexity requiring someone with a Masters of Science degree to understand. Does the EIS meet NEPA's regulatory requirements? If not, what steps can the agency take to rectify the problem?

6. An EIS analysis concludes that a proposed herbicide application project could result in significant harmful effects to wildlife and workers. The project manager is concerned that public reaction will upset plans to pursue the project. This manager orders that the EIS staff not disclose this fact in the EIS. Does this direction violate any of NEPA's requirements? Justify your response.
7. Flooding is a common problem in the surrounding region. It has resulted in substantial economic damages and loss of life. An agency prepares an EIS for a flood control program. The statement of purpose and need (SPN) states: "The underlying purpose and need for the proposed flood control program is to prepare an EIS that satisfies NEPA and its regulatory requirements." Do you believe this SPN is correctly written? If not, how would you rewrite it?
8. Imagine that a proposed microbiology laboratory would perform research on dangerous viruses. The EIS analyzes the potential impacts, including an accident scenario involving an accidental release of a hazardous virus into the biosphere. However, there is no available information on what the potentially significant effects would be if there was an accidental release. In light of this unknown information, what would you recommend the EIS manager do?
9. The main body of a final EIS is 941 pages long. Does this meet NEPA page limit direction? Explain your answer. List some means by which the agency might reduce the page length while complying with NEPA's requirements to rigorously assess potentially significant impacts.
10. An EIS is prepared to analyze impacts and alternatives for logging an undeveloped area in a national forest. The EIS uses verbiage such as "The proposed project would involve…." The agency's program manager reads the EIS and complains that the decision has already been made. She demands that all references to the "proposed" action be dropped because the project will definitely be implemented following the record of decision. She also wants the verbiage about potential actions that would take place changed from "would" to "will." Is this acceptable NEPA practice? If not, what problems do you see in terms of NEPA's requirements?

Notes

1. 40 Code of Federal Regulations (CFR) Parts 1500–1508.
2. Eccleston C. *NEPA and Environmental Planning: Tools, Techniques, and Approaches*. Introduction. CRC Press, Boca Raton, FL (2008).
3. Eccleston C.H. *The NEPA Planning Process: A Comprehensive Guide with Emphasis on Efficiency*. John Wiley & Sons, pp. 2–3 (1999).
4. *Oregon Envtl. Council v. Kunzman*, United States District Court for the District of Oregon, 636 F. Supp 632 (1986).

5. *Presidential Memorandum on Plain Language* (dated June 1, 1998).
6. Eccleston C.H. *Environmental Impact Statements: A Comprehensive Guide to Project and Strategic Planning*.
7. DOE. *Management of Spent Nuclear Fuel From the k-Basins at the Hanford Site, Richland, Washington*, Final, DOE/EIS-0245, (61 FR 3922) (1996).
8. Gerber E. Personal communications (1999).
9. Eccleston C. *NEPA and Environmental Planning: Tools, Techniques, and Approaches*. CRC Press, Boca Raton, FL (2008).
10. CEQ. *Council on Environmental Quality—Forty Most Asked Questions Concerning CEQ's National Environmental Policy Act Regulations (40 CFR 1500–1508), Federal Register*, Vol. 46, No. 55, 18026–18038, Number 26a (March 23, 1981), question number 2a.
11. CEQ. *Council on Environmental Quality—Forty Most Asked Questions Concerning CEQ's National Environmental Policy Act Regulations (40 CFR 1500–1508), Federal Register*, Vol. 46, No. 55, 18026–18038, Number 26a (March 23, 1981), question number 4a.
12. *Hughes River Watershed Conservancy v. Glickman*, 81 F.3d 437 (4th Cir. April 12, 1996).
13. Miller G.A. The magical number seven, plus or minus two: Some limits on our capacity for processing information. *Psychological Review* 63(2): 81–97 (1956).
14. CEQ. Preamble to Final CEQ NEPA Regulations, 43 *Fed. Reg*. 55978, Section 3 (November 29, 1978).
15. Eccleston C. *NEPA and Environmental Planning: Tools, Techniques, and Approaches*. CRC Press, Boca Raton, FL (2008).
16. Eccleston C. *Preparing NEPA Environmental Assessments: A User's Guide to Best Professional Practices*. Introduction. CRC Press, Boca Raton, FL (2012).
17. United States Court of Appeals, District of Columbia Circuit (June 8, 2012).
18. CEQ. Public memorandum titled "Talking Points on CEQ's Oversight of Agency Compliance with the NEPA Regulations" (1980).
19. CEQ. *Council on Environmental Quality—Forty Most Asked Questions Concerning CEQ's National Environmental Policy Act Regulations (40 CFR 1500–1508), Federal Register*, Vol. 46, No. 55, 18026–18038, Number 7 (March 23, 1981).
20. CEQ. *Council on Environmental Quality—Forty Most Asked Questions Concerning CEQ's National Environmental Policy Act Regulations (40 CFR 1500–1508), Federal Register*, Vol. 46, No. 55, 18026–18038, Number 2b (March 23, 1981), Questions 1 and 2.
21. CEQ. *Council on Environmental Quality—Forty Most Asked Questions Concerning CEQ's National Environmental Policy Act Regulations (40 CFR 1500–1508), Federal Register*, Vol. 46, No. 55, 18026–18038, Number 2b (March 23, 1981), Question 1b.
22. CEQ. *Council on Environmental Quality—Forty Most Asked Questions Concerning CEQ's National Environmental Policy Act Regulations (40 CFR 1500–1508), Federal Register*, Vol. 46, No. 55, 18026–18038, Number 2b (March 23, 1981).
23. CEQ, *Council on Environmental Quality—Forty Most Asked Questions Concerning CEQ's National Environmental Policy Act Regulations (40 CFR 1500–1508), Federal Register*, Vol. 46, No. 55, 18026–18038, Number 1a (March 23, 1981).
24. CEQ, *Council on Environmental Quality—Forty Most Asked Questions Concerning CEQ's National Environmental Policy Act Regulations (40 CFR 1500–1508), Federal Register*, Vol. 46, No. 55, 18026–18038, Number 4b (March 23, 1981).

25. CEQ, *Council on Environmental Quality—Forty Most Asked Questions Concerning CEQ's National Environmental Policy Act Regulations (40 CFR 1500–1508), Federal Register*, Vol. 46, No. 55, 18026–18038, Number 6a (March 23, 1981).
26. Yost N.C. & Rubin J.W. *The National Environmental Policy Act*. Unpublished.
27. CEQ, *Council on Environmental Quality—Forty Most Asked Questions Concerning CEQ's National Environmental Policy Act Regulations (40 CFR 1500–1508), Federal Register*, Vol. 46, No. 55, 18026–18038 (March 23, 1981), Question 2b.
28. CEQ, *Council on Environmental Quality—Forty Most Asked Questions Concerning CEQ's National Environmental Policy Act Regulations (40 CFR 1500–1508), Federal Register*, Vol. 46, No. 55, 18026–18038, Number 19a (March 23, 1981).
29. CEQ, *Council on Environmental Quality—Forty Most Asked Questions Concerning CEQ's National Environmental Policy Act Regulations (40 CFR 1500–1508), Federal Register*, Vol. 46, No. 55, 18026–18038, Number 19b (March 23, 1981).
30. *The Steamboaters v. Federal Energy Regulatory Commission*, 759 F. 2d 1382 (9th Cir. 1985); *Northwest Indian Cemetery Protective Association v. Peterson*, 795 F. 2d 688 (9th Cir. 1986).
31. Eccleston C. *NEPA and Environmental Planning: Tools, Techniques, and Approaches for Practitioners*. CRC Press, Boca Raton, FL, pp. 124–125 (2008). Chapter 10 discusses the general topic of accident analysis. Chapter 10.7 describes the use of the decision tool for determining when an accident analysis should be assessed in an EIS.
32. Eccleston C. *Environmental Impact Assessment: A Guide to Best Professional Practices*. CRC Press, Boca Raton, FL (2011).
33. Executive Order No. 12898, *Federal Actions to Address Environmental Justice in Minority and Low-Income Populations* (February 11, 1994).
34. Presidential Memorandum for the Heads of all Departments and Agencies, which was released concurrently with Executive Order No. 12898 (1994).
35. CEQ. Environmental Justice: Guidance Under the National Environmental Policy Act, (December 10, 1997). Available at: http://www.whitehouse.gov/CEQ/.
36. Executive Order 13045, Protection of Children from Environmental Health Risks and Safety Risks, 62 *Federal Register* [FR] (1985).
37. 36 CFR 60.2.
38. 36 CFR 60.2 and 48 *Federal Register* (FR) 44723.
39. 43 CFR 7.
40. Final rule for "Determining Conformity of General Federal Actions to State or Federal Implementation Plans," 58 *Federal Register* 63214, No. 228 (November 30, 1993, and took effect on January 31, 1944) (40 CFR parts 6, 51, and 93).
41. Air quality thresholds listed in the CEQA Environmental Checklist, Appendix G of the CEQA Guidelines and the Imperial County Air Pollution Control District (ICAPCD) CEQA Air Quality Handbook.
42. Executive Orders 11988 Floodplain Management (May 24, 1977) and 11990 Protection of Wetlands (May 24, 1977).
43. Section 102(2) of NEPA.
44. *Webster's New Twentieth Century Dictionary Unabridged*. Second ed. Simon & Schuster, 1983.
45. Superfund Act, 43 CFR 11, Section 107(a)(4)(C).

46. Superfund Act, 43 CFR 11, Section 107.
47. Superfund Act, 43 CFR 11, Section 107 (f)(1).
48. US Senate. Senate Report Number 848, 96th Congress, 2nd session (1980).
49. *The National Environmental Policy Act of 1969*, Purpose, Sec. 2, 42 USC §4321.
50. *The National Environmental Policy Act of 1969*, Sec. 101(b), 42 USC §4321.
51. CEQ. Preamble to Final CEQ NEPA Regulations, 43 *Fed. Reg.* 55978, Section 4 (November 29, 1978).
52. CEQ. *Council on Environmental Quality—Forty Most Asked Questions Concerning CEQ's National Environmental Policy Act Regulations (40 CFR 1500–1508), Federal Register*, Vol. 46, No. 55, 18026–18038, Number 23a (March 23, 1981).
53. CEQ. *Council on Environmental Quality—Forty Most Asked Questions Concerning CEQ's National Environmental Policy Act Regulations (40 CFR 1500–1508), Federal Register*, Vol. 46, No. 55, 18026–18038, Number 23b (March 23, 1981).
54. CEQ. *Council on Environmental Quality—Forty Most Asked Questions Concerning CEQ's National Environmental Policy Act Regulations (40 CFR 1500–1508), Federal Register*, Vol. 46, No. 55, 18026–18038, Number 27c (March 23, 1981).
55. CEQ. *Council on Environmental Quality—Forty Most Asked Questions Concerning CEQ's National Environmental Policy Act Regulations (40 CFR 1500–1508), Federal Register*, Vol. 46, No. 55, 18026–18038, Number 27a (March 23, 1981).
56. CEQ. *Council on Environmental Quality—Forty Most Asked Questions Concerning CEQ's National Environmental Policy Act Regulations (40 CFR 1500–1508), Federal Register*, Vol. 46, No. 55, 18026–18038, Number 27b (March 23, 1981).
57. CEQ. *Council on Environmental Quality—Forty Most Asked Questions Concerning CEQ's National Environmental Policy Act Regulations (40 CFR 1500–1508), Federal Register*, Vol. 46, No. 55, 18026–18038, Number 26b (March 23, 1981).
58. CEQ. *Council on Environmental Quality—Forty Most Asked Questions Concerning CEQ's National Environmental Policy Act Regulations (40 CFR 1500–1508), Federal Register*, Vol. 46, No. 55, 18026–18038, Number 26a (March 23, 1981).
59. CEQ. *Council on Environmental Quality—Forty Most Asked Questions Concerning CEQ's National Environmental Policy Act Regulations (40 CFR 1500–1508), Federal Register*, Vol. 46, No. 55, 18026–18038, Number 25a (March 23, 1981).
60. CEQ. *Council on Environmental Quality—Forty Most Asked Questions Concerning CEQ's National Environmental Policy Act Regulations (40 CFR 1500–1508), Federal Register*, Vol. 46, No. 55, 18026–18038, Number 26a (March 23, 1981), Question Number 25a.
61. CEQ. *Council on Environmental Quality—Forty Most Asked Questions Concerning CEQ's National Environmental Policy Act Regulations (40 CFR 1500–1508), Federal Register*, Vol. 46, No. 55, 18026–18038, Number 25b (March 23, 1981).
62. CEQ. *Council on Environmental Quality—Forty Most Asked Questions Concerning CEQ's National Environmental Policy Act Regulations (40 CFR 1500–1508), Federal Register*, Vol. 46, No. 55, 18026–18038, Number 23c (March 23, 1981).
63. CEQ. *Council on Environmental Quality—Forty Most Asked Questions Concerning CEQ's National Environmental Policy Act Regulations (40 CFR 1500–1508), Federal Register*, Vol. 46, No. 55, 18026–18038, Number 26a (March 23, 1981), Question Number 34c.
64. CEQ. *Council on Environmental Quality—Forty Most Asked Questions Concerning CEQ's National Environmental Policy Act Regulations (40 CFR 1500–1508), Federal Register*, Vol. 46, No. 55, 18026–18038, Number 26a (March 23, 1981) Question Number 19b.

65. CEQ. *Council on Environmental Quality—Forty Most Asked Questions Concerning CEQ's National Environmental Policy Act Regulations (40 CFR 1500–1508)*, *Federal Register*, Vol. 46, No. 55, 18026–18038, Number 14b (March 23, 1981).
66. CEQ. *Council on Environmental Quality—Forty Most Asked Questions Concerning CEQ's National Environmental Policy Act Regulations (40 CFR 1500–1508)*, *Federal Register*, Vol. 46, No. 55, 18026–18038, Number 34d (March 23, 1981).

Closing thoughts

> Sometimes I wonder whether the world is being run by smart people who are putting us on, or by imbeciles who really mean it.
>
> **Mark Twain**

Being somewhat of an optimistic soul, Mark Twain later pondered:

> In the beginning there was nothing. God said, "Let there be light!" And there was light. There was still nothing, but you could see it a whole lot better.

Given such sentiments, Mark Twain might well have supported adoption of the National Environmental Policy Act (NEPA). NEPA's environmental impact statement (EIS) process opened the federal project and decision-making cycle to public review, and cast public "light" on its decision-making process. Indeed, it was Congress's intent that NEPA would force decision makers to consider the consequences of future decisions a "whole lot better." While we can now see the consequences of decisions a "whole lot better," this by itself does not mean that wise and rational decisions are being made. Nor does it guarantee that an EIS has provided an objective and accurate assessment of such consequences to the decision maker and public.

The purpose of this book has been to provide a comprehensive guide for preparing EISs that advance NEPA's purpose of rational and informed decision making. To this end, Chapters 3 and 4 presented a step-by-step approach for navigating the complexities of the EIS process. The purpose of these chapters was to present the reader with all pertinent EIS procedural requirements (process requirements) from issuing the notice of intent, through public scoping, to preparing the EIS, and cumulating with the issuance of the record of decision (ROD). This was followed by Chapter 5, which presented the analytical requirements (analysis requirements), including guidance and direction for preparing an objective and rigorous analysis of impacts; tools, techniques, and best professional practices for performing a systematic and rigorous analysis were also introduced.

The book concluded with Chapter 6, which details all key EIS documentation requirements. The intent of this chapter is to synthesize a large and diverse body of guidance and requirements to describe all requirements that a legally sufficient EIS document must satisfy. These requirements and guidance included the Council on Environmental Quality's (CEQ's) NEPA implementing regulations, CEQ guidance, Environmental Protection Agency (EPA) direction and guidance, presidential executive orders, best professional practices, and lessons from case law.

By now the reader should have a firm grasp of the step-by-step process for preparing an EIS, including all key regulatory requirements that a legally sufficient EIS document must satisfy. But as we have seen, this in and of itself, is not sufficient to ensure either a quality analysis or informed decision making. Nor does it necessarily ensure that the public has been presented with objective facts and information. Some agencies are making diligent efforts to prepare fair, objective, and impartial scientific assessments that can truly contribute to rational and informed decision making. Regrettably, a few others are mired in planning and decision-making quagmires; this is indeed ironic given that the avoidance of such quagmires was the very impetus that led Congress to enact NEPA in the first place.

Recall the "three laws of the environmental movement," presented in the Introduction to this book. The first law requires a "top-level commitment to environmental quality"; the second law states that the force an EIS brings to bear on environmental protection is equal to effort expended on planning multiplied by the decision maker's commitment to environmental quality; the third law tells us that for every agency official or manager attempting to side step the EIS process, there is an equal and opposite adversary waiting to contest the project. The third law is a forewarning that a disingenuous EIS process can fester not only public discontent and mistrust, but can also lead to an embarrassing legal challenge.

As exemplified in Chapter 1, we continue to witness examples of haphazard EISs that present defective, skewed, and sometimes even deceptive assessments to the public and decision maker. Chapter 1 documented how many NEPA stumbling blocks can be traced directly back to faulty management, direction, and oversight. As we saw, Pham and Holian's own project managers voiced disturbing concerns such as

- "Poor management decisions" are being made.
- "Managers don't listen—they act like know-it-alls."
- Managers are "bypassing the regulatory process and compromising the safety mission to impress upper management."
- Managers have "dominant personalities"—they place pressure on project managers to shortcut the process.
- Managers are "sacrificing quality for schedule."

This is a troubling critique, particularly given that these charges were lodged by the very project managers responsible for preparing the Nuclear Regulatory Commission's EISs for renewing the operating licenses of aging nuclear reactors. One of the key objectives of this book has been to show the reader how to avoid repeating similar mistakes. But avoiding such mistakes is not as straightforward as it sounds. It necessitates that an agency adhere to the first law of the environmental movement—a true and top-level "commitment to environmental quality and excellence." Achieving this objective requires a serious commitment by management and agency officials to prepare accurate, fair, and objective assessments that truly contribute to the goal of transparency and quality-based decision making. Lacking such a commitment, an agency's EIS process may amount to little more than a futile paper chase.

How to avoid the pitfalls of flawed planning and decision making

Over the last several decades, we have witnessed many preventable disasters: Japanese Fukushima, Chernobyl, and Three Mile Island Nuclear Power disasters; lack if preparedness for the Mount St. Helen's volcanic eruption; 9/11 terrorist attack; Hurricane Katrina; and the Columbia space shuttle disaster, to name just a few. In the aftermath of such disasters, there has been a surge of interest in attempting to dissect the systemic and psychological root-cause factors that contribute to poor and sometimes even deceptive planning and decision making. This begs the question, "Given the information available at the time of these events, why didn't responsible officials act to avoid such disasters?" MacLean details some of the root-cause factors that contribute to flawed planning and decision making.[1] The author has revised his assessment in terms of the NEPA decision-making process. Some of the key factors that contribute to flawed planning and decision making include the following:

1. Critical information never reaches the decision makers. A defective EIS is a sure-fire method for ensuring that key information regarding the alternatives and consequences is misrepresented or obscured as it passes through different organizational and public review levels.
2. As detailed in Chapter 1, some federal NEPA officials may not be skilled or experienced in managing complex analyses.
3. As illustrated in Chapter 1, technical and quality issues can be strongly influenced or even misrepresented in response to factors such as: reaction or alarm from the public if they were to learn the truth about potential consequences and how it could affect their livelihood; how public opposition could affect the agency's mission; potential reaction from regulatory organizations; and the fact that project schedules may trump quality.
4. Management often resists change and stubbornly holds onto their preconceived belief system.
5. Some managers suffer from the "Why rock the boat?" syndrome.
6. Management dynamics such as "group think" can hinder unconventional thinking or prevent an organization from "thinking outside the box." This can also lead to poor judgment, failure to address critical safety and environmental concerns, lackluster quality commitments, and ultimately, flawed decisions that may threaten society and the environment.
7. Individual differences in risk tolerance. For example, the public and stakeholders, environmental organizations, and analysts and risk assessment professionals tend to be more conscious of risk than many federal managers.

Note

1. MacLean R. Ignoring impending disasters: Why do the warning signs go unheeded? *EM Magazine* 30–31 (January 2008).

Capstone problems

Upon completing this book, you should be able to prepare an environmental impact statement (EIS) using the concepts, requirements, and tools detailed in this book. This book ends with three capstone problems. Collectively, these problems involve consideration of the key requirements, principles, and concepts presented in this book. It is important to note that all three problems require professional judgment and there is no completely correct response to these problems; however, some responses and approaches can be viewed as being superior. The instructor is encouraged to reformulate these problems as necessary to fit the needs of the students and the class schedule.

Capstone problem no. 1: Port and harbor development

Divide the class into several groups. The class will represent the interdisciplinary team (IDT) of the NEPA Office of the US Department of Everything for Everybody (DEE). Your mission is to perform an EIS prescoping effort (see Chapter 3) for an EIS on a proposed port and harbor development project. Your project is controversial and hotly contested by the Citizen's Committee against the Department of Everything for Everybody (CCADEE). The CCADEE has already stated that it plans to "… sue the pants of DEE for their rabid, poisonous, death-promoting, anti-human, anti-environmental, anti-dolphin, and anti-fish practices."

The class will use the Internet and other sources to research real port and harbor areas within the United States and choose an area for the setting of this hypothetical project. The instructor will designate an EIS manager that will be coordinating the tasks performed by four different groups. Students are encouraged to exercise creativity. Each respective group will be responsible for performing the following exercises:

1. Groups 1–4: All four groups will collaborate together in an effort to research real port and harbor development projects. On the basis of these other projects, they will prepare a brief description of their proposed port and harbor development project. Prepare a simple map of an actual area in which your project would take place. Prepare a three- to five-page synopsis of the hypothetical project with emphasis on describing what activities would take place.
2. Group 1: Members of Group 1 will prepare a management action plan (MAP) that provides a "road map" for preparing the EIS (i.e., Chapters 3 through 6). At a minimum, it will include:
   - Outline of the EIS process, and how it will be implemented and coordinated
   - Annotated outline of the EIS
   - Roles and responsibilities (functional roles and responsibility matrix) for preparing the EIS (i.e., class member assignments)
   - Brief description of the proposal
   - Schedule outlining significant milestones
   - Outline of a process including quality control measures for ensuring that the assessment fairly considers the needs and interests of all potential stakeholders
3. Group 2: Group 2 will prepare a public involvement strategy with emphasis on incorporating concerns of the CCADEE. It will also identify and describe potential barriers to the proposal.
4. Group 3: Group 3 will prepare a scoping plan including how scoping will be conducted. Prepare at least one draft advertisement on the DEE's plan to prepare the EIS, and invite the public to participate in the public scoping process.
5. Group 4: Group 4 will develop a data requirements document that identifies the types of data that your group believes will be required to prepare the affected environment and environmental consequences section of the EIS. The content will vary with the description of the proposal.

Each individual group will present the results of their respective task in front of the class. The rest of the class will act as project peer reviewers who will ask questions and critique the work of that group.

Capstone problem no. 2: Mining application

Divide the class into five groups. The class represents staff members for the NEPA Office of the Department of Mining Everywhere and Anytime (DMEA). Assume that a mining application has been submitted to the DMEA for approval. The agency must prepare an EIS for the application. Your project is controversial and viciously opposed by the United Citizens Against Government, and Rabid and Poisonous Mining Projects (UCAGRPMP). The UCAGRPMP has held a rally and press conference in which it vows to "... blow up DMEA projects... and rejects the hateful, discriminatory, capitalist–fascist, noxious, lethal, and generally death-promoting practices of the DMEA." The EIS manager has already indicated that he/she is going to increase his life insurance policy. UCAGRPMP will certainly sue DMEA's EIS.

The class represents the IDT that will be in charge of the EIS process. The class will use the Internet to research mining areas in the United States and choose an area for the setting of their proposed project (again, exercise creativity). All data for this exercise will be generated by the groups designated below and based on Internet research of a real area and other similar mining projects. The instructor will designate an EIS manager that will be in charge of coordinating five groups. Each respective group will be responsible for performing the following exercises:

1. Groups 1–5: All five groups will collaborate in an effort to locate an area that is actually being mined for some mineral (e.g., iron, bauxite). On the basis of these other projects, prepare a brief description and scope of the proposed area. Prepare a simple map of the project area and delineate key features (mountains, deserts, forests, water resources, etc.).
2. Group 2: Group 2 will identify and describe at least two alternatives (in addition to no action) based on factors such as different mining technologies, practices, or alternative sites. Describe potential mitigation measures for reducing impacts. Each alternative should be a minimum of several pages long.
3. Group 3: Prepare an annotated schedule identifying all key steps (scoping, notice of intent, preparation of draft and final statements, issuing notice of availability, issuing record of decision, etc.) that must be accomplished to prepare the EIS (see Chapter 4).
4. Group 4: Prepare a description of the affected environment (minimum five pages long, based on research of the environmental resources in the area that was chosen for the site of the proposal).
5. Group 5: Use a Leopold Matrix (Chapter 5) to identify potentially significant impacts. Provide a brief description of each impact. Identify adverse effects that cannot be mitigated.

Each individual group will present the results of their respective task in front of the class. The rest of the class will act as peer reviewers who will ask questions and critique the work of that group.

Capstone problem no. 3: Recreational facility

Divide the class into five groups. The class will represent staff members for the NEPA Office of an agency that has recently changed its name from "Department of Parks and Recreation" to the Department of "Lets Park and Re-create" (LPR). Your mission is to prepare an EIS (see Chapter 4) for a new recreational area in an environmentally sensitive area. Your project is controversial and vigorously opposed by the Mother's Committee Against Parking and Re-creating (MCAPR). MCAPR will certainly sue the LPR's EIS.

The class represents the IDT that will be in charge of the EIS process. The class will use the Internet to research parks and recreational areas in the United States and choose an area for the setting of this project (again, exercise creativity). All material for this exercise will be generated by the groups based on Internet research of a real area park or recreational (federal or state) and other similar projects. The IDT will designate an EIS manager that will be in charge of coordinating five groups. Students are encouraged to exercise creativity. Each respective group will be responsible for performing the following exercises:

1. Groups 1–5: All five groups will collaborate together in an effort to locate an area for development of a park or recreational area. On the basis of these other projects, prepare a brief description and scope of the proposed park or recreational area project. Delineate key features (mountains, deserts, forests, water resources, etc.).
2. Group 1: Group 1 will use Schmidt's model (Chapter 3) to develop a statement of the purpose and need for the proposal.
3. Group 2: Prepare a draft notice of intent for the EIS.
4. Group 3: Use an environmental checklist (Chapter 5) to identify potentially significant impacts. Provide a brief description of each impact. Identify adverse effects that cannot be mitigated.
5. Group 4: On the basis of your research and the results of the environmental checklist, prepare an annotated outline of the EIS document.
6. Group 5: Choose one potentially significant impact and describe it and how it would affect the environment.

Each individual group will present the results of their respective task in front of the class. The rest of the class will act as stakeholders, public members, and opponents who will ask questions and critique the work of that group.

Glossary

Act: A synonym used in the Council on Environmental Quality regulations to refer to the National Environmental Policy Act, as amended (42 U.S.C. 4321, et seq.).

actions: The Council on Environmental Quality's NEPA regulations define three types of actions, other than unconnected single actions, which must be taken into consideration during a NEPA analysis. These three actions are (1) connected, (2) cumulative, and (3) similar actions.

Administrative Procedures Act: A law that specifies the requirements and procedures that must be followed in issuing regulations.

alternatives: The term "alternatives," as used in the Council on Environmental Quality's NEPA regulations, refers to other reasonable options that would meet the need of a proposed action. There are three types of alternatives: (1) no-action alternative, (2) other reasonable courses of actions, and (3) mitigation measures (not in the proposed action).

analyzed alternatives: Alternatives that are both reasonable and are also examined in detail.

applicant: An applicant is a nonfederal party that has filed an application with a federal agency, and that is subject to a NEPA review before the agency may approve the application. Such applications normally involve required federal approvals or permits that must be obtained before the applicant may proceed with a specified action.

categorical exclusions: Categorical exclusions (CATX) are class of actions under NEPA that do not have a significance, either individually or cumulatively, on the human environment, and therefore does not require preparation of an environmental assessment or environmental impact statement.

CEQ: See Council on Environmental Quality.

Commission: Abbreviation for the US Nuclear Regulatory Commission.

connected actions: The term "connected action," as defined by the Council on Environmental Quality's NEPA regulations, means actions that are closely related and therefore should be discussed in the same impact statement. Actions are connected if they (1) automatically trigger other actions that may require environmental impact statements, (2) cannot or will not proceed unless other actions are taken previously or simultaneously, or (3) are interdependent parts of a larger action and depend on the larger action for their justification.

context: The term "context," as used in the Council on Environmental Quality's NEPA regulations, refers to a factor that must be considered in making a determination regarding the significance of an impact. In making a determination regarding the significance of an action, the impacts must be analyzed in several contexts, such as society as a whole (human, national), the affected region, the affected interests, and the locality. Significance varies with the setting of the proposed action. For instance, in the case of a site-specific action, significance would usually depend on the effects in the locale rather than in the world as a whole. Both short- and long-term effects are relevant.

cooperating agency: A federal agency other than a lead agency that has jurisdiction by law or special expertise with respect to any environmental impact involved in a proposal (or a reasonable alternative) for legislation or other major federal action significantly affecting the quality of the human environment.

Council: A synonym used for the Council on Environmental Quality.

Council on Environmental Quality (CEQ): The council created by Title II of the NEPA Act to oversee the NEPA process.

Council on Environmental Quality regulations: The regulations issued by the Council on Environmental Quality (40 CFR parts 1500–1508) for implementing the procedural aspects of NEPA.

cumulative actions: The term "connected action," as defined by the Council on Environmental Quality's NEPA regulations, to mean actions that when viewed with other proposed actions have cumulatively significant impacts and should therefore be discussed in the same impact statement.

cumulative impact: The impact on the environment that results from the incremental impact of an action when it is added to other past, present, and reasonably foreseeable future actions, regardless of what agency (federal or nonfederal) or person has undertaken these other actions. This is an important concept because individually minor but collectively significant impacts can take place over a period of time.

direct impacts: Effects that are caused by the action and occur at the same time and place as the action.

EA: See environmental assessment.

effects: The term effects and impacts as used in the NEPA regulations are synonymous. Effects may include impacts of an action on ecological (such as the effects on natural resources and on the components, structures, and functioning of affected ecosystems), aesthetic, historic, economic, social, health, and cultural resources. The concept of effects includes direct, indirect, and cumulative effects, and includes both beneficial and detrimental impacts. There are three types of impacts: (1) direct, (2) indirect, and (3) cumulative.

EIS: See environmental impact statement.

emission: A pollution discharge into the atmosphere from smokestacks, vents, and other sources.

endangered species: Organisms that are threatened with extinction by man-made or natural changes in the environment.

environment: See human environment.

environmental assessment: An environmental assessment (EA) is a concise public document that is used to briefly provide sufficient evidence and analysis for determining whether to prepare an environmental impact statement or a finding of no significant impact for a proposed action. An EA may also be used to assist an agency in compliance with the NEPA act when no environmental impact statement is necessary. An EA may also be used to facilitate preparation of an EIS when one is necessary. An EA must include brief discussions of the need for the proposal, alternatives, environmental impacts of the proposed action and alternatives, and a listing of agencies and persons consulted.

environmental document: As defined in the Council on Environmental Quality's NEPA regulations, this document includes environmental assessment, environmental impact statement, finding of no significant impact, and the notice of intent.

environmental impact statement: A detailed document that is required to be prepared under the Council on Environmental Quality's NEPA regulations for a proposed action that may result in a significant environmental impact.

federal agency: As defined by the Council on Environmental Quality's NEPA regulations, a federal agency means all agencies of the federal government. This term does not include Congress, the judiciary, or the president, including the performance of staff functions for the president in his executive office.

finding of no significant impact: The term "finding of no significant impact" (FONSI) means a document by a federal agency that briefly presents the reasons why an action that has not already been categorically excluded will not have a significant effect on the human environment and, therefore, for which an environmental impact statement will not be required. It must include the environmental assessment or a summary of it. The FONSI must also note any other environmental documents related to it.

FONSI: See finding of no significant impact.

habitat: The location and surroundings where a population of plants or animals live.

hazardous waste: Waste that can pose a hazard to human health and the environment according to the Resource Conservation and Recovery Act. To be designated as a hazardous, the waste must possess one of the following four characteristics: (1) reactivity, (2) corrosivity, (3) ignitability, or (4) toxicity. A waste may also be designated hazardous if it is listed by the Environmental Protection Agency (EPA) as a hazardous waste.

human environment: The term "human environment," as defined by the Council on Environmental Quality's NEPA regulations, is interpreted comprehensively to include the natural and physical environment, and the relationship of people with that environment. This means that economic or social effects are not intended by themselves to require preparation of an environmental impact statement. When an environmental impact statement is prepared, and economic or social and natural or physical environmental effects are interrelated, then the environmental impact statement will discuss all of these effects on the human environment.

impacts: See effects.

implementation plan (IP): A document used by many federal agencies to record the results of the EIS scoping process. The IP also provides a plan for preparing the EIS.

indirect impacts: Reasonably foreseeable impacts that are caused by an action but occur at a later time or that are removed in distance from the action. Indirect impacts may include growth-inducing effects and other effects related to induced changes in the pattern of land use, population density or growth rate, and related effects on air and water and other natural systems, including ecosystems.

intensity: The term "intensity," as used in the Council on Environmental Quality's NEPA regulations, refers to a factor that must be considered in making a determination regarding the significance of an impact.

In making a determination regarding the significance of an action, the impacts must be analyzed in terms of its intensity. The intensity is the degree to which the impact would affect the environment.

interim action: An action within the scope of a proposal that is the subject of an ongoing EIS and that an agency proposes to pursue before the ROD is issued, and that is permissible under 40 CFR 1506.1.

IP: See implementation plan.

jurisdiction by law: The term "jurisdiction by law," as used in NEPA, means agency authority to approve, veto, or finance all or part of the proposal.

land use plans: With respect to NEPA, the term "land use plans" includes any formally adopted documents for land use planning or zoning, including proposed plans that have been formally proposed by a government body and are under active consideration (see Council on Environmental Quality's 40 Questions, Question No. 23b).

land use policies: The term "land use policies," as used in reference to NEPA, includes formally adopted statements of land use policy embodied in laws or regulations. It also includes land use policies that have been formally proposed but have not yet been adopted (see Council on Environmental Quality's 40 Questions, Question No. 23b).

lead agency: The term "lead agency," as used in the Council on Environmental Quality's NEPA regulations, means the agency or agencies preparing or having taken primary responsibility for preparing the environmental impact statement.

legislation: The term "legislation" includes a bill or legislative proposal to Congress developed by or with the significant cooperation and support of a federal agency, but does not include requests for appropriations. The test for significant cooperation is whether the proposal is in fact predominantly that of the agency rather than another source. Drafting does not by itself constitute significant cooperation. Proposals for legislation include requests for ratification of treaties. Only the agency that has primary responsibility for the subject matter involved will prepare a legislative environmental impact statement.

major federal action: The term "major federal action," as used in the Council on Environmental Quality's NEPA regulations, includes actions with effects that may be major and which are potentially subject to federal control and responsibility. "Major" reinforces but does not have a meaning independent of the term "significantly." Actions include the circumstance where the responsible officials fail to act and that failure to act is reviewable by courts or administrative tribunals under the Administrative Procedure Act or other applicable law as agency action.

mitigation: Measures that may be taken to avoid, minimize, rectify, reduce, or compensate the adverse impacts of an action on the environment.

mitigation action plan: Refers to a document describing the plan for implementing commitments made in an EIS/ROD or EA/FONSI.

monitoring: The process of observing and measuring environmental impacts on environmental resources to verify compliance with the description of the proposed action and any mitigation factors that were cited in a NEPA document.

National Environmental Policy Act (NEPA): Federal statute passed by Congress in 1969, establishing the basic environmental policy for protection of the environment (42 U.S.C. 4321 et seq.). It provides a systematic and interdisciplinary process that agencies are required to follow to reduce or prevent environment degradation. The Act contains "action-forcing" procedures that must be followed by federal agencies to ensure federal decision makers take environmental factors before making a final decision regarding a proposed action.

NEPA: See National Environmental Policy Act.

NEPA process: The term "NEPA process" refers to all measures that are necessary for compliance with the requirements of section 2 and Title I of NEPA.

NEPA review: The process followed in complying with section 102(2) of NEPA.

NOA: See notice of availability.

NOI: See notice of intent.

notice of availability: A formal notice as defined in 40 CFR 1508.22, published in the *Federal Register,* announcing the issuance and public availability of a draft or final EIS.

notice of intent: A formal notice, published in the *Federal Register,* announcing the issuance and public availability of a draft or final EIS.

program: For the purposes of NEPA, a program can be defined as a sequence of connected or related actions as discussed in 40 CFR 1508.18(b)(3) and 1508.25(a).

programmatic EA/EIS: A broadly scoped EA or EIS prepared to evaluate an agency program and/or including a sequence of connected or related agency actions or projects as discussed in 40 CFR 1508.18(b)(3) and 1508.25(a).

project: For the purposes of NEPA, a "project" refers to a specific agency effort, including actions approved by a permit or regulatory decision, federal and federally assisted activities, or similar activities, as described in 40 CFR 1508.18(b)(4).

proposal: A "proposal" as used in the Council on Environmental Quality's NEPA regulations exists at that stage in the development of an action when an agency subject to the Act has a goal and is actively preparing to make a decision on one or more alternative means of accomplishing that goal and the effects can be meaningfully evaluated. Preparation of an environmental impact statement on a proposal should be timed so that the final statement may be completed in time for the statement to be included in any recommendation or report on the proposal. A proposal may exist in fact as well as by agency declaration that one exists.

proposed action: The alternative that the agency is proposing to implement.

public scoping: Refers to that portion of the scoping process where the public is invited to participate, as described in 40 CFR 1501.7(a)(1) and (b)(4).

record of decision (ROD): A public document that is prepared on completion of an EIS. This document records the agency's final decision and rationale for making the decision, and any commitments to monitoring and mitigation.

referring agency: A "referring agency" as used in the Council on Environmental Quality's NEPA regulations means the federal agency that has referred any matter to the Council after a determination that the matter is unsatisfactory from the standpoint of public health or welfare or environmental quality.

Regulations: As used in this book, the term "Regulations" refers to NEPA regulations that were issued by the Council on Environmental Quality (40 CFR parts 1500–1508).

resources: With respect to NEPA, environmental resources include all physical (e.g., geological, biological, atmospheric), socioeconomic, and other related aspects of the environment that may be potentially affected by the agency's action.

risk: As used in this text, "risk" is defined as the probability that the accident would occur, multiplied by the consequences of the accident.

ROD: See record of decision.

S-EIS: See supplemental EIS.

scope: The term "referring agency," as used in the Council on Environmental Quality's NEPA regulations, is defined to consist of the range of actions, alternatives, and impacts to be considered in an environmental impact statement. The scope of an individual statement may depend on its relationships to other statements. To determine the scope of environmental impact statements, agencies must consider three types of actions, three types of alternatives, and three types of impacts.

significance: The degree to which an impact may affect the human environment. The term, as used in the Council on Environmental Quality's NEPA regulations, requires consideration of both context and intensity of an impact.

similar actions: The term "connected action," as defined by the Council on Environmental Quality's NEPA regulations, means actions that when viewed with other reasonably foreseeable or proposed agency actions, have similarities that provide a basis for evaluating their environmental consequences together, such as common timing or geography. An agency may wish to analyze these actions in the same impact statement. It should do so when the best way to assess adequately the combined impacts of similar actions or reasonable alternatives to such actions is to treat them in a single impact statement.

special expertise: The term "special expertise," as defined by the Council on Environmental Quality's NEPA regulations, means statutory responsibility, agency mission, or related program experience.

supplemental EIS (S-EIS): An EIS prepared to supplement an existing EIS as described in 40 CFR 1502.9(c). A supplemental EIS is prepared when a substantial change to the proposed action or when important new information is acquired regarding the action.

tiering: The term "tiering" refers to the coverage of general matters in broader environmental impact statements (such as national program or policy statements) with subsequent narrower statements or environmental analyses (such as regional or basin-wide program statements or ultimately site-specific statements) incorporating by reference the general discussions and concentrating solely on the issues specific to the statement subsequently prepared.

tribal lands: The area of "Indian country," as defined in 18 U.S.C. 1151, that is under the tribe's jurisdiction.

wetlands: An area that is saturated or partially saturated. An area need only be saturated during a small portion of the year to be designated a wetlands. To be designated a wetlands, the area must exhibit certain soil, hydrological, and vegetative characteristics.

Appendix A

The National Environmental Policy Act of 1969

A.1 The National Environmental Policy Act of 1969, as amended

(Pub. L. 91-190, 42 U.S.C. 4321-4347, January 1, 1970, as amended by Pub. L. 94-52, July 3, 1975, Pub. L. 94-83, August 9, 1975, and Pub. L. 97-258, § 4(b), Sept. 13, 1982)

An Act to establish a national policy for the environment, to provide for the establishment of a Council on Environmental Quality, and for other purposes.

Be it enacted by the Senate and House of Representatives of the United States of America in Congress assembled, That this Act may be cited as the "National Environmental Policy Act of 1969."

A.2 Purpose

A.2.1 Sec. 2 [42 USC § 4321].

The purposes of this Act are: To declare a national policy which will encourage productive and enjoyable harmony between man and his environment; to promote efforts which will prevent or eliminate damage to the environment and biosphere and stimulate the health and welfare of man; to enrich the understanding of the ecological systems and natural resources important to the Nation; and to establish a Council on Environmental Quality.

A.3 Title I

A.3.1 Congressional declaration of National Environmental Policy

A.3.1.1 Sec. 101 [42 USC § 4331].

(a) The Congress, recognizing the profound impact of man's activity on the interrelations of all components of the natural environment, particularly the profound influences of population growth, high-density urbanization, industrial expansion, resource exploitation, and new and expanding technological advances and recognizing further the critical importance of restoring and maintaining environmental quality to the overall welfare and development of man, declares that it is the continuing policy of the Federal Government, in cooperation with State and local governments, and other concerned public and private organizations, to use all practicable means and measures, including financial and technical assistance, in a manner calculated to foster and promote the general welfare, to create and maintain conditions under which man and nature can exist in productive harmony, and fulfill the social, economic, and other requirements of present and future generations of Americans.

(b) In order to carry out the policy set forth in this Act, it is the continuing responsibility of the Federal Government to use all practicable means, consist with other essential considerations of national policy, to improve and coordinate Federal plans, functions, programs, and resources to the end that the Nation may—

1. fulfill the responsibilities of each generation as trustee of the environment for succeeding generations;
2. assure for all Americans safe, healthful, productive, and aesthetically and culturally pleasing surroundings;
3. attain the widest range of beneficial uses of the environment without degradation, risk to health or safety, or other undesirable and unintended consequences;
4. preserve important historic, cultural, and natural aspects of our national heritage, and maintain, wherever possible, an environment which supports diversity, and variety of individual choice;
5. achieve a balance between population and resource use which will permit high standards of living and a wide sharing of life's amenities; and
6. enhance the quality of renewable resources and approach the maximum attainable recycling of depletable resources.

(c) The Congress recognizes that each person should enjoy a healthful environment and that each person has a responsibility to contribute to the preservation and enhancement of the environment.

A.3.1.2 Sec. 102 [42 USC § 4332].

The Congress authorizes and directs that, to the fullest extent possible: (1) the policies, regulations, and public laws of the United States shall be interpreted and administered in accordance with the policies set forth in this Act, and (2) all agencies of the Federal Government shall—

(A) utilize a systematic, interdisciplinary approach which will insure the integrated use of the natural and social sciences and the environmental design arts in planning and in decisionmaking which may have an impact on man's environment;

(B) identify and develop methods and procedures, in consultation with the Council on Environmental Quality established by title II of this Act, which will insure that presently unquantified environmental amenities and values may be given appropriate consideration in decisionmaking along with economic and technical considerations;

(C) include in every recommendation or report on proposals for legislation and other major Federal actions significantly affecting the quality of the human environment, a detailed statement by the responsible official on—

  (i) the environmental impact of the proposed action,

  (ii) any adverse environmental effects which cannot be avoided should the proposal be implemented,

  (iii) alternatives to the proposed action,

  (iv) the relationship between local short-term uses of man's environment and the maintenance and enhancement of long-term productivity, and

  (v) any irreversible and irretrievable commitments of resources which would be involved in the proposed action should it be implemented.

Prior to making any detailed statement, the responsible Federal official shall consult with and obtain the comments of any Federal agency which has jurisdiction by law or special expertise with respect to any environmental impact involved. Copies of such statement and the comments and views of the appropriate Federal, State, and local agencies, which are authorized to develop and enforce environmental standards, shall be made available to the President, the Council on Environmental Quality and to the public as provided by section 552 of title 5, United States Code, and shall accompany the proposal through the existing agency review processes;

(D) Any detailed statement required under subparagraph (C) after January 1, 1970, for any major Federal action funded under a program of grants to States shall not be deemed to be legally insufficient solely by reason of having been prepared by a State agency or official, if:

(i) the State agency or official has statewide jurisdiction and has the responsibility for such action,

(ii) the responsible Federal official furnishes guidance and participates in such preparation,

(iii) the responsible Federal official independently evaluates such statement prior to its approval and adoption, and

(iv) after January 1, 1976, the responsible Federal official provides early notification to, and solicits the views of, any other State or any Federal land management entity of any action or any alternative thereto which may have significant impacts upon such State or affected Federal land management entity and, if there is any disagreement on such impacts, prepares a written assessment of such impacts and views for incorporation into such detailed statement.

The procedures in this subparagraph shall not relieve the Federal official of his responsibilities for the scope, objectivity, and content of the entire statement or of any other responsibility under this Act; and further, this subparagraph does not affect the legal sufficiency of statements prepared by State agencies with less than statewide jurisdiction.

(E) study, develop, and describe appropriate alternatives to recommended courses of action in any proposal which involves unresolved conflicts concerning alternative uses of available resources;

(F) recognize the worldwide and long-range character of environmental problems and, where consistent with the foreign policy of the United States, lend appropriate support to initiatives, resolutions, and programs designed to maximize international cooperation in anticipating and preventing a decline in the quality of mankind's world environment;

(G) make available to States, counties, municipalities, institutions, and individuals, advice and information useful in restoring, maintaining, and enhancing the quality of the environment;

(H) initiate and utilize ecological information in the planning and development of resource-oriented projects; and

(I) assist the Council on Environmental Quality established by title II of this Act.

A.3.1.3 Sec. 103 [42 USC § 4333].

All agencies of the Federal Government shall review their present statutory authority, administrative regulations, and current policies and procedures for the purpose of determining whether there are any deficiencies or inconsistencies therein which prohibit full compliance with the purposes and provisions of this Act and shall propose to the President not later than July 1, 1971, such measures as may be necessary to bring their authority and policies into conformity with the intent, purposes, and procedures set forth in this Act.

A.3.1.4 Sec. 104 [42 USC § 4334].

Nothing in section 102 [42 USC § 4332] or 103 [42 USC § 4333] shall in any way affect the specific statutory obligations of any Federal agency (1) to comply with criteria or standards of environmental quality, (2) to coordinate or consult with any other Federal or State agency, or (3) to act, or refrain from acting contingent upon the recommendations or certification of any other Federal or State agency.

A.3.1.5 Sec. 105 [42 USC § 4335].

The policies and goals set forth in this Act are supplementary to those set forth in existing authorizations of Federal agencies.

A.4 Title II

A.4.1 Council on Environmental Quality

A.4.1.1 Sec. 201 [42 USC § 4341].

The President shall transmit to the Congress annually beginning July 1, 1970, an Environmental Quality Report (hereinafter referred to as the "report") which shall set forth (1) the status and condition of the major natural, manmade, or altered environmental classes of the Nation, including, but not limited to, the air, the aquatic, including marine, estuarine, and fresh water, and the terrestrial environment, including, but not limited to, the forest, dryland, wetland, range, urban, suburban an rural environment; (2) current and foreseeable trends in the quality, management and utilization of such environments and the effects of those trends on the social, economic, and other requirements of the Nation; (3) the adequacy of available natural resources for fulfilling human and economic requirements of the Nation in the light of expected population pressures; (4) a review of the programs and activities (including regulatory activities) of the Federal Government, the State and local governments, and nongovernmental entities or individuals with particular reference to their effect on the environment and on the conservation, development and utilization of natural resources; and (5) a program for remedying the deficiencies of existing programs and activities, together with recommendations for legislation.

A.4.1.2 Sec. 202 [42 USC § 4342].

There is created in the Executive Office of the President a Council on Environmental Quality (hereinafter referred to as the "Council"). The Council shall be composed of three members who shall be appointed by the President to serve at his pleasure, by and with the advice and consent of the Senate. The President shall designate one of the members of the Council to serve as Chairman. Each member shall be a person who, as a result of his training, experience, and attainments, is exceptionally well qualified to analyze and interpret environmental trends and information of all kinds; to appraise programs and activities of the Federal Government in the light of the policy set forth in title I of this Act; to be conscious of and responsive to the scientific, economic, social, aesthetic, and cultural needs and interests of the Nation; and to formulate and recommend national policies to promote the improvement of the quality of the environment.

A.4.1.3 Sec. 203 [42 USC § 4343].

(a) The Council may employ such officers and employees as may be necessary to carry out its functions under this Act. In addition, the Council may employ and fix the compensation of such experts and consultants as may be necessary for the carrying out of its functions under this Act, in accordance with section 3109 of title 5, United States Code (but without regard to the last sentence thereof).
(b) Notwithstanding section 1342 of title 31, the Council may accept and employ voluntary and uncompensated services in furtherance of the purposes of the Council.

A.4.1.4 Sec. 204 [42 USC § 4344].

It shall be the duty and function of the Council—

1. to assist and advise the President in the preparation of the Environmental Quality Report required by section 201 [42 USC § 4341] of this title;
2. to gather timely and authoritative information concerning the conditions and trends in the quality of the environment both current and prospective, to analyze and interpret such information for the purpose of determining whether such conditions and trends are interfering, or are likely to interfere, with the achievement of the policy set forth in title I of this Act, and to compile and submit to the President studies relating to such conditions and trends;
3. to review and appraise the various programs and activities of the Federal Government in the light of the policy set forth in title I of this Act for the purpose of determining the extent to which such programs and activities are contributing to the achievement of such policy, and to make recommendations to the President with respect thereto;
4. to develop and recommend to the President national policies to foster and promote the improvement of environmental quality to meet

the conservation, social, economic, health, and other requirements and goals of the Nation;

5. to conduct investigations, studies, surveys, research, and analyses relating to ecological systems and environmental quality;
6. to document and define changes in the natural environment, including the plant and animal systems, and to accumulate necessary data and other information for a continuing analysis of these changes or trends and an interpretation of their underlying causes;
7. to report at least once each year to the President on the state and condition of the environment; and
8. to make and furnish such studies, reports thereon, and recommendations with respect to matters of policy and legislation as the President may request.

A.4.1.5 Sec. 205 [42 USC § 4345].

In exercising its powers, functions, and duties under this Act, the Council shall—

1. consult with the Citizens' Advisory Committee on Environmental Quality established by Executive Order No. 11472, dated May 29, 1969, and with such representatives of science, industry, agriculture, labor, conservation organizations, State and local governments and other groups, as it deems advisable; and
2. utilize, to the fullest extent possible, the services, facilities and information (including statistical information) of public and private agencies and organizations, and individuals, in order that duplication of effort and expense may be avoided, thus assuring that the Council's activities will not unnecessarily overlap or conflict with similar activities authorized by law and performed by established agencies.

A.4.1.6 Sec. 206 [42 USC § 4346].

Members of the Council shall serve full time and the Chairman of the Council shall be compensated at the rate provided for Level II of the Executive Schedule Pay Rates [5 USC §5313]. The other members of the Council shall be compensated at the rate provided for Level IV of the Executive Schedule Pay Rates [5 USC § 5315].

A.4.1.7 Sec. 207 [42 USC § 4346a].

The Council may accept reimbursements from any private nonprofit organization or from any department, agency, or instrumentality of the Federal Government, any State, or local government, for the reasonable travel expenses incurred by an officer or employee of the Council in connection with his attendance at any conference, seminar, or similar meeting conducted for the benefit of the Council.

A.4.1.8 Sec. 208 [42 USC § 4346b].

The Council may make expenditures in support of its international activities, including expenditures for: (1) international travel; (2) activities in implementation of international agreements; and (3) the support of international exchange programs in the United States and in foreign countries.

A.4.1.9 Sec. 209 [42 USC § 4347].

There are authorized to be appropriated to carry out the provisions of this chapter not to exceed $300,000 for fiscal year 1970, $700,000 for fiscal year 1971, and $1,000,000 for each fiscal year thereafter.

Appendix B

The CEQ NEPA Implementing Regulations

Council on Environmental Quality
Executive Office of the President

REGULATIONS

For Implementing The Procedural Provisions Of The

NATIONAL ENVIRONMENTAL POLICY ACT

Reprint
40 CFR Parts 1500-1508
(2005)

TABLE OF CONTENTS

PART 1503—COMMENTING

Sec.
1503.1 Inviting comments.
1503.2 Duty to comment.
1503.3 Specificity of comments.
1503.4 Response to comments.

PART 1504—PREDECISION REFERRALS TO THE COUNCIL OF PROPOSED FEDERAL ACTIONS DETERMINED TO BE ENVIRONMENTALLY UNSATISFACTORY

Sec.
1504.1 Purpose.
1504.2 Criteria for referral.
1504.3 Procedure for referrals and response.

PART 1505—NEPA AND AGENCY DECISIONMAKING

Sec.
1505.1 Agency decisionmaking procedures.
1505.2 Record of decision in cases requiring environmental impact statements.
1505.3 Implementing the decision.

PART 1506—OTHER REQUIREMENTS OF NEPA

Sec.
1506.1 Limitations on actions during NEPA process.
1506.2 Elimination of duplication with State and local procedures.
1506.3 Adoption.
1506.4 Combining documents.
1506.5 Agency responsibility.
1506.6 Public involvement.
1506.7 Further guidance.
1506.8 Proposals for legislation.
1506.9 Filing requirements.
1506.10 Timing of agency action.
1506.11 Emergencies.
1506.12 Effective date.

PART 1507—AGENCY COMPLIANCE

Sec.
1507.1 Compliance.
1507.2 Agency capability to comply.
1507.3 Agency procedures.

PART 1508—TERMINOLOGY AND INDEX

PART 1500—PURPOSE, POLICY, AND MANDATE

AUTHORITY: NEPA, the Environmental Quality Improvement Act of 1970, as amended (42 U.S.C. 4371 et seq.), sec. 309 of the Clean Air Act, as amended (42 U.S.C. 7609) and E.O. 11514, Mar. 5, 1970, as amended by E.O. 11991, May 24, 1977).

SOURCE: 43 FR 55990, Nov. 28, 1978, unless otherwise noted.

§1500.1 Purpose.

(a) The National Environmental Policy Act (NEPA) is our basic national charter for protection of the environment. It establishes policy, sets goals (section 101), and provides means (section 102) for carrying out the policy. Section 102(2) contains "action-forcing" provisions to make sure that federal agencies act according to the letter and spirit of the Act. The regulations that follow implement section 102(2). Their purpose is to tell federal agencies what they must do to comply with the procedures and achieve the goals of the Act. The President, the federal agencies, and the courts share responsibility for enforcing the Act so as to achieve the substantive requirements of section 101.

(b) NEPA procedures must insure that environmental information is available to public officials and citizens before decisions are made and before actions are taken. The information must be of high quality. Accurate scientific analysis, expert agency comments, and public scrutiny are essential to implementing NEPA. Most important, NEPA documents must concentrate on the issues that are truly significant to the action in question, rather than amassing needless detail.

(c) Ultimately, of course, it is not better documents but better decisions that count. NEPA's purpose is not to generate paperwork—even excellent paperwork—but to foster excellent action. The NEPA process is intended to help public officials make decisions that are based on understanding of environmental consequences, and take actions that protect, restore, and enhance the environment. These regulations provide the direction to achieve this purpose.

§1500.2 Policy.

Federal agencies shall to the fullest extent possible:

(a) Interpret and administer the policies, regulations, and public laws of the United States in accordance with the policies set forth in the Act and in these regulations.

(b) Implement procedures to make the NEPA process more useful to decisionmakers and the public; to reduce paperwork and the accumulation of extraneous background data; and to emphasize real environmental issues and alternatives. Environmental impact statements shall be concise, clear, and to the point, and shall be supported by evidence that agencies have made the necessary environmental analyses.

(c) Integrate the requirements of NEPA with other planning and environmental review procedures required by law or by agency practice so that all such procedures run concurrently rather than consecutively.
(d) Encourage and facilitate public involvement in decisions which affect the quality of the human environment.
(e) Use the NEPA process to identify and assess the reasonable alternatives to proposed actions that will avoid or minimize adverse effects of these actions upon the quality of the human environment.
(f) Use all practicable means, consistent with the requirements of the Act and other essential considerations of national policy, to restore and enhance the quality of the human environment and avoid or minimize any possible adverse effects of their actions upon the quality of the human environment.

§1500.3 Mandate.

Parts 1500 through 1508 of this title provide regulations applicable to and binding on all federal agencies for implementing the procedural provisions of the National Environmental Policy Act of 1969, as amended (Pub. L. 91–190, 42 U.S.C. 4321 *et seq.*) (NEPA or the Act) except where compliance would be inconsistent with other statutory requirements. These regulations are issued pursuant to NEPA, the Environmental Quality Improvement Act of 1970, as amended (42 U.S.C. 4371 *et seq.*) section 309 of the Clean Air Act, as amended (42 U.S.C. 7609) and Executive Order 11514, Protection and Enhancement of Environmental Quality (March 5, 1970, as amended by Executive Order 11991, May 24, 1977). These regulations, unlike the predecessor guidelines, are not confined to sec. 102(2)(C) (environmental impact statements). The regulations apply to the whole of section 102(2). The provisions of the Act and of these regulations must be read together as a whole in order to comply with the spirit and letter of the law. It is the Council's intention that judicial review of agency compliance with these regulations not occur before an agency has filed the final environmental impact statement, or has made a final finding of no significant impact (when such a finding will result in action affecting the environment), or takes action that will result in irreparable injury. Furthermore, it is the Council's intention that any trivial violation of these regulations not give rise to any independent cause of action.

§1500.4 Reducing paperwork.

Agencies shall reduce excessive paperwork by:

(a) Reducing the length of environmental impact statements (§1502.2(c)), by means such as setting appropriate page limits (§§1501.7(b)(1) and 1502.7).

(b) Preparing analytic rather than encyclopedic environmental impact statements (§1502.2(a)).
(c) Discussing only briefly issues other than significant ones (§1502.2(b)).
(d) Writing environmental impact statements in plain language (§1502.8).
(e) Following a clear format for environmental impact statements (§1502.10).
(f) Emphasizing the portions of the environmental impact statement that are useful to decisionmakers and the public (§§1502.14 and 1502.15) and reducing emphasis on background material (§1502.16).
(g) Using the scoping process, not only to identify significant environmental issues deserving of study, but also to deemphasize insignificant issues, narrowing the scope of the environmental impact statement process accordingly (§1501.7).
(h) Summarizing the environmental impact statement (§1502.12) and circulating the summary instead of the entire environmental impact statement if the latter is unusually long (§1502.19).
(i) Using program, policy, or plan environmental impact statements and tiering from statements of broad scope to those of narrower scope, to eliminate repetitive discussions of the same issues (§§1502.4 and 1502.20).
(j) Incorporating by reference (§1502.21).
(k) Integrating NEPA requirements with other environmental review and consultation requirements (§1502.25).
(l) Requiring comments to be as specific as possible (§1503.3).
(m) Attaching and circulating only changes to the draft environmental impact statement, rather than rewriting and circulating the entire statement when changes are minor (§1503.4(c)).
(n) Eliminating duplication with state and local procedures, by providing for joint preparation (§1506.2), and with other federal procedures, by providing that an agency may adopt appropriate environmental documents prepared by another agency (§1506.3).
(o) Combining environmental documents with other documents (§1506.4).
(p) Using categorical exclusions to define categories of actions which do not individually or cumulatively have a significant effect on the human environment and which are therefore exempt from requirements to prepare an environmental impact statement (§1508.4).
(q) Using a finding of no significant impact when an action not otherwise excluded will not have a significant effect on the human environment and is therefore exempt from requirements to prepare an environmental impact statement (§1508.13).

[43 FR 55990, Nov. 29, 1978; 44 FR 873, Jan. 3, 1979]

§1500.5 Reducing delay.

Agencies shall reduce delay by:

(a) Integrating the NEPA process into early planning (§1501.2).
(b) Emphasizing interagency cooperation before the environmental impact statement is prepared, rather than submission of adversary comments on a completed document (§1501.6).
(c) Insuring the swift and fair resolution of lead agency disputes (§1501.5).
(d) Using the scoping process for an early identification of what are and what are not the real issues (§1501.7).
(e) Establishing appropriate time limits for the environmental impact statement process (§§1501.7(b)(2) and 1501.8).
(f) Preparing environmental impact statements early in the process (§1502.5).
(g) Integrating NEPA requirements with other environmental review and consultation requirements (§1502.25).
(h) Eliminating duplication with state and local procedures by providing for joint preparation (§1506.2), and with other federal procedures by providing that an agency may adopt appropriate environmental documents prepared by another agency (§1506.3).
(i) Combining environmental documents with other documents (§1506.4).
(j) Using accelerated procedures for proposals for legislation (§1506.8).
(k) Using categorical exclusions to define categories of actions which do not individually or cumulatively have a significant effect on the human environment (§1508.4) and which are therefore exempt from requirements to prepare an environmental impact statement.
(l) Using a finding of no significant impact when an action not otherwise excluded will not have a significant effect on the human environment (§1508.13) and is therefore exempt from requirements to prepare an environmental impact statement.

§1500.6 Agency authority.

Each agency shall interpret the provisions of the Act as a supplement to its existing authority and as a mandate to view traditional policies and missions in the light of the Act's national environmental objectives. Agencies shall review their policies, procedures, and regulations accordingly and revise them as necessary to insure full compliance with the purposes and provisions of the Act. The phrase "to the fullest extent possible" in section 102 means that each agency of the federal government shall comply with that section unless existing law applicable to the agency's operations expressly prohibits or makes compliance impossible.

PART 1501—NEPA AND AGENCY PLANNING

AUTHORITY: NEPA, the Environmental Quality Improvement Act of 1970, as amended (42 U.S.C. 4371 *et seq*.), sec. 309 of the Clean Air Act, as amended (42 U.S.C. 7609, and E.O. 11514 (Mar. 5, 1970, as amended by E.O. 11991, May 24, 1977).
SOURCE: 43 FR 55992, Nov. 29, 1978, unless otherwise noted.

§1501.1 Purpose.

The purposes of this part include:

(a) Integrating the NEPA process into early planning to insure appropriate consideration of NEPA's policies and to eliminate delay.
(b) Emphasizing cooperative consultation among agencies before the environmental impact statement is prepared rather than submission of adversary comments on a completed document.
(c) Providing for the swift and fair resolution of lead agency disputes.
(d) Identifying at an early stage the significant environmental issues deserving of study and deemphasizing insignificant issues, narrowing the scope of the environmental impact statement accordingly.
(e) Providing a mechanism for putting appropriate time limits on the environmental impact statement process.

§1501.2 Apply NEPA early in the process.

Agencies shall integrate the NEPA process with other planning at the earliest possible time to insure that planning and decisions reflect environmental values, to avoid delays later in the process, and to head off potential conflicts. Each agency shall:

(a) Comply with the mandate of section 102(2)(A) to "utilize a systematic, interdisciplinary approach which will insure the integrated use of the natural and social sciences and the environmental design arts in planning and in decisionmaking which may have an impact on man's environment," as specified by §1507.2.

(b) Identify environmental effects and values in adequate detail so they can be compared to economic and technical analyses. Environmental documents and appropriate analyses shall be circulated and reviewed at the same time as other planning documents.
(c) Study, develop, and describe appropriate alternatives to recommended courses of action in any proposal which involves unresolved conflicts concerning alternative uses of available resources as provided by section 102(2)(E) of the Act.
(d) Provide for cases where actions are planned by private applicants or other non-federal entities before federal involvement so that:
    (1) Policies or designated staff are available to advise potential applicants of studies or other information foreseeably required for later federal action.
    (2) The federal agency consults early with appropriate state and local agencies and Indian tribes and with interested private persons and organizations when its own involvement is reasonably foreseeable.
    (3) The federal agency commences its NEPA process at the earliest possible time.

§1501.3 When to prepare an environmental assessment.

(a) Agencies shall prepare an environmental assessment (§1508.9) when necessary under the procedures adopted by individual agencies to supplement these regulations as described in §1507.3. An assessment is not necessary if the agency has decided to prepare an environmental impact statement.
(b) Agencies may prepare an environmental assessment on any action at any time in order to assist agency planning and decision making.

§1501.4 Whether to prepare an environmental impact statement.

In determining whether to prepare an environmental impact statement the federal agency shall:

(a) Determine under its procedures supplementing these regulations (described in §1507.3) whether the proposal is one which:
    (1) Normally requires an environmental impact statement, or
    (2) Normally does not require either an environmental impact statement or an environmental assessment (categorical exclusion).
(b) If the proposed action is not covered by paragraph (a) of this section, prepare an environmental assessment (§1508.9). The agency shall involve environmental agencies, applicants, and the public, to the extent practicable, in preparing assessments required by §1508.9(a)(1).
(c) Based on the environmental assessment make its determination whether to prepare an environmental impact statement.

(d) Commence the scoping process (§1501.7), if the agency will prepare an environmental impact statement.
(e) Prepare a finding of no significant impact (§1508.13), if the agency determines on the basis of the environmental assessment not to prepare a statement.
   (1) The agency shall make the finding of no significant impact available to the affected public as specified in §1506.6.
   (2) In certain limited circumstances, which the agency may cover in its procedures under §1507.3, the agency shall make the finding of no significant impact available for public review (including state and areawide clearinghouses) for 30 days before the agency makes its final determination whether to prepare an environmental impact statement and before the action may begin. The circumstances are:
      (i) The proposed action is, or is closely similar to, one which normally requires the preparation of an environmental impact statement under the procedures adopted by the agency pursuant to §1507.3, or
      (ii) The nature of the proposed action is one without precedent.

§1501.5 Lead agencies.

(a) A lead agency shall supervise the preparation of an environmental impact statement if more than one federal agency either:
   (1) Proposes or is involved in the same action; or
   (2) Is involved in a group of actions directly related to each other because of their functional interdependence or geographical proximity.
(b) Federal, state, or local agencies, including at least one federal agency, may act as joint lead agencies to prepare an environmental impact statement (§1506.2).
(c) If an action falls within the provisions of paragraph (a) of this section the potential lead agencies shall determine by letter or memorandum which agency shall be the lead agency and which shall be cooperating agencies. The agencies shall resolve the lead agency question so as not to cause delay. If there is disagreement among the agencies, the following factors (which are listed in order of descending importance) shall determine lead agency designation:
   (1) Magnitude of agency's involvement.
   (2) Project approval/disapproval authority.
   (3) Expertise concerning the action's environmental effects.
   (4) Duration of agency's involvement.
   (5) Sequence of agency's involvement.
(d) Any federal agency, or any state or local agency or private person substantially affected by the absence of lead agency designation,

may make a written request to the potential lead agencies that a lead agency be designated.

(e) If federal agencies are unable to agree on which agency will be the lead agency or if the procedure described in paragraph (c) of this section has not resulted within 45 days in a lead agency designation, any of the agencies or persons concerned may file a request with the Council asking it to determine which Federal agency shall be the lead agency. A copy of the request shall be transmitted to each potential lead agency. The request shall consist of:
  (1) A precise description of the nature and extent of the proposed action.
  (2) A detailed statement of why each potential lead agency should or should not be the lead agency under the criteria specified in paragraph (c) of this section.

(f) A response may be filed by any potential lead agency concerned within 20 days after a request is filed with the Council. The Council shall determine as soon as possible but not later than 20 days after receiving the request and all responses to it which federal agency shall be the lead agency and which other federal agencies shall be cooperating agencies.

[43 FR 55992, Nov. 29, 1978; 44 FR 873, Jan. 3, 1979]

§1501.6 Cooperating agencies.

The purpose of this section is to emphasize agency cooperation early in the NEPA process. Upon request of the lead agency, any other federal agency which has jurisdiction by law shall be a cooperating agency. In addition any other federal agency which has special expertise with respect to any environmental issue, which should be addressed in the statement may be a cooperating agency upon request of the lead agency. An agency may request the lead agency to designate it a cooperating agency.

(a) The lead agency shall:
  (1) Request the participation of each cooperating agency in the NEPA process at the earliest possible time.
  (2) Use the environmental analysis and proposals of cooperating agencies with jurisdiction by law or special expertise, to the maximum extent possible consistent with its responsibility as lead agency.
  (3) Meet with a cooperating agency at the latter's request.

(b) Each cooperating agency shall:
  (1) Participate in the NEPA process at the earliest possible time.
  (2) Participate in the scoping process (described below in §1501.7).
  (3) Assume on request of the lead agency responsibility for developing information and preparing environmental analyses including

portions of the environmental impact statement concerning which the cooperating agency has special expertise.
(4) Make available staff support at the lead agency's request to enhance the latter's interdisciplinary capability.
(5) Normally use its own funds. The lead agency shall, to the extent available funds permit, fund those major activities or analyses it requests from cooperating agencies. Potential lead agencies shall include such funding requirements in their budget requests.

(c) A cooperating agency may in response to a lead agency's request for assistance in preparing the environmental impact statement (described in paragraph (b) (3), (4), or (5) of this section) reply that other program commitments preclude any involvement or the degree of involvement requested in the action that is the subject of the environmental impact statement. A copy of this reply shall be submitted to the Council.

§1501.7 Scoping.

There shall be an early and open process for determining the scope of issues to be addressed and for identifying the significant issues related to a proposed action. This process shall be termed scoping. As soon as practicable after its decision to prepare an environmental impact statement and before the scoping process the lead agency shall publish a notice of intent (§1508.22) in the FEDERAL REGISTER except as provided in §1507.3(e).

(a) As part of the scoping process the lead agency shall:
(1) Invite the participation of affected federal, state, and local agencies, any affected Indian tribe, the proponent of the action, and other interested persons (including those who might not be in accord with the action on environmental grounds), unless there is a limited exception under §1507.3(c). An agency may give notice in accordance with §1506.6.
(2) Determine the scope (§1508.25) and the significant issues to be analyzed in depth in the environmental impact statement.
(3) Identify and eliminate from detailed study the issues which are not significant or which have been covered by prior environmental review (§1506.3), narrowing the discussion of these issues in the statement to a brief presentation of why they will not have a significant effect on the human environment or providing a reference to their coverage elsewhere.
(4) Allocate assignments for preparation of the environmental impact statement among the lead and cooperating agencies, with the lead agency retaining responsibility for the statement.

(5) Indicate any public environmental assessments and other environmental impact statements which are being or will be prepared that are related to but are not part of the scope of the impact statement under consideration.
(6) Identify other environmental review and consultation requirements so the lead and cooperating agencies may prepare other required analyses and studies concurrently with, and integrated with, the environmental impact statement as provided in §1502.25.
(7) Indicate the relationship between the timing of the preparation of environmental analyses and the agency's tentative planning and decisionmaking schedule.

(b) As part of the scoping process the lead agency may:
(1) Set page limits on environmental documents (§1502.7).
(2) Set time limits (§1501.8).
(3) Adopt procedures under §1507.3 to combine its environmental assessment process with its scoping process.
(4) Hold an early scoping meeting or meetings which may be integrated with any other early planning meeting the agency has. Such a scoping meeting will often be appropriate when the impacts of a particular action are confined to specific sites.

(c) An agency shall revise the determinations made under paragraphs (a) and (b) of this section if substantial changes are made later in the proposed action, or if significant new circumstances or information arise which bear on the proposal or its impacts.

§1501.8 Time limits.

Although the Council has decided that prescribed universal time limits for the entire NEPA process are too inflexible, federal agencies are encouraged to set time limits appropriate to individual actions (consistent with the time intervals required by §1506.10). When multiple agencies are involved the reference to agency below means lead agency.

(a) The agency shall set time limits if an applicant for the proposed action requests them: *Provided,* That the limits are consistent with the purposes of NEPA and other essential considerations of national policy.

(b) The agency may:
(1) Consider the following factors in determining time limits:
(i) Potential for environmental harm.
(ii) Size of the proposed action.
(iii) State of the art of analytic techniques.
(iv) Degree of public need for the proposed action, including the consequences of delay.
(v) Number of persons and agencies affected.

(vi) Degree to which relevant information is known and if not known the time required for obtaining it.
(vii) Degree to which the action is controversial.
(viii) Other time limits imposed on the agency by law, regulations, or executive order.

(2) Set overall time limits or limits for each constituent part of the NEPA process, which may include:
(i) Decision on whether to prepare an environmental impact statement (if not already decided).
(ii) Determination of the scope of the environmental impact statement.
(iii) Preparation of the draft environmental impact statement.
(iv) Review of any comments on the draft environmental impact statement from the public and agencies.
(v) Preparation of the final environmental impact statement.
(vi) Review of any comments on the final environmental impact statement.
(vii) Decision on the action based in part on the environmental impact statement.

(3) Designate a person (such as the project manager or a person in the agency's office with NEPA responsibilities) to expedite the NEPA process.

(c) State or local agencies or members of the public may request a federal agency to set time limits.

PART 1502—ENVIRONMENTAL IMPACT STATEMENT

Sec.

AUTHORITY: NEPA, the Environmental Quality Improvement Act of 1970, as amended (42 U.S.C. 4371 *et seq.*), sec. 309 of the Clean Air Act, as amended (42 U.S.C. 7609), and E.O. 11514 (Mar. 5, 1970, as amended by E.O. 11991, May 24, 1977).

SOURCE: 43 FR 55994, Nov. 29, 1978, unless otherwise noted.

§1502.1 Purpose.

The primary purpose of an environmental impact statement is to serve as an action-forcing device to insure that the policies and goals defined in the Act are infused into the ongoing programs and actions of the federal government. It shall provide full and fair discussion of significant environmental impacts and shall inform decisionmakers and the public of the reasonable alternatives which would avoid or minimize adverse impacts or enhance the quality of the human environment. Agencies shall focus on significant environmental issues and alternatives and shall reduce paperwork and the accumulation of extraneous background data. Statements shall be concise, clear, and to the point, and shall be supported by evidence that the agency has made the necessary environmental analyses. An environmental impact statement is more than a disclosure document. It shall be used by federal officials in conjunction with other relevant material to plan actions and make decisions.

§1502.2 Implementation.

To achieve the purposes set forth in §1502.1 agencies shall prepare environmental impact statements in the following manner:

(a) Environmental impact statements shall be analytic rather than encyclopedic.
(b) Impacts shall be discussed in proportion to their significance. There shall be only brief discussion of other than significant issues. As in

a finding of no significant impact, there should be only enough discussion to show why more study is not warranted.

(c) Environmental impact statements shall be kept concise and shall be no longer than absolutely necessary to comply with NEPA and with these regulations. Length should vary first with potential environmental problems and then with project size.

(d) Environmental impact statements shall state how alternatives considered in it and decisions based on it will or will not achieve the requirements of sections 101 and 102(1) of the Act and other environmental laws and policies.

(e) The range of alternatives discussed in environmental impact statements shall encompass those to be considered by the ultimate agency decisionmaker.

(f) Agencies shall not commit resources prejudicing selection of alternatives before making a final decision (§1506.1).

(g) Environmental impact statements shall serve as the means of assessing the environmental impact of proposed agency actions, rather than justifying decisions already made.

§1502.3 Statutory requirements for statements.

As required by sec. 102(2)(C) of NEPA environmental impact statements (§1508.11) are to be included in every recommendation or report.

On proposals (§1508.23).
For legislation and (§1508.17).
Other major federal actions (§1508.18).
Significantly (§1508.27).
Affecting (§§1508.3, 1508.8).
The quality of the human environment (§1508.14).

§1502.4 Major Federal actions requiring the preparation of environmental impact statements.

(a) Agencies shall make sure the proposal which is the subject of an environmental impact statement is properly defined. Agencies shall use the criteria for scope (§1508.25) to determine which proposal(s) shall be the subject of a particular statement. Proposals or parts of proposals which are related to each other closely enough to be, in effect, a single course of action shall be evaluated in a single impact statement.

(b) Environmental impact statements may be prepared, and are sometimes required, for broad federal actions such as the adoption of new agency programs or regulations (§1508.18). Agencies shall prepare statements on broad actions so that they are relevant to policy and are timed to coincide with meaningful points in agency planning and decisionmaking.

(c) When preparing statements on broad actions (including proposals by more than one agency), agencies may find it useful to evaluate the proposal(s) in one of the following ways:
   (1) Geographically, including actions occurring in the same general location, such as body of water, region, or metropolitan area.
   (2) Generically, including actions which have relevant similarities, such as common timing, impacts, alternatives, methods of implementation, media, or subject matter.
   (3) By stage of technological development including federal or federally assisted research, development or demonstration programs for new technologies which, if applied, could significantly affect the quality of the human environment. Statements shall be prepared on such programs and shall be available before the program has reached a stage of investment or commitment to implementation likely to determine subsequent development or restrict later alternatives.

(d) Agencies shall as appropriate employ scoping (§1501.7), tiering (§1502.20), and other methods listed in §§1500.4 and 1500.5 to relate broad and narrow actions and to avoid duplication and delay.

§1502.5 Timing.

An agency shall commence preparation of an environmental impact statement as close as possible to the time the agency is developing or is presented with a proposal (§1508.23) so that preparation can be completed in time for the final statement to be included in any recommendation or report on the proposal. The statement shall be prepared early enough so that it can serve practically as an important contribution to the decisionmaking process and will not be used to rationalize or justify decisions already made (§§1500.2(c), 1501.2, and 1502.2). For instance:

(a) For projects directly undertaken by federal agencies the environmental impact statement shall be prepared at the feasibility analysis (go–no go) stage and may be supplemented at a later stage if necessary.

(b) For applications to the agency appropriate environmental assessments or statements shall be commenced no later than immediately after the application is received. Federal agencies are encouraged to begin preparation of such assessments or statements earlier, preferably jointly with applicable state or local agencies.

(c) For adjudication, the final environmental impact statement shall normally precede the final staff recommendation and that portion of the public hearing related to the impact study. In appropriate circumstances the statement may follow preliminary hearings designed to gather information for use in the statements.
(d) For informal rulemaking the draft environmental impact statement shall normally accompany the proposed rule.

§1502.6 Interdisciplinary preparation.

Environmental impact statements shall be prepared using an interdisciplinary approach which will insure the integrated use of the natural and social sciences and the environmental design arts (section 102(2)(A) of the Act). The disciplines of the preparers shall be appropriate to the scope and issues identified in the scoping process (§1501.7).

§1502.7 Page limits.

The text of final environmental impact statements (e.g., paragraphs (d) through (g) of §1502.10) shall normally be less than 150 pages and for proposals of unusual scope or complexity shall normally be less than 300 pages.

§1502.8 Writing.

Environmental impact statements shall be written in plain language and may use appropriate graphics so that decisionmakers and the public can readily understand them. Agencies should employ writers of clear prose or editors to write, review, or edit statements, which will be based upon the analysis and supporting data from the natural and social sciences and the environmental design arts.

§1502.9 Draft, final, and supplemental statements.

Except for proposals for legislation as provided in §1506.8 environmental impact statements shall be prepared in two stages and may be supplemented.

(a) Draft environmental impact statements shall be prepared in accordance with the scope decided upon in the scoping process. The lead agency shall work with the cooperating agencies and shall obtain comments as required in part 1503 of this chapter. The draft statement must fulfill and satisfy to the fullest extent possible the requirements established for final statements in section 102(2)(C) of the Act. If a draft statement is so inadequate as to preclude meaningful analysis, the agency shall prepare and circulate a revised draft of the appropriate portion. The agency shall make every effort to disclose and discuss at appropriate points in the draft statement all major points of view on the environmental impacts of the alternatives including the proposed action.

(b) Final environmental impact statements shall respond to comments as required in part 1503 of this chapter. The agency shall discuss at appropriate points in the final statement any responsible opposing view which was not adequately discussed in the draft statement and shall indicate the agency's response to the issues raised.

(c) Agencies:

  (1) Shall prepare supplements to either draft or final environmental impact statements if:

    (i) The agency makes substantial changes in the proposed action that are relevant to environmental concerns; or

    (ii) There are significant new circumstances or information relevant to environmental concerns and bearing on the proposed action or its impacts.

  (2) May also prepare supplements when the agency determines that the purposes of the Act will be furthered by doing so.

  (3) Shall adopt procedures for introducing a supplement into its formal administrative record, if such a record exists.

  (4) Shall prepare, circulate, and file a supplement to a statement in the same fashion (exclusive of scoping) as a draft and final statement unless alternative procedures are approved by the Council.

§1502.10 Recommended format.

Agencies shall use a format for environmental impact statements which will encourage good analysis and clear presentation of the alternatives including the proposed action. The following standard format for environmental impact statements should be followed unless the agency determines that there is a compelling reason to do otherwise:

(a) Cover sheet.
(b) Summary.
(c) Table of contents.
(d) Purpose of and need for action.
(e) Alternatives including proposed action (sections 102(2)(C)(iii) and 102(2)(E) of the Act).
(f) Affected environment.
(g) Environmental consequences (especially sections 102(2)(C)(i), (ii), (iv), and (v) of the Act).
(h) List of preparers.
(i) List of agencies, organizations, and persons to whom copies of the statement are sent.

(j) Index.
(k) Appendices (if any).

If a different format is used, it shall include paragraphs (a), (b), (c), (h), (i), and (j), of this section and shall include the substance of paragraphs (d), (e), (f), (g), and (k) of this section, as further described in §§1502.11 through 1502.18, in any appropriate format.

§1502.11 Cover sheet.

The cover sheet shall not exceed one page. It shall include:
(a) A list of the responsible agencies including the lead agency and any cooperating agencies.
(b) The title of the proposed action that is the subject of the statement (and if appropriate the titles of related cooperating agency actions), together with the state(s) and county(ies) (or other jurisdiction if applicable) where the action is located.
(c) The name, address, and telephone number of the person at the agency who can supply further information.
(d) A designation of the statement as a draft, final, or draft or final supplement.
(e) A one paragraph abstract of the statement.
(f) The date by which comments must be received (computed in cooperation with EPA under §1506.10). The information required by this section may be entered on Standard Form 424 (in items 4, 6, 7, 10, and 18).

§1502.12 Summary.

Each environmental impact statement shall contain a summary which adequately and accurately summarizes the statement. The summary shall stress the major conclusions, areas of controversy (including issues raised by agencies and the public), and the issues to be resolved (including the choice among alternatives). The summary will normally not exceed 15 pages.

§1502.13 Purpose and need.

The statement shall briefly specify the underlying purpose and need to which the agency is responding in proposing the alternatives including the proposed action.

§1502.14 Alternatives including the proposed action.

This section is the heart of the environmental impact statement. Based on the information and analysis presented in the sections on the Affected

Environment (§1502.15) and the Environmental Consequences (§1502.16), it should present the environmental impacts of the proposal and the alternatives in comparative form, thus sharply defining the issues and providing a clear basis for choice among options by the decisionmaker and the public. In this section agencies shall:

(a) Rigorously explore and objectively evaluate all reasonable alternatives, and for alternatives which were eliminated from detailed study, briefly discuss the reasons for their having been eliminated.
(b) Devote substantial treatment to each alternative considered in detail including the proposed action so that reviewers may evaluate their comparative merits.
(c) Include reasonable alternatives not within the jurisdiction of the lead agency.
(d) Include the alternative of no action.
(e) Identify the agency's preferred alternative or alternatives, if one or more exists, in the draft statement and identify such alternative in the final statement unless another law prohibits the expression of such a preference.
(f) Include appropriate mitigation measures not already included in the proposed action or alternatives.

§1502.15 Affected environment.

The environmental impact statement shall succinctly describe the environment of the area(s) to be affected or created by the alternatives under consideration. The description shall be no longer than is necessary to understand the effects of the alternatives. Data and analyses in a statement shall be commensurate with the importance of the impact, with less important material summarized, consolidated, or simply referenced. Agencies shall avoid useless bulk in statements and shall concentrate effort and attention on important issues. Verbose descriptions of the affected environment are themselves no measure of the adequacy of an environmental impact statement.

§1502.16 Environmental consequences.

This section forms the scientific and analytic basis for the comparisons under §1502.14. It shall consolidate the discussions of those elements required by sections 102(2)(C)(i), (ii), (iv), and (v) of NEPA which are within the scope of the statement and as much of section 102(2)(C)(iii) as is necessary to support the comparisons. The discussion will include the environmental impacts of the alternatives including the proposed action, any adverse environmental effects which cannot be avoided should the proposal be implemented, the relationship between short-term uses of man's environment and the maintenance and enhancement of long-term productivity, and any irreversible or irretrievable commitments of resources which would be involved in the proposal should it be implemented. This section should not duplicate discussions in §1502.14. It shall include discussions of:

(a) Direct effects and their significance (§1508.8).
(b) Indirect effects and their significance (§1508.8).
(c) Possible conflicts between the proposed action and the objectives of federal, regional, state, and local (and in the case of a reservation, Indian tribe) land use plans, policies and controls for the area concerned. (See §1506.2(d).)
(d) The environmental effects of alternatives including the proposed action. The comparisons under §1502.14 will be based on this discussion.
(e) Energy requirements and conservation potential of various alternatives and mitigation measures.
(f) Natural or depletable resource requirements and conservation potential of various alternatives and mitigation measures.
(g) Urban quality, historic and cultural resources, and the design of the built environment, including the reuse and conservation potential of various alternatives and mitigation measures.
(h) Means to mitigate adverse environmental impacts (if not fully covered under §1502.14(f)).

[43 FR 55994, Nov. 29, 1978; 44 FR 873, Jan. 3, 1979]

§1502.17 List of preparers.

The environmental impact statement shall list the names, together with their qualifications (expertise, experience, professional disciplines), of the persons who were primarily responsible for preparing the environmental impact statement or significant background papers, including basic components of the statement (§§1502.6 and 1502.8). Where possible the persons who are responsible for a particular analysis, including analyses in background papers, shall be identified. Normally the list will not exceed two pages.

§1502.18 Appendix.

If an agency prepares an appendix to an environmental impact statement the appendix shall:

(a) Consist of material prepared in connection with an environmental impact statement (as distinct from material which is not so prepared and which is incorporated by reference (§1502.21)).

(b) Normally consist of material which substantiates any analysis fundamental to the impact statement.
(c) Normally be analytic and relevant to the decision to be made.
(d) Be circulated with the environmental impact statement or be readily available on request.

§1502.19 Circulation of the environmental impact statement.

Agencies shall circulate the entire draft and final environmental impact statements except for certain appendices as provided in §1502.18(d) and unchanged statements as provided in §1503.4(c). However, if the statement is unusually long, the agency may circulate the summary instead, except that the entire statement shall be furnished to:

(a) Any federal agency which has jurisdiction by law or special expertise with respect to any environmental impact involved and any appropriate federal, state or local agency authorized to develop and enforce environmental standards.
(b) The applicant, if any.
(c) Any person, organization, or agency requesting the entire environmental impact statement.
(d) In the case of a final environmental impact statement any person, organization, or agency which submitted substantive comments on the draft. If the agency circulates the summary and thereafter receives a timely request for the entire statement and for additional time to comment, the time for that requestor only shall be extended by at least 15 days beyond the minimum period.

§1502.20 Tiering.

Agencies are encouraged to tier their environmental impact statements to eliminate repetitive discussions of the same issues and to focus on the actual issues ripe for decision at each level of environmental review (§1508.28). Whenever a broad environmental impact statement has been prepared (such as a program or policy statement) and a subsequent statement or environmental assessment is then prepared on an action included within the entire program or policy (such as a site specific action) the subsequent statement or environmental assessment need only summarize the issues discussed in the broader statement and incorporate discussions from the broader statement by reference and shall concentrate on the issues specific to the subsequent action. The subsequent document shall state where the earlier document is available. Tiering may also be appropriate for different stages of actions. (§1508.28).

§1502.21 Incorporation by reference.

Agencies shall incorporate material into an environmental impact statement by reference when the effect will be to cut down on bulk without impeding agency and public review of the action. The incorporated material shall be cited in the statement and its content briefly described. No material may be incorporated by reference unless it is reasonably available for inspection by potentially interested persons within the time allowed for comment. Material based on proprietary data which is itself not available for review and comment shall not be incorporated by reference.

§1502.22 Incomplete or unavailable information.

When an agency is evaluating reasonably foreseeable significant adverse effects on the human environment in an environmental impact statement and there is incomplete or unavailable information, the agency shall always make clear that such information is lacking.

(a) If the incomplete information relevant to reasonably foreseeable significant adverse impacts is essential to a reasoned choice among alternatives and the overall costs of obtaining it are not exorbitant, the agency shall include the information in the environmental impact statement.

(b) If the information relevant to reasonably foreseeable significant adverse impacts cannot be obtained because the overall costs of obtaining it are exorbitant or the means to obtain it are not known, the agency shall include within the environmental impact statement: (1) A statement that such information is incomplete or unavailable; (2) a statement of the relevance of the incomplete or unavailable information to evaluating reasonably foreseeable significant adverse impacts on the human environment; (3) a summary of existing credible scientific evidence which is relevant to evaluating the reasonably foreseeable significant adverse impacts on the human environment; and (4) the agency's evaluation of such impacts based upon theoretical approaches or research methods generally accepted in the scientific community. For the purposes of this section, "reasonably foreseeable" includes impacts which have catastrophic consequences, even if their probability of occurrence is low, provided that the analysis of the impacts is supported by credible scientific evidence, is not based on pure conjecture, and is within the rule of reason.

(c) The amended regulation will be applicable to all environmental impact statements for which a Notice of Intent (40 CFR 1508.22) is published in the FEDERAL REGISTER on or after May 27, 1986. For environmental impact statements in progress, agencies may choose to comply with the requirements of either the original or amended regulation.

[51 FR 15625, Apr. 25, 1986]

§1502.23 Cost–benefit analysis.

If a cost–benefit analysis relevant to the choice among environmentally different alternatives is being considered for the proposed action, it shall be incorporated by reference or appended to the statement as an aid in evaluating the environmental consequences. To assess the adequacy of compliance with section 102(2)(B) of the Act the statement shall, when a cost–benefit analysis is prepared, discuss the relationship between that analysis and any analyses of unquantified environmental impacts, values, and amenities. For purposes of complying with the Act, the weighing of the merits and drawbacks of the various alternatives need not be displayed in a monetary cost–benefit analysis and should not be when there are important qualitative considerations. In any event, an environmental impact statement should at least indicate those considerations, including factors not related to environmental quality, which are likely to be relevant and important to a decision.

§1502.24 Methodology and scientific accuracy.

Agencies shall insure the professional integrity, including scientific integrity, of the discussions and analyses in environmental impact statements. They shall identify any methodologies used and shall make explicit reference by footnote to the scientific and other sources relied upon for conclusions in the statement. An agency may place discussion of methodology in an appendix.

§1502.25 Environmental review and consultation requirements.

(a) To the fullest extent possible, agencies shall prepare draft environmental impact statements concurrently with and integrated with environmental impact analyses and related surveys and studies required by the Fish and Wildlife Coordination Act (16 U.S.C. 661 *et seq.*), the National Historic Preservation Act of 1966 (16 U.S.C. 470 *et seq.*), the Endangered Species Act of 1973 (16 U.S.C. 1531 *et seq.*), and other environmental review laws and executive orders.

(b) The draft environmental impact statement shall list all federal permits, licenses, and other entitlements which must be obtained in implementing the proposal. If it is uncertain whether a federal permit, license, or other entitlement is necessary, the draft environmental impact statement shall so indicate.

PART 1503—COMMENTING

AUTHORITY: NEPA, the Environmental Quality Improvement Act of 1970, as amended (42 U.S.C. 4371 *et seq.*), sec. 309 of the Clean Air Act, as amended (42 U.S.C. 7609), and E.O. 11514 (Mar. 5, 1970, as amended by E.O. 11991, May 24, 1977).
SOURCE: 43 FR 55997, Nov. 29, 1978, unless otherwise noted.

§1503.1 Inviting comments.

(a) After preparing a draft environmental impact statement and before preparing a final environmental impact statement the agency shall:
   (1) Obtain the comments of any federal agency which has jurisdiction by law or special expertise with respect to any environmental impact involved or which is authorized to develop and enforce environmental standards.
   (2) Request the comments of:
      (i) Appropriate state and local agencies which are authorized to develop and enforce environmental standards;
      (ii) Indian tribes, when the effects may be on a reservation; and
      (iii) Any agency which has requested that it receive statements on actions of the kind proposed. Office of Management and Budget Circular A-95 (Revised), through its system of clearinghouses, provides a means of securing the views of state and local environmental agencies. The clearinghouses may be used, by mutual agreement of the lead agency and the clearinghouse, for securing state and local reviews of the draft environmental impact statements.
   (3) Request comments from the applicant, if any.
   (4) Request comments from the public, affirmatively soliciting comments from those persons or organizations who may be interested or affected.
(b) An agency may request comments on a final environmental impact statement before the decision is finally made. In any case other agencies or persons may make comments before the final decision unless a different time is provided under §1506.10.

§1503.2 Duty to comment.

Federal agencies with jurisdiction by law or special expertise with respect to any environmental impact involved and agencies which are authorized to develop and enforce environmental standards shall comment on statements within their jurisdiction, expertise, or authority. Agencies shall comment within the time period specified for comment in §1506.10. A Federal agency may reply that it has no comment. If a cooperating agency is satisfied that its views are adequately reflected in the environmental impact statement, it should reply that it has no comment.

§1503.3 Specificity of comments.

(a) Comments on an environmental impact statement or on a proposed action shall be as specific as possible and may address either the adequacy of the statement or the merits of the alternatives discussed or both.
(b) When a commenting agency criticizes a lead agency's predictive methodology, the commenting agency should describe the alternative methodology which it prefers and why.
(c) A cooperating agency shall specify in its comments whether it needs additional information to fulfill other applicable environmental reviews or consultation requirements and what information it needs. In particular, it shall specify any additional information it needs to comment adequately on the draft statement's analysis of significant site-specific effects associated with the granting or approving by that cooperating agency of necessary federal permits, licenses, or entitlements.
(d) When a cooperating agency with jurisdiction by law objects to or expresses reservations about the proposal on grounds of environmental impacts, the agency expressing the objection or reservation shall specify the mitigation measures it considers necessary to allow the agency to grant or approve applicable permit, license, or related requirements or concurrences.

§1503.4 Response to comments.

(a) An agency preparing a final environmental impact statement shall assess and consider comments both individually and collectively, and shall respond by one or more of the means listed below, stating its response in the final statement. Possible responses are to:
    (1) Modify alternatives including the proposed action.

(2) Develop and evaluate alternatives not previously given serious consideration by the agency.
(3) Supplement, improve, or modify its analyses.
(4) Make factual corrections.
(5) Explain why the comments do not warrant further agency response, citing the sources, authorities, or reasons which support the agency's position and, if appropriate, indicate those circumstances which would trigger agency reappraisal or further response.

(b) All substantive comments received on the draft statement (or summaries thereof where the response has been exceptionally voluminous), should be attached to the final statement whether or not the comment is thought to merit individual discussion by the agency in the text of the statement.

(c) If changes in response to comments are minor and are confined to the responses described in paragraphs (a)(4) and (5) of this section, agencies may write them on errata sheets and attach them to the statement instead of rewriting the draft statement. In such cases only the comments, the responses, and the changes and not the final statement need be circulated (§1502.19). The entire document with a new cover sheet shall be filed as the final statement (§1506.9).

PART 1504—PREDECISION REFERRALS TO THE COUNCIL OF PROPOSED FEDERAL ACTIONS DETERMINED TO BE ENVIRONMENTALLY UNSATISFACTORY

AUTHORITY: NEPA, the Environmental Quality Improvement Act of 1970, as amended (42 U.S.C. 4371 *et seq.*), sec. 309 of the Clean Air Act, as amended (42 U.S.C. 7609), and E.O. 11514 (Mar. 5, 1970, as amended by E.O. 11991, May 24, 1977).
Source: 43FR 55998, Nov. 29, 1978 unless otherwise noted.

§1504.1 Purpose.

(a) This part establishes procedures for referring to the Council federal interagency disagreements concerning proposed major federal actions that might cause unsatisfactory environmental effects. It provides means for early resolution of such disagreements.

(b) Under section 309 of the Clean Air Act (42 U.S.C. 7609), the Administrator of the Environmental Protection Agency is directed to review and comment publicly on the environmental impacts of federal activities, including actions for which environmental impact statements are prepared. If after this review the Administrator determines that the matter is "unsatisfactory from the standpoint of public health or welfare or environmental quality," section 309 directs that the matter be referred to the Council (hereafter "environmental referrals").

(c) Under section 102(2)(C) of the Act other federal agencies may make similar reviews of environmental impact statements, including judgments on the acceptability of anticipated environmental impacts. These reviews must be made available to the President, the Council and the public.

[43 FR 55998, Nov. 29, 1978]

§1504.2 Criteria for referral.

Environmental referrals should be made to the Council only after concerted, timely (as early as possible in the process), but unsuccessful attempts to resolve differences with the lead agency. In determining what environmental objections to the matter are appropriate to refer to the Council, an agency should weigh potential adverse environmental impacts, considering:

(a) Possible violation of national environmental standards or policies.
(b) Severity.
(c) Geographical scope.
(d) Duration.
(e) Importance as precedents.
(f) Availability of environmentally preferable alternatives.

[43 FR 55998, Nov. 29, 1978]

§1504.3 Procedure for referrals and response.

(a) A federal agency making the referral to the Council shall:
  (1) Advise the lead agency at the earliest possible time that it intends to refer a matter to the Council unless a satisfactory agreement is reached.
  (2) Include such advice in the referring agency's comments on the draft environmental impact statement, except when the statement does not contain adequate information to permit an assessment of the matter's environmental acceptability.
  (3) Identify any essential information that is lacking and request that it be made available at the earliest possible time.
  (4) Send copies of such advice to the Council.

(b) The referring agency shall deliver its referral to the Council not later than twenty-five (25) days after the final environmental impact statement has been made available to the Environmental Protection Agency, commenting agencies, and the public. Except when an extension of this period has been granted by the lead agency, the Council will not accept a referral after that date.

(c) The referral shall consist of:

(1) A copy of the letter signed by the head of the referring agency and delivered to the lead agency informing the lead agency of the referral and the reasons for it, and requesting that no action be taken to implement the matter until the Council acts upon the referral. The letter shall include a copy of the statement referred to in (c)(2) of this section.

(2) A statement supported by factual evidence leading to the conclusion that the matter is unsatisfactory from the standpoint of public health or welfare or environmental quality. The statement shall:

(i) Identify any material facts in controversy and incorporate (by reference if appropriate) agreed upon facts,

(ii) Identify any existing environmental requirements or policies which would be violated by the matter,

(iii) Present the reasons why the referring agency believes the matter is environmentally unsatisfactory,

(iv) Contain a finding by the agency whether the issue raised is of national importance because of the threat to national environmental resources or policies or for some other reason,

(v) Review the steps taken by the referring agency to bring its concerns to the attention of the lead agency at the earliest possible time, and

(vi) Give the referring agency's recommendations as to what mitigation alternative, further study, or other course of action (including abandonment of the matter) are necessary to remedy the situation.

(d) Not later than twenty-five (25) days after the referral to the Council the lead agency may deliver a response to the Council, and the referring agency. If the lead agency requests more time and gives assurance that the matter will not go forward in the interim, the Council may grant an extension. The response shall:

(1) Address fully the issues raised in the referral.

(2) Be supported by evidence.

(3) Give the lead agency's response to the referring agency's recommendations.

(e) Interested persons (including the applicant) may deliver their views in writing to the Council. Views in support of the referral should be delivered not later than the referral. Views in support of the response shall be delivered not later than the response.

(f) Not later than twenty-five (25) days after receipt of both the referral and any response or upon being informed that there will be no response (unless the lead agency agrees to a longer time), the Council may take one or more of the following actions:

(1) Conclude that the process of referral and response has successfully resolved the problem.

(2) Initiate discussions with the agencies with the objective of mediation with referring and lead agencies.

(3) Hold public meetings or hearings to obtain additional views and information.

(4) Determine that the issue is not one of national importance and request the referring and lead agencies to pursue their decision process.

(5) Determine that the issue should be further negotiated by the referring and lead agencies and is not appropriate for Council consideration until one or more heads of agencies report to the Council that the agencies' disagreements are irreconcilable.

(6) Publish its findings and recommendations (including where appropriate a finding that the submitted evidence does not support the position of an agency).

(7) When appropriate, submit the referral and the response together with the Council's recommendation to the President for action.

(g) The Council shall take no longer than 60 days to complete the actions specified in paragraph (f)(2), (3), or (5) of this section.

(h) When the referral involves an action required by statute to be determined on the record after opportunity for agency hearing, the referral shall be conducted in a manner consistent with 5 U.S.C. 557(d) (Administrative Procedure Act).

[43 FR 55998, Nov. 29, 1978; 44 FR 873, Jan. 3, 1979]

PART 1505—NEPA AND AGENCY DECISIONMAKING

Sec.

1505.1 Agency decisionmaking procedures.

1505.2 Record of decision in cases requiring environmental impact statements.

1505.3 Implementing the decision.

AUTHORITY: NEPA, the Environmental Quality Improvement Act of 1970, as amended (42 U.S.C. 4371 *et seq.*), sec. 309 of the Clean Air Act, as amended (42 U.S.C. 7609), and E.O. 11514 (Mar. 5, 1970, as amended by E.O. 11991, May 24, 1977).

SOURCE: 43 FR 55999, Nov. 29, 1978, unless otherwise noted.

§1505.1 Agency decisionmaking procedures.

Agencies shall adopt procedures (§1507.3) to ensure that decisions are made in accordance with the policies and purposes of the Act. Such procedures shall include but not be limited to:

(a) Implementing procedures under section 102(2) to achieve the requirements of sections 101 and 102(1).
(b) Designating the major decision points for the agency's principal programs likely to have a significant effect on the human environment and assuring that the NEPA process corresponds with them.
(c) Requiring that relevant environmental documents, comments, and responses be part of the record in formal rulemaking or adjudicatory proceedings.
(d) Requiring that relevant environmental documents, comments, and responses accompany the proposal through existing agency review processes so that agency officials use the statement in making decisions.
(e) Requiring that the alternatives considered by the decisionmaker are encompassed by the range of alternatives discussed in the relevant environmental documents and that the decisionmaker consider the alternatives described in the environmental impact statement. If another decision document accompanies the relevant environmental documents to the decisionmaker, agencies are encouraged to make available to the public before the decision is made any part of that document that relates to the comparison of alternatives.

§1505.2 Record of decision in cases requiring environmental impact statements.

At the time of its decision (§1506.10) or, if appropriate, its recommendation to Congress, each agency shall prepare a concise public record of decision. The record, which may be integrated into any other record prepared by the agency, including that required by OMB Circular A-95 (Revised), part I, sections 6(c) and (d), and part II, section 5(b)(4), shall:

(a) State what the decision was.
(b) Identify all alternatives considered by the agency in reaching its decision, specifying the alternative or alternatives which were considered to be environmentally preferable. An agency may discuss preferences among alternatives based on relevant factors including economic and technical considerations and agency statutory missions. An agency shall identify and discuss all such factors including any essential considerations of national policy which were balanced

by the agency in making its decision and state how those considerations entered into its decision.

(c) State whether all practicable means to avoid or minimize environmental harm from the alternative selected have been adopted, and if not, why they were not. A monitoring and enforcement program shall be adopted and summarized where applicable for any mitigation.

§1505.3 Implementing the decision.

Agencies may provide for monitoring to assure that their decisions are carried out and should do so in important cases. Mitigation (§1505.2(c)) and other conditions established in the environmental impact statement or during its review and committed as part of the decision shall be implemented by the lead agency or other appropriate consenting agency. The lead agency shall:

(a) Include appropriate conditions in grants, permits or other approvals.
(b) Condition funding of actions on mitigation.
(c) Upon request, inform cooperating or commenting agencies on progress in carrying out mitigation measures which they have proposed and which were adopted by the agency making the decision.
(d) Upon request, make available to the public the results of relevant monitoring.

PART 1506—OTHER REQUIREMENTS OF NEPA

Sec.
1506.1 Limitations on actions during NEPA process.
1506.2 Elimination of duplication with state and local procedures.
1506.3 Adoption.
1506.4 Combining documents.
1506.5 Agency responsibility.
1506.6 Public involvement.
1506.7 Further guidance.
1506.8 Proposals for legislation.
1506.9 Filing requirements.
1506.10 Timing of agency action.
1506.11 Emergencies.
1506.12 Effective date.

AUTHORITY: NEPA, the Environmental Quality Improvement Act of 1970, as amended (42 U.S.C. 4371 *et seq.*), sec. 309 of the Clean Air Act, as amended (42 U.S.C. 7609), and E.O. 11514 (Mar. 5, 1970, as amended by E.O. 11991, May 24, 1977).
SOURCE: 43 FR 56000, Nov. 29, 1978, unless otherwise noted.

§1506.1 Limitations on actions during NEPA process.

(a) Until an agency issues a record of decision as provided in §1505.2 (except as provided in paragraph (c) of this section), no action concerning the proposal shall be taken which would:
   (1) Have an adverse environmental impact; or
   (2) Limit the choice of reasonable alternatives.

(b) If any agency is considering an application from a non-federal entity, and is aware that the applicant is about to take an action within the agency's jurisdiction that would meet either of the criteria in paragraph (a) of this section, then the agency shall promptly notify the applicant that the agency will take appropriate action to insure that the objectives and procedures of NEPA are achieved.

(c) While work on a required program environmental impact statement is in progress and the action is not covered by an existing program statement, agencies shall not undertake in the interim any major federal action covered by the program which may significantly affect the quality of the human environment unless such action:
   (1) Is justified independently of the program;
   (2) Is itself accompanied by an adequate environmental impact statement; and
   (3) Will not prejudice the ultimate decision on the program. Interim action prejudices the ultimate decision on the program when it tends to determine subsequent development or limit alternatives.

(d) This section does not preclude development by applicants of plans or designs or performance of other work necessary to support an application for federal, state or local permits or assistance. Nothing in this section shall preclude Rural Electrification Administration approval of minimal expenditures not affecting the environment (e.g. long leadtime equipment and purchase options) made by non-governmental entities seeking loan guarantees from the Administration.

§1506.2 Elimination of duplication with state and local procedures.

(a) Agencies authorized by law to cooperate with state agencies of statewide jurisdiction pursuant to section 102(2)(D) of the Act may do so.

(b) Agencies shall cooperate with state and local agencies to the fullest extent possible to reduce duplication between NEPA and state and local requirements, unless the agencies are specifically barred from doing so by some other law. Except for cases covered by paragraph (a) of this section, such cooperation shall to the fullest extent possible include:
   (1) Joint planning processes.
   (2) Joint environmental research and studies.

(3) Joint public hearings (except where otherwise provided by statute).

(4) Joint environmental assessments.

(c) Agencies shall cooperate with state and local agencies to the fullest extent possible to reduce duplication between NEPA and comparable State and local requirements, unless the agencies are specifically barred from doing so by some other law. Except for cases covered by paragraph (a) of this section, such cooperation shall to the fullest extent possible include joint environmental impact statements. In such cases one or more federal agencies and one or more state or local agencies shall be joint lead agencies. Where state laws or local ordinances have environmental impact statement requirements in addition to but not in conflict with those in NEPA, federal agencies shall cooperate in fulfilling these requirements as well as those of federal laws so that one document will comply with all applicable laws.

(d) To better integrate environmental impact statements into state or local planning processes, statements shall discuss any inconsistency of a proposed action with any approved state or local plan and laws (whether or not federally sanctioned). Where an inconsistency exists, the statement should describe the extent to which the agency would reconcile its proposed action with the plan or law.

§1506.3 Adoption.

(a) An agency may adopt a federal draft or final environmental impact statement or portion thereof provided that the statement or portion thereof meets the standards for an adequate statement under these regulations.

(b) If the actions covered by the original environmental impact statement and the proposed action are substantially the same, the agency adopting another agency's statement is not required to recirculate it except as a final statement. Otherwise the adopting agency shall treat the statement as a draft and recirculate it (except as provided in paragraph (c) of this section).

(c) A cooperating agency may adopt without recirculating the environmental impact statement of a lead agency when, after an independent review of the statement, the cooperating agency concludes that its comments and suggestions have been satisfied.

(d) When an agency adopts a statement which is not final within the agency that prepared it, or when the action it assesses is the subject of a referral under part 1504, or when the statement's adequacy is the subject of a judicial action which is not final, the agency shall so specify.

§1506.4 Combining documents.

Any environmental document in compliance with NEPA may be combined with any other agency document to reduce duplication and paperwork.

§1506.5 Agency responsibility.

(a) Information. If an agency requires an applicant to submit environmental information for possible use by the agency in preparing an environmental impact statement, then the agency should assist the applicant by outlining the types of information required. The agency shall independently evaluate the information submitted and shall be responsible for its accuracy. If the agency chooses to use the information submitted by the applicant in the environmental impact statement, either directly or by reference, then the names of the persons responsible for the independent evaluation shall be included in the list of preparers (§1502.17). It is the intent of this paragraph that acceptable work not be redone, but that it be verified by the agency.

(b) Environmental assessments. If an agency permits an applicant to prepare an environmental assessment, the agency, besides fulfilling the requirements of paragraph (a) of this section, shall make its own evaluation of the environmental issues and take responsibility for the scope and content of the environmental assessment.

(c) Environmental impact statements. Except as provided in §§1506.2 and 1506.3 any environmental impact statement prepared pursuant to the requirements of NEPA shall be prepared directly by or by a contractor selected by the lead agency or where appropriate under §1501.6(b), a cooperating agency. It is the intent of these regulations that the contractor be chosen solely by the lead agency, or by the lead agency in cooperation with cooperating agencies, or where appropriate by a cooperating agency to avoid any conflict of interest. Contractors shall execute a disclosure statement prepared by the lead agency, or where appropriate the cooperating agency, specifying that they have no financial or other interest in the outcome of the project. If the document is prepared by contract, the responsible federal official shall furnish guidance and participate in the preparation and shall independently evaluate the statement prior to its approval and take responsibility for its scope and contents. Nothing in this section is intended to prohibit any agency from requesting any person to submit information to it or to prohibit any person from submitting information to any agency.

§1506.6 Public involvement.

Agencies shall:

(a) Make diligent efforts to involve the public in preparing and implementing their NEPA procedures.
(b) Provide public notice of NEPA-related hearings, public meetings, and the availability of environmental documents so as to inform those persons and agencies who may be interested or affected.
  (1) In all cases the agency shall mail notice to those who have requested it on an individual action.
  (2) In the case of an action with effects of national concern notice shall include publication in the FEDERAL REGISTER and notice by mail to national organizations reasonably expected to be interested in the matter and may include listing in the 102 Monitor. An agency engaged in rule-making may provide notice by mail to national organizations who have requested that notice regularly be provided. Agencies shall maintain a list of such organizations.
  (3) In the case of an action with effects primarily of local concern the notice may include:
    (i) Notice to state and areawide clearinghouses pursuant to OMB Circular A-95 (Revised).
    (ii) Notice to Indian tribes when effects may occur on reservations.
    (iii) Following the affected state's public notice procedures for comparable actions.
    (iv) Publication in local newspapers (in papers of general circulation rather than legal papers).
    (v) Notice through other local media.
    (vi) Notice to potentially interested community organizations including small business associations.
    (vii) Publication in newsletters that may be expected to reach potentially interested persons.
    (viii) Direct mailing to owners and occupants of nearby or affected property.
    (ix) Posting of notice on and off site in the area where the action is to be located.
(c) Hold or sponsor public hearings or public meetings whenever appropriate or in accordance with statutory requirements applicable to the agency. Criteria shall include whether there is:
  (1) Substantial environmental controversy concerning the proposed action or substantial interest in holding the hearing.

(2) A request for a hearing by another agency with jurisdiction over the action supported by reasons why a hearing will be helpful. If a draft environmental impact statement is to be considered at a public hearing, the agency should make the statement available to the public at least 15 days in advance (unless the purpose of the hearing is to provide information for the draft environmental impact statement).

(d) Solicit appropriate information from the public.

(e) Explain in its procedures where interested persons can get information or status reports on environmental impact statements and other elements of the NEPA process.

(f) Make environmental impact statements, the comments received, and any underlying documents available to the public pursuant to the provisions of the Freedom of Information Act (5 U.S.C. 552), without regard to the exclusion for interagency memoranda where such memoranda transmit comments of Federal agencies on the environmental impact of the proposed action. Materials to be made available to the public shall be provided to the public without charge to the extent practicable, or at a fee which is not more than the actual costs of reproducing copies required to be sent to other federal agencies, including the Council.

§1506.7 Further guidance.

The Council may provide further guidance concerning NEPA and its procedures including:

(a) A handbook which the Council may supplement from time to time, which shall in plain language provide guidance and instructions concerning the application of NEPA and these regulations.

(b) Publication of the Council's Memoranda to Heads of Agencies.

(c) In conjunction with the Environmental Protection Agency and the publication of the 102 Monitor, notice of:

(1) Research activities;

(2) Meetings and conferences related to NEPA; and

(3) Successful and innovative procedures used by agencies to implement NEPA.

§1506.8 Proposals for legislation.

(a) The NEPA process for proposals for legislation (§1508.17) significantly affecting the quality of the human environment shall be integrated with the legislative process of the Congress. A legislative environmental impact statement is the detailed statement required by law to be included in a recommendation or report on a legislative proposal to Congress. A legislative environmental impact statement shall be considered part of the formal transmittal of a legislative proposal to Congress; however, it may be transmitted to Congress up to 30 days later in order to allow time for completion of an accurate statement which can serve as the basis for public and Congressional debate. The statement must be available in time for Congressional hearings and deliberations.

(b) Preparation of a legislative environmental impact statement shall conform to the requirements of these regulations except as follows:
  (1) There need not be a scoping process.
  (2) The legislative statement shall be prepared in the same manner as a draft statement, but shall be considered the "detailed statement" required by statute; Provided, That when any of the following conditions exist both the draft and final environmental impact statement on the legislative proposal shall be prepared and circulated as provided by §§1503.1 and 1506.10.
    (i) A Congressional committee with jurisdiction over the proposal has a rule requiring both draft and final environmental impact statements.
    (ii) The proposal results from a study process required by statute (such as those required by the Wild and Scenic Rivers Act (16 U.S.C. 1271 *et seq.*) and the Wilderness Act (16 U.S.C. 1131 *et seq.*)).
    (iii) Legislative approval is sought for federal or federally assisted construction or other projects which the agency recommends be located at specific geographic locations. For proposals requiring an environmental impact statement for the acquisition of space by the General Services Administration, a draft statement shall accompany the Prospectus or the 11(b) Report of Building Project Surveys to the Congress, and a final statement shall be completed before site acquisition.
    (iv) The agency decides to prepare draft and final statements.

(c) Comments on the legislative statement shall be given to the lead agency which shall forward them along with its own responses to the Congressional committees with jurisdiction.

§1506.9 Filing requirements.

Environmental impact statements together with comments and responses shall be filed with the Environmental Protection Agency, attention Office of Federal Activities (MC2252-A), 1200 Pennsylvania Ave., NW, Washington, DC 20460. Statements shall be filed with EPA no earlier than they are also transmitted to commenting agencies and made available to the public. EPA shall deliver one copy of each statement to the Council, which shall satisfy the requirement of availability to the President. EPA may issue guidelines to agencies to implement its responsibilities under this section and §1506.10.

§1506.10 Timing of agency action.

(a) The Environmental Protection Agency shall publish a notice in the FEDERAL REGISTER each week of the environmental impact statements filed during the preceding week. The minimum time periods set forth in this section shall be calculated from the date of publication of this notice.

(b) No decision on the proposed action shall be made or recorded under §1505.2 by a federal agency until the later of the following dates:
   (1) Ninety (90) days after publication of the notice described above in paragraph (a) of this section for a draft environmental impact statement.
   (2) Thirty (30) days after publication of the notice described above in paragraph (a) of this section for a final environmental impact statement.

An exception to the rules on timing may be made in the case of an agency decision which is subject to a formal internal appeal. Some agencies have a formally established appeal process which allows other agencies or the public to take appeals on a decision and make their views known, after publication of the final environmental impact statement. In such cases, where a real opportunity exists to alter the decision, the decision may be made and recorded at the same time the environmental impact statement is published. This means that the period for appeal of the decision and the 30-day period prescribed in paragraph (b)(2) of this section may run concurrently. In such cases the environmental impact statement shall explain the timing and the public's right of appeal. An agency engaged in rulemaking under the Administrative Procedure Act or other statute for the purpose of protecting the public health or safety, may waive the time period in paragraph (b)(2) of this section and publish a decision on the final rule simultaneously with publication of the notice of the availability of the final environmental impact statement as described in paragraph (a) of this section.

(c) If the final environmental impact statement is filed within ninety (90) days after a draft environmental impact statement is filed with the Environmental Protection Agency, the minimum thirty (30) day period and the minimum ninety (90) day period may run concurrently. However, subject to paragraph (d) of this section agencies shall allow not less than 45 days for comments on draft statements.

(d) The lead agency may extend prescribed periods. The Environmental Protection Agency may upon a showing by the lead agency of compelling reasons of national policy reduce the prescribed periods and may upon a showing by any other Federal agency of compelling reasons of national policy also extend prescribed periods, but only after consultation with the lead agency. (Also see §1507.3(d).) Failure to file timely comments shall not be a sufficient reason for extending a period. If the lead agency does not concur with the extension of time, EPA may not extend it for more than 30 days. When the Environmental Protection Agency reduces or extends any period of time it shall notify the Council.

[43 FR 56000, Nov. 29, 1978; 44 FR 874, Jan. 3, 1979]

§1506.11 Emergencies.

Where emergency circumstances make it necessary to take an action with significant environmental impact without observing the provisions of these regulations, the federal agency taking the action should consult with the Council about alternative arrangements. Agencies and the Council will limit such arrangements to actions necessary to control the immediate impacts of the emergency. Other actions remain subject to NEPA review.

§1506.12 Effective date.

The effective date of these regulations is July 30, 1979, except that for agencies that administer programs that qualify under section 102(2)(D) of the Act or under section 104(h) of the Housing and Community Development Act of 1974 an additional four months shall be allowed for the state or local agencies to adopt their implementing procedures.

(a) These regulations shall apply to the fullest extent practicable to ongoing activities and environmental documents begun before the effective date. These regulations do not apply to an environmental impact statement or supplement if the draft statement was filed before the effective date of these regulations. No completed environmental documents need be redone by reasons of these regulations. Until these regulations are applicable, the Council's guidelines published in the FEDERAL REGISTER of August 1, 1973, shall continue to be applicable. In cases where these regulations are applicable the guidelines are superseded. However, nothing shall prevent an agency from proceeding under these regulations at an earlier time.

(b) NEPA shall continue to be applicable to actions begun before January 1, 1970, to the fullest extent possible.

PART 1507—AGENCY COMPLIANCE

Sec.
1507.1 Compliance.
1507.2 Agency capability to comply.
1507.3 Agency procedures.

AUTHORITY: NEPA, the Environmental Quality Improvement Act of 1970, as amended (42 U.S.C. 4371 *et seq.*), sec. 309 of the Clean Air Act, as amended (42 U.S.C. 7609), and E.O. 11514 (Mar. 5, 1970, as amended by E.O. 11991, May 24, 1977).
SOURCE: 43 FR 56002, Nov. 29, 1978, unless otherwise noted.

§1507.1 Compliance.

All agencies of the federal government shall comply with these regulations. It is the intent of these regulations to allow each agency flexibility in adapting its implementing procedures authorized by §1507.3 to the requirements of other applicable laws.

§1507.2 Agency capability to comply.

Each agency shall be capable (in terms of personnel and other resources) of complying with the requirements enumerated below. Such compliance may include use of other's resources, but the using agency shall itself have sufficient capability to evaluate what others do for it. Agencies shall:

(a) Fulfill the requirements of section 102(2)(A) of the Act to utilize a systematic, interdisciplinary approach which will insure the integrated use of the natural and social sciences and the environmental design arts in planning and in decisionmaking which may have an impact on the human environment. Agencies shall designate a person to be responsible for overall review of agency NEPA compliance.
(b) Identify methods and procedures required by section 102(2)(B) to insure that presently unquantified environmental amenities and values may be given appropriate consideration.
(c) Prepare adequate environmental impact statements pursuant to section 102(2)(C) and comment on statements in the areas where the agency has jurisdiction by law or special expertise or is authorized to develop and enforce environmental standards.
(d) Study, develop, and describe alternatives to recommended courses of action in any proposal which involves unresolved conflicts concerning alternative uses of available resources. This requirement of section 102(2)(E) extends to all such proposals, not just the more limited scope of section 102(2)(C)(iii) where the discussion of alternatives is confined to impact statements.

(e) Comply with the requirements of section 102(2)(H) that the agency initiate and utilize ecological information in the planning and development of resource-oriented projects.
(f) Fulfill the requirements of sections 102(2)(F), 102(2)(G), and 102(2)(I), of the Act and of Executive Order 11514, Protection and Enhancement of Environmental Quality, Sec. 2.

§1507.3 Agency procedures.

(a) Not later than eight months after publication of these regulations as finally adopted in the FEDERAL REGISTER, or five months after the establishment of an agency, whichever shall come later, each agency shall as necessary adopt procedures to supplement these regulations. When the agency is a department, major subunits are encouraged (with the consent of the department) to adopt their own procedures. Such procedures shall not paraphrase these regulations. They shall confine themselves to implementing procedures. Each agency shall consult with the Council while developing its procedures and before publishing them in the FEDERAL REGISTER for comment. Agencies with similar programs should consult with each other and the Council to coordinate their procedures, especially for programs requesting similar information from applicants. The procedures shall be adopted only after an opportunity for public review and after review by the Council for conformity with the Act and these regulations. The Council shall complete its review within 30 days. Once in effect they shall be filed with the Council and made readily available to the public. Agencies are encouraged to publish explanatory guidance for these regulations and their own procedures. Agencies shall continue to review their policies and procedures and in consultation with the Council to revise them as necessary to ensure full compliance with the purposes and provisions of the Act.
(b) Agency procedures shall comply with these regulations except where compliance would be inconsistent with statutory requirements and shall include:
    (1) Those procedures required by §§1501.2(d), 1502.9(c)(3), 1505.1, 1506.6(e), and 1508.4.
    (2) Specific criteria for and identification of those typical classes of action:
        (i) Which normally do require environmental impact statements.
        (ii) Which normally do not require either an environmental impact statement or an environmental assessment (categorical exclusions (§1508.4)).
        (iii) Which normally require environmental assessments but not necessarily environmental impact statements.

(c) Agency procedures may include specific criteria for providing limited exceptions to the provisions of these regulations for classified proposals. They are proposed actions which are specifically authorized under criteria established by an Executive Order or statute to be kept secret in the interest of national defense or foreign policy and are in fact properly classified pursuant to such Executive Order or statute. Environmental assessments and environmental impact statements which address classified proposals may be safeguarded and restricted from public dissemination in accordance with agencies' own regulations applicable to classified information. These documents may be organized so that classified portions can be included as annexes, in order that the unclassified portions can be made available to the public.

(d) Agency procedures may provide for periods of time other than those presented in §1506.10 when necessary to comply with other specific statutory requirements.

(e) Agency procedures may provide that where there is a lengthy period between the agency's decision to prepare an environmental impact statement and the time of actual preparation, the notice of intent required by §1501.7 may be published at a reasonable time in advance of preparation of the draft statement.

PART 1508—TERMINOLOGY AND INDEX

Sec.

AUTHORITY: NEPA, the Environmental Quality Improvement Act of 1970, as amended (42 U.S.C. 4371 *et seq.*), sec. 309 of the Clean Air Act, as amended (42 U.S.C. 7609), and E.O. 11514 (Mar. 5, 1970, as amended by E.O. 11991, May 24, 1977).
SOURCE: 43 FR 56003, Nov. 29, 1978, unless otherwise noted.

§1508.1 Terminology.

The terminology of this part shall be uniform throughout the federal government.

§1508.2 Act.

"Act" means the National Environmental Policy Act, as amended (42 U.S.C. 4321, *et seq.*) which is also referred to as "NEPA."

§1508.3 Affecting.

"Affecting" means will or may have an effect on.

§1508.4 Categorical exclusion.

"Categorical exclusion" means a category of actions which do not individually or cumulatively have a significant effect on the human environment and which have been found to have no such effect in procedures adopted by a federal agency in implementation of these regulations (§1507.3) and for which, therefore, neither an environmental assessment nor an environmental impact statement is required. An agency may decide in its procedures or otherwise, to prepare environmental assessments for the reasons stated in §1508.9 even though it is not required to do so. Any procedures under this section shall provide for extraordinary circumstances in which a normally excluded action may have a significant environmental effect.

§1508.5 Cooperating agency.

"Cooperating agency" means any federal agency other than a lead agency which has jurisdiction by law or special expertise with respect to any environmental impact involved in a proposal (or a reasonable alternative) for legislation or other major federal action significantly affecting the quality of the human environment. The selection and responsibilities of a cooperating agency are described in §1501.6. A state or local agency of similar qualifications or, when the effects are on a reservation, an Indian tribe, may by agreement with the lead agency become a cooperating agency.

§1508.6 Council.

"Council" means the Council on Environmental Quality established by title II of the Act.

§1508.7 Cumulative impact.

"Cumulative impact" is the impact on the environment which results from the incremental impact of the action when added to other past, present, and reasonably foreseeable future actions regardless of what agency (federal or non-federal) or person undertakes such other actions. Cumulative impacts can result from individually minor but collectively significant actions taking place over a period of time.

§1508.8 Effects.

"Effects" include:

(a) Direct effects, which are caused by the action and occur at the same time and place.
(b) Indirect effects, which are caused by the action and are later in time or farther removed in distance, but are still reasonably foreseeable. Indirect effects may include growth inducing effects and other effects related to induced changes in the pattern of land use, population density or growth rate, and related effects on air and water and other natural systems, including ecosystems.

Effects and impacts as used in these regulations are synonymous. Effects includes ecological (such as the effects on natural resources and on the components, structures, and functioning of affected ecosystems), aesthetic, historic, cultural, economic, social, or health, whether direct, indirect, or cumulative. Effects may also include those resulting from actions which may have both beneficial and detrimental effects, even if on balance the agency believes that the effect will be beneficial.

§1508.9 Environmental assessment.

"Environmental assessment":

(a) Means a concise public document for which a federal agency is responsible that serves to:
  (1) Briefly provide sufficient evidence and analysis for determining whether to prepare an environmental impact statement or a finding of no significant impact.
  (2) Aid an agency's compliance with the Act when no environmental impact statement is necessary.
  (3) Facilitate preparation of a statement when one is necessary.

(b) Shall include brief discussions of the need for the proposal, of alternatives as required by section 102(2)(E), of the environmental impacts of the proposed action and alternatives, and a listing of agencies and persons consulted.

§1508.10 Environmental document.

"Environmental document" includes the documents specified in §1508.9 (environmental assessment), §1508.11 (environmental impact statement), §1508.13 (finding of no significant impact), and §1508.22 (notice of intent).

§1508.11 Environmental impact statement.

"Environmental impact statement" means a detailed written statement as required by section 102(2)(C) of the Act.

§1508.12 Federal agency.

"Federal agency" means all agencies of the federal government. It does not mean the Congress, the Judiciary, or the President, including the performance of staff functions for the President in his Executive Office. It also includes for purposes of these regulations states and units of general local government and Indian tribes assuming NEPA responsibilities under section 104(h) of the Housing and Community Development Act of 1974.

§1508.13 Finding of no significant impact.

"Finding of no significant impact" means a document by a federal agency briefly presenting the reasons why an action, not otherwise excluded

(§1508.4), will not have a significant effect on the human environment and for which an environmental impact statement therefore will not be prepared. It shall include the environmental assessment or a summary of it and shall note any other environmental documents related to it (§1501.7(a)(5)). If the assessment is included, the finding need not repeat any of the discussion in the assessment but may incorporate it by reference.

§1508.14 Human environment.

"Human environment" shall be interpreted comprehensively to include the natural and physical environment and the relationship of people with that environment. (See the definition of "effects" (§1508.8).) This means that economic or social effects are not intended by themselves to require preparation of an environmental impact statement. When an environmental impact statement is prepared and economic or social and natural or physical environmental effects are interrelated, then the environmental impact statement will discuss all of these effects on the human environment.

§1508.15 Jurisdiction by law.

"Jurisdiction by law" means agency authority to approve, veto, or finance all or part of the proposal.

§1508.16 Lead agency.

"Lead agency" means the agency or agencies preparing or having taken primary responsibility for preparing the environmental impact statement.

§1508.17 Legislation.

"Legislation" includes a bill or legislative proposal to Congress developed by or with the significant cooperation and support of a federal agency, but does not include requests for appropriations. The test for significant cooperation is whether the proposal is in fact predominantly that of the agency rather than another source. Drafting does not by itself constitute significant cooperation. Proposals for legislation include requests for ratification of treaties. Only the agency which has primary responsibility for the subject matter involved will prepare a legislative environmental impact statement.

§1508.18 Major federal action.

"Major federal action" includes actions with effects that may be major and which are potentially subject to federal control and responsibility. Major reinforces but does not have a meaning independent of significantly (§1508.27). Actions include the circumstance where the responsible officials fail to act and that failure to act is reviewable by courts or administrative tribunals under the Administrative Procedure Act or other applicable law as agency action.

(a) Actions include new and continuing activities, including projects and programs entirely or partly financed, assisted, conducted, regulated, or approved by federal agencies; new or revised agency rules, regulations, plans, policies, or procedures; and legislative proposals (§§1506.8, 1508.17). Actions do not include funding assistance solely in the form of general revenue sharing funds, distributed under the State and Local Fiscal Assistance Act of 1972, 31 U.S.C. 1221 *et seq.*, with no federal agency control over the subsequent use of such funds. Actions do not include bringing judicial or administrative civil or criminal enforcement actions.

(b) Federal actions tend to fall within one of the following categories:

  (1) Adoption of official policy, such as rules, regulations, and interpretations adopted pursuant to the Administrative Procedure Act, 5 U.S.C. 551 *et seq.*; treaties and international conventions or agreements; formal documents establishing an agency's policies which will result in or substantially alter agency programs.

  (2) Adoption of formal plans, such as official documents prepared or approved by federal agencies which guide or prescribe alternative uses of federal resources, upon which future agency actions will be based.

  (3) Adoption of programs, such as a group of concerted actions to implement a specific policy or plan; systematic and connected agency decisions allocating agency resources to implement a specific statutory program or executive directive.

  (4) Approval of specific projects, such as construction or management activities located in a defined geographic area. Projects include actions approved by permit or other regulatory decision as well as federal and federally assisted activities.

§1508.19 Matter.

"Matter" includes for purposes of part 1504:

(a) With respect to the Environmental Protection Agency, any proposed legislation, project, action or regulation as those terms are used in section 309(a) of the Clean Air Act (42 U.S.C. 7609).

(b) With respect to all other agencies, any proposed major federal action to which section 102(2)(C) of NEPA applies.

§1508.20 Mitigation.

"Mitigation" includes:

(a) Avoiding the impact altogether by not taking a certain action or parts of an action.
(b) Minimizing impacts by limiting the degree or magnitude of the action and its implementation.
(c) Rectifying the impact by repairing, rehabilitating, or restoring the affected environment.
(d) Reducing or eliminating the impact over time by preservation and maintenance operations during the life of the action.
(e) Compensating for the impact by replacing or providing substitute resources or environments.

§1508.21 NEPA process.

"NEPA process" means all measures necessary for compliance with the requirements of section 2 and title I of NEPA.

§1508.22 Notice of intent.

"Notice of intent" means a notice that an environmental impact statement will be prepared and considered. The notice shall briefly:

(a) Describe the proposed action and possible alternatives.
(b) Describe the agency's proposed scoping process including whether, when, and where any scoping meeting will be held.
(c) State the name and address of a person within the agency who can answer questions about the proposed action and the environmental impact statement.

§1508.23 Proposal.

"Proposal" exists at that stage in the development of an action when an agency subject to the Act has a goal and is actively preparing to make a decision on one or more alternative means of accomplishing that goal and the effects can be meaningfully evaluated. Preparation of an environmental impact statement on a proposal should be timed (§1502.5) so that the final statement may be completed in time for the statement to be included in any recommendation or report on the proposal. A proposal may exist in fact as well as by agency declaration that one exists.

§1508.24 Referring agency.

"Referring agency" means the federal agency which has referred any matter to the Council after a determination that the matter is unsatisfactory from the standpoint of public health or welfare or environmental quality.

§1508.25 Scope.

"Scope" consists of the range of actions, alternatives, and impacts to be considered in an environmental impact statement. The scope of an individual statement may depend on its relationships to other statements (§§1502.20 and 1508.28). To determine the scope of environmental impact statements, agencies shall consider 3 types of actions, 3 types of alternatives, and 3 types of impacts. They include:

- (a) Actions (other than unconnected single actions) which may be:
  - (1) Connected actions, which means that they are closely related and therefore should be discussed in the same impact statement. Actions are connected if they:
    - (i) Automatically trigger other actions which may require environmental impact statements.
    - (ii) Cannot or will not proceed unless other actions are taken previously or simultaneously.
    - (iii) Are interdependent parts of a larger action and depend on the larger action for their justification.
  - (2) Cumulative actions, which when viewed with other proposed actions have cumulatively significant impacts and should therefore be discussed in the same impact statement.
  - (3) Similar actions, which when viewed with other reasonably foreseeable or proposed agency actions, have similarities that provide a basis for evaluating their environmental consequences together, such as common timing or geography. An agency may wish to analyze these actions in the same impact statement. It should do so when the best way to assess adequately the combined impacts of similar actions or reasonable alternatives to such actions is to treat them in a single impact statement.
- (b) Alternatives, which include:
  - (1) No action alternative.
  - (2) Other reasonable courses of actions.
  - (3) Mitigation measures (not in the proposed action).
- (c) Impacts, which may be: (1) direct; (2) indirect; (3) cumulative.

§1508.26 Special expertise.

"Special expertise" means statutory responsibility, agency mission, or related program experience.

§1508.27 Significantly.

"Significantly" as used in NEPA requires considerations of both context and intensity:

(a) Context. This means that the significance of an action must be analyzed in several contexts such as society as a whole (human, national), the affected region, the affected interests, and the locality. Significance varies with the setting of the proposed action. For instance, in the case of a site-specific action, significance would usually depend upon the effects in the locale rather than in the world as a whole. Both short and long-term effects are relevant.

(b) Intensity. This refers to the severity of impact. Responsible officials must bear in mind that more than one agency may make decisions about partial aspects of a major action. The following should be considered in evaluating intensity:

  (1) Impacts that may be both beneficial and adverse. A significant effect may exist even if the federal agency believes that on balance the effect will be beneficial.
  (2) The degree to which the proposed action affects public health or safety.
  (3) Unique characteristics of the geographic area such as proximity to historic or cultural resources, park lands, prime farmlands, wetlands, wild and scenic rivers, or ecologically critical areas.
  (4) The degree to which the effects on the quality of the human environment are likely to be highly controversial.
  (5) The degree to which the possible effects on the human environment are highly uncertain or involve unique or unknown risks.
  (6) The degree to which the action may establish a precedent for future actions with significant effects or represents a decision in principle about a future consideration.
  (7) Whether the action is related to other actions with individually insignificant but cumulatively significant impacts. Significance exists if it is reasonable to anticipate a cumulatively significant impact on the environment. Significance cannot be avoided by terming an action temporary or by breaking it down into small component parts.
  (8) The degree to which the action may adversely affect districts, sites, highways, structures, or objects listed in or eligible for listing in the

National Register of Historic Places or may cause loss or destruction of significant scientific, cultural, or historical resources.

(9) The degree to which the action may adversely affect an endangered or threatened species or its habitat that has been determined to be critical under the Endangered Species Act of 1973.

(10) Whether the action threatens a violation of federal, state, or local law or requirements imposed for the protection of the environment.

[43 FR 56003, Nov. 29, 1978; 44 FR 874, Jan. 3, 1979]

§1508.28 Tiering.

"Tiering" refers to the coverage of general matters in broader environmental impact statements (such as national program or policy statements) with subsequent narrower statements or environmental analyses (such as regional or basinwide program statements or ultimately site-specific statements) incorporating by reference the general discussions and concentrating solely on the issues specific to the statement subsequently prepared. Tiering is appropriate when the sequence of statements or analyses is:

(a) From a program, plan, or policy environmental impact statement to a program, plan, or policy statement or analysis of lesser scope or to a site-specific statement or analysis.

(b) From an environmental impact statement on a specific action at an early stage (such as need and site selection) to a supplement (which is preferred) or a subsequent statement or analysis at a later stage (such as environmental mitigation). Tiering in such cases is appropriate when it helps the lead agency to focus on the issues which are ripe for decision and exclude from consideration issues already decided or not yet ripe.

Index to Parts 1500 Through 1508

Editorial Note: This listing is provided for information purposes only. It is compiled and kept up-to-date by the Council on Environmental Quality.

Appendix C
Environmental impact statement checklists

A set of check-off lists are provided in Tables C.1 through C.21 to assist the reader in preparing and reviewing an environmental impact statement (EIS) for adequacy.[1] Not all questions will apply in all circumstances. Professional experience must be exercised in responding to each question. The reader should note that these lists address many, but not necessarily all, of the regulatory requirements that an EIS must meet. Each and every proposal also has specific nuisances that must be addressed individually. As such, these checklists should not be relied upon as the sole method for ensuring quality and compliance with regulatory requirements. Prudence should be exercised in applying the checklists as

- No checklist can be prepared that is universally applicable to all circumstances.
- The checklists cannot guarantee that the EIS will be adequate or in full compliance with NEPA and other related regulations, guidance or laws.

C.1 Acronyms used in the EIS Checklists

CEQ President's Council on Environmental Quality
CERCLA Comprehensive Environmental Response, Compensation, and Liability Act
CFR United States Code of Federal Regulations
EA environmental assessment
EIS environmental impact statement
EO executive order
FONSI finding of no significant impact
FR Federal Register

N/A not applicable
NEPA National Environmental Policy Act
NMFS United States National Marine Fisheries Service
RCRA Resource Conservation and Recovery Act
SHPO State Historic Preservation Officer
US United States
U.S.C. United States Code
USFWS United States Fish and Wildlife Service

COVER SHEET					
	Yes	No	N/A	EIS Page Num.	Adequacy Evaluation and Comments
1.1 Does the cover sheet include: – A list of responsible agencies, including the lead agency and any cooperating agencies:					
– The title of the proposal and its location (state[s], other jurisdiction)?					
– The name(s), address(es), and telephone number(s) of a person (or persons) to contact for further information (on the general NEPA process or on the specific EIS)?					
– The EIS designation as draft, final, or supplemental?					
– A one-paragraph abstract of the EIS?					
– The date (for a draft EIS) by which comments must be received? [40 CFR 1502.11]					
1.2 Is the cover sheet one page length? [40 CFR 1502.11]					

Table C.1 Summary

	Yes	No	N/A	EIS Page Num.	Adequacy Evaluation and Comments
1.1 Does the summary describe the underlying purpose and need for agency action? – The proposed action?					
– Each or the alternatives?					
– The preferred alternative, if any?					
– The principal environmental issues analyzed and the results?					
1.2 Does the summary highlight key differences among the alternatives?					
1.3 Does the summary stress – The major conclusions?					
– Areas of controversy (including issues raised by agencies and the public)?					
– The issues to be resolved (including the choice among alternatives)? [40 CFR 1502.12]					
1.4 Are the discussions in the summary consistent with the EIS text and appendices?					
1.5 Does the summary adequately and accurately summarize the EIS? [40 CFR 1502.12]					

Source: Eccleston C.H., The EIS Book, CRC Press (2013).

Table C.2 Purpose and Need for Taking Action

	Yes	No	N/A	EIS Page Num.	Adequacy Evaluation and Comments
2.1 Does the EIS specify the underlying purpose and need to which the agency is responding in proposing the alternatives including the proposed action? [40 CFR 1502.13]					
2.2 Does the statement of purpose and need relate to the broad requirement or desire for action, and not to the need for one specific proposal or the need for the EIS?					
2.3 Does the statement of purpose and need adequately explain the problem or opportunity to which the agency is responding?					
2.4 Is the statement of purpose and need written so that it (a) does not inappropriately narrow the range of reasonable alternatives, nor (b) is too broadly defined as to make the number of alternatives virtually limitless?					

Source: Eccleston C.H., The EIS Book, CRC Press (2013).

Table C.3 Description of the Proposed Action and Alternatives

	Yes	No	N/A	EIS Page Num.	Adequacy Evaluation and Comments
3.1 Does the EIS clearly describe the proposed action and alternatives?					
3.2 Is the proposed action described in terms of the actions to be taken (even a private action that has been federalized or enabled by funding)?					
3.3 Does the proposed action exclude elements that are more appropriate to the statement of purpose and need?					
3.4 Does the EIS identify the range of reasonable alternatives that satisfy the agency's purpose and need?					
3.5 Does the EIS "rigorously explore and objectively evaluate" all reasonable alternatives that encompass the range to be considered by the decision maker? [40 CFR 1502.14(a)]					
3.6a For a draft EIS, does the document indicate whether a preferred alternative(s) exist, and, if so, is it identified? [40 CFR 1502.14(e)]					
3.4.6b For a final EIS , is the preferred alternative identified? [40 CFR 1502.14(e)]					
3.7 Does the EIS include the no-action alternative? [40 CFR 1502.14(d)]					
3.8 Is the no-action alternative described in sufficient detail so that its scope is clear and potential impacts can be identified?					

(*continued*)

Table C.3 (Continued) Description of the Proposed Action and Alternatives

	Yes	No	N/A	EIS Page Num.	Adequacy Evaluation and Comments
3.9 Does the non-action alternative include a discussion of the legal ramification of taking no action, if appropriate?					
3.10 As appropriate, does the EIS identify and analyze reasonable technology, transportation, and siting alternatives, including those that could occur offsite?					
3.11 Does the EIS include reasonable alternatives outside the agency's jurisdiction? [40 CFR 1502.14(c)]					
3.12 For alternatives that were eliminated from detailed study (including those that appear obvious) does the EIS fully and objectively explain why they were found to be unreasonable? [40 CFR 1502.14(a)]					
3.13 For each alternative analyzed in detail (including the no-action alternative), is the depth of analysis approximately the same, allowing reviewers to evaluate their comparative merits? [40 CFR 1502.14(b)]					
3.14 Are the proposed action/alternatives described in sufficient detail so that potential impacts can be identified?					
3.15 Are all phases of the proposed action/alternatives described (e.g., construction, operation, and post-operation/decommissioning)?					
3.16 Are environmental releases associated with the proposed action and alternatives quantified, including both the rates and durations?					
3.17 As appropriate, are mitigation measures included in the description of the proposed action and alternatives? [40 CFR 1502.14(f)]					

(continued)

Table C.3 (Continued) Description of the Proposed Action and Alternatives

	Yes	No	N/A	EIS Page Num.	Adequacy Evaluation and Comments
3.18 Are cost-effective waste minimization and pollution prevention activities included in the description of the proposed action and alternatives?					
3.19 As appropriate, are environmentally and economically beneficial landscape practices included in the description of the proposed action and alternatives?					
3.20 Are the descriptions of the proposed action and alternatives written broadly enough to encompass future modifications?					
3.21 Does the proposed action comply with CEQ regulations for interim actions? [40 CFR 1506.1]					
34.22 Does the EIS take into account relationships between the proposed action and other actions to be taken by the agency in order to avoid improper segmentation?					

Source: Eccleston C.H., The EIS Book, CRC Press (2013).

Table C.4 Description of the Affected Environment

	Yes	No	N/A	EIS Page Num.	Adequacy Evaluation and Comments
4.1 Does the EIS succinctly describe the environment of the area(s) to be affected or created by the proposed action and alternatives? [40 CFR 1502.15]					
4.2 Does the EIS identify either the presence or absence of the following within the area potentially affected by the proposed action and alternatives: – Floodplains? [EO 11988; 10 CFR 1022] – Wetlands? [EO 11990; 10 CFR 1022; 40 CFR 1508.27(b)(3)]					
– Threatened, endangered, or candidate species and/or their critical habitat, and other special status (e.g., state-listed) species? [16 U.S.C. 1531; 40 CFR 1508.27(b)(9)]					
– Prime or unique farmland? [7 U.S.C. 4201; 7 CFR 658; 40 CFR 1508.27(b)(3)]					
– State or national parks, forests, conservation areas, or other areas of recreational, ecological, scenic, or aesthetic importance? [40 CFR 1508.27(b)(3)]					
– Wild and scenic rivers? [16U.S.C. 1271; 40 CFR 1508.27(b)(3)]					
– Natural resources (e.g., timber, range, soils, minerals, fish, migratory birds, wildlife, water bodies, aquifers)? [40 CFR 1508.8]					
– Property of historic, archaeological, or architectural significance (including sites on or eligible for the National Registry of Natural Landmarks)? [OE 11593; 16 U.S.C. 470; 36 CFR 800; 40 CFR 1508.27(b)(3) and(8)]					
– Native Americans' concerns? [EO 13007; 25 U.S.C. 3001; 16 U.S.C. 470; 42 U.S.C. 1996]					
– Minority and low-income populations (including a description of their use and consumption of environmental resources)? [EO 12898]					

(continued)

Table C.4 (Continued) Description of the Affected Environment

	Yes	No	N/A	EIS Page Num.	Adequacy Evaluation and Comments
4.3 Does the description of the affected environment provide the necessary information to support the impact analysis, including cumulative impact analysis? [40 CFR 1502.15] 4.4 Are the descriptions of the affected environment substantially consistent with current baseline studies (e.g., descriptions of plant communities, wildlife habitat, and cultural resources)?					
4.5 Is the discussion appropriately limited to information that is directly related to the scope of the proposed action and alternatives? [40 CFR 1502.15]					
4.6 Does the EIS concentrate on important issues, avoiding useless bulk and verbose descriptions of the affected environment? [40 CFR 1502.15]					

Source: Eccleston C.H., The EIS Book, CRC Press (2013).

Table C.5 Environmental Effects

	Yes	No	N/A	EIS Page Num.	Adequacy Evaluation and Comments
5.1 Does the EIS adequately identify the direct and the indirect impacts of the proposed action/alternatives and discuss their significance? [40 CFR 1502.16(a) and (b)]					
5.2 Does the EIS adequately analyze both short-term and long-term effects?					
5.3 Does the EIS analyze both beneficial and adverse impacts? [40 CFR 1508.27(b)(1)]					
5.4 Does the EIS discuss reasonably foreseeable impacts of cumulative *actions* with regard to both the proposed action/alternatives? [40 CFR 1508.25(a)(2)]					
5.5 Does the EIS discuss the potential direct, indirect, and cumulative *effects* to the following, as identified in question 5.2: – Floodplains? [OE 11988; 10 CFR 1022]					
– Wetlands? [EO 11990; 10 CFR 1022; 40 CFR 1508.27(b)(3)]					
– Threatened, endangered, or candidate species and/or their critical habitat, and other special status (e.g., state-listed) species? [16 U.S.C. 1531; 40 CFR 1508.27(b)(9)]					
– Prime or unique farmland? [7 U.S.C. 4201; 7 CFR 658; 40 CFR 1508.27(b)(3)]					
– State or national parks, forests, conservation areas, or other areas of recreational, ecological, scenic, or aesthetic importance? [40 CFR 1508.27(b)(3)]					
– Wild and scenic rivers? [16 U.S.C. 1271; 40 CFR 1508.27(b)(3)]					
– Natural resources (e.g., timber, range, soils, minerals, fish, migratory birds, wildlife, water bodies, aquifers)? [40 CFR 1508.8]					

(continued)

Table C.5 (Continued) Environmental Effects

	Yes	No	N/A	EIS Page Num.	Adequacy Evaluation and Comments
– Property of historic, archaeological, or architectural significance (including sites on or eligible for the National Registry of Natural Landmarks)? [OE 11593; 16 U.S.C. 470; 36 CFR 800; 40 CFR 1508.27(b)(3) and (8)]					
– Native Americans' concerns? [OE 13007; 25 U.S.C. 3001; 16 U.S.C. 470; 42 U.S.C. 1996]					
– Minority and low-income populations to the extent that such effects are disproportionately high and adverse? [OE 12898]					
5.6 Does the EIS discuss: – Possible conflicts with land use plans, policies, or controls? [40 CFR 1502.16(c)					
– Energy requirements and conservation potential of various alternatives and mitigation measures? [40 CFR 1502.16(e)]					
– Natural or depletable resource requirements and conservation potential of the proposed action and alternatives? [40 CFR 1502.16(f)]					
– Urban quality, historic, and cultural resources, and the design of the built environment, including the reuse and conservation potential of the proposed action and alternatives? [40 CFR 1502.16(g)]					
– The means to mitigate adverse impacts? [40 CFR 1502.16(h)]					
5.7 Does the EIS discuss: – Any unavoidable, adverse environmental effects?					
– The relationship between short-term uses of the environment and long-term productivity?					
– Any irreversible or irretrievable commitments of resources? [40 CFR 1502.16]					

(*continued*)

Table C.5 (Continued) Environmental Effects

	Yes	No	N/A	EIS Page Num.	Adequacy Evaluation and Comments
5.8 Do the discussions of environmental impacts include (as appropriate):					
– Human health effects?					
– Effects of accidents?					
– Transportation effects?					
5.9 Does the EIS discuss the potential effects of released pollutants, rather than just identifying the releases?					
5.10 Does the EIS avoid presenting a description of severe impacts (e.g., from accidents), without also describing the likelihood/probability of such impacts occurring?					
5.11 Are the methodologies used for impact assessment generally accepted/recognized in the scientific community? [40 CFR 1502.22 and 1504.24]					
5.12 Does the EIS quantify environmental impacts where practical?					
5.13 Are impacts analyzed using a sliding-scale approach, as appropriate; i.e., proportional to their potential significance?					
5.14 Does the EIS avoid presenting bounding impact estimates that obscure differences among alternatives.					
5.15 Are sufficient data and references presented to allow validation of analysis methods and results?					
5.16a If information related to significant adverse effects is incomplete or unavailable, does the EIS state that such information is lacking?					
5.16b If this information is essential to a choice among alternatives and the cost of obtaining it is not exorbitant, is the information included?					
5.16c If this information cannot be obtained, does the EIS include: (1) a statement that the information is incomplete or unavailable, (2) the relevance of the information in evaluating significant effects, (3) a summary of credible scientific evidence, and (4) an evaluation based on theoretical approaches? [40 CFR 1502.22]					
5.17 As appropriate, does the EIS identify important sources of uncertainty in the analyses and conclusions?					

Source: Eccleston C.H., The EIS Book, CRC Press (2013).

Table C.6 Overall Considerations and Incorporation of Nepa Values

	Yes	No	N/A	EIS Page Num.	Adequacy Evaluation and Comments
6.1 Does the EIS identify all reasonably foreseeable impacts? [40 CFR 1508.8]					
6.2 Do the conclusions regarding potential impacts follow from the information and analyses presented in the EIS?					
6.3 Does the EIS avoid the implication that compliance with regulatory requirements demonstrates the absence of environmental effects?					
6.4 To the extent possible, does the EIS assess reasonable alternatives and identify measures to restore and enhance the environment and avoid or minimize potential adverse effects? [40 CFR 1500.2(f)]					
6.5 Does the EIS identify best management practices associated with the proposed action or with mitigation measures that would help avoid or minimize environmental disturbance, emissions, and other adverse effects?					
6.6 Does the EIS avoid (including the appearance) justifying decisions that have already been made? [40 CFR 1502.5]					
6.7 Are all assumptions conservative, and are the analyses and methodologies generally accepted/recognized by the scientific community? [40 CFR 1502.22 and 1502.24]					
6.8 Does the EIS indicate that the agency "has taken a 'hard look' at environmental consequences"? [*Kleppe V. Sierra Club*, 427 US 390, 410 (1976)]					
6.9 Does the EIS present the potential environmental effects of the proposal and the alternatives in comparative form, sharply defining the issues and providing a clear basis for choice? [40 CFR 1502.14]					

Source: Eccleston C.H., The EIS Book, CRC Press (2013).

Table C.7 Format, General Document Quality, User-Friendliness

	Yes	No	N/A	EIS Page Num.	Adequacy Evaluation and Comments
7.1 Is the EIS written precisely and concisely, using plain language, and defining any technical terms that must be used?					
7.2 Is information in tables and figures consistent with information in the text and appendices?					
7.3 As appropriate, is the metric system of units used (with English units in parentheses)?					
7.4 Are the units consistent throughout the document?					
7.5 Are technical terms defined, using plain language?					
7.6 If scientific notation is used, is an explanation provided?					
7.7 If regulatory terms are used, are they consistent with their regulatory definitions?					
7.8 Does the EIS use conditional language (e.g., "would" rather than "will") in describing the proposed action and alternatives and their potential consequences?					
7.9 Are graphics and other visual aids used whenever possible to simplify the EIS?					
7.10 Are abbreviations and acronyms defined the first time they are used?					
7.11 Is the use of abbreviations and acronyms minimized to the extent practical?					
7.12 Does the EIS make appropriate use of appendices (e.g., for material prepared in connection with the EIS and related environmental reviews, substantiating material, official communications, and descriptions of methodologies)? [40 CFR 1502.18 and 1502.24]					
7.13 Do the appendices support the content and conclusions contained in the main body of the EIS?					
7.14 Is there a discussion of the relationship between this EIS and related NEPA documents?					
7.15 Is the issue date (month and year of approval) on the cover?					

Source: Eccleston C.H., The EIS Book, CRC Press (2013).

Table C.8 Other Regulatory Requirements

	Yes	No	N/A	EIS Page Num.	Adequacy Evaluation and Comments
8.1 Unless there is a compelling reason to do otherwise, does the EIS include a: – Table of contents?					
– Index?					
– List of agencies, organizations, and persons to whom copies of the EIS were sent? [40 CFR 1502.19]					
8.2 Does the EIS identify all federal permits, licenses, and other entitlement that must be obtained in implementing the proposal? [40 CFR 1502.25(b)]					
8.3 Does the EIS identify methodologies used in the analyses, include references to sources relied upon for conclusions, supporting material, and methodologies? [40 CFR 1502.24]					
8.4 If a cost–benefit analysis has been prepared, has it been incorporated by reference or appended to the EIS? [40 CFR 1502.23]					
8.5 If this EIS adopts, in whole or in part, a NEPA document prepared by another federal agency, has the agency independently evaluated this information? [40 CFR 1506.3]					
8.6 Does the EIS appropriately use incorporation by reference, i.e.: – Is the information up to date?					
– Is the information summarized in EIS?					
– Are cited references publicly available? [40 CFR 1502.21]					
8.7 Does the EIS contain a list of preparers and their qualifications? [40 CFR 1502.17]					
8.8 If an EIS contractor has been used, was a disclosure statement prepared? [40 CFR 1506.5(c)]					
8.9 If the EIS was prepared by a contractor, is the agency's name listed as the preparer on the title page of the EIS and has the agency evaluated all information and accepted responsibility for the contents? [40 CFR 1506.5]					

Source: Eccleston C.H., The EIS Book, CRC Press (2013).

Table C.9 Procedural Considerations

	Yes	No	N/A	EIS Page Num.	Adequacy Evaluation and Comments
9.1 If appropriate, did the agency notify the host state and host tribe, and other affected states and tribes, of the determination to prepare the EIS					
9.2 Did the agency publish a Notice of Intent in the *Federal Register*, allowing reasonable time for public comment? [40 CFR 1501.7]					
9.3 Is a floodplain/wetlands assessment required, and if so has a notice of involvement been published in the Federal Register?					
9.4 In addition to EPA's notice of availability, has the agency otherwise publicize the availability of the draft EIS, focusing on potentially interested or affected persons? [40 CFR 1506.6]					
9.5 Has the agency actively sought the participation of low-income and minority communities in the preparation and review of the EIS? [EO 12898; Effective Public Participation guidance, p.11]					
9.6 Is the EIS administrative record being maintained contemporaneously, and does it provide evidence that the agency considered all relevant issues?					
9.7 To the fullest extent possible, have other environmental review and consultation requirements been integrated with NEPA requirements? [40 CFR 1502.25]					

Source: Eccleston C.H., The EIS Book, CRC Press (2013).

Table C.10 Draft EIS Considerations

	Yes	No	N/A	EIS Page Num.	Adequacy Evaluation and Comments
10.1 Has the agency considered scoping comments from other agencies and the public?					
10.2 Does the draft EIS demonstrate that the agency considered possible connected actions, cumulative actions, and similar actions? [40 CFR 1508.25(a)]					
10.3 If the draft EIS identifies a preferred alternative(s), does the document present the criteria and selection process? [40 CFR 1502.14(e)]					
10.4a Does the draft EIS demonstrate adequate consultation with appropriate agencies to ensure compliance with sensitive resource laws and regulations? 10.4b Does the document contain a list of agencies and persons consulted? 10.4c Are letters of consultation (e.g., SHPO, USFWS) appended? [40 CFR 1502.25]					

Source: Eccleston C.H., The EIS Book, CRC Press (2013).

Table C.11 Final EIS Considerations

	Yes	No	N/A	EIS Page Num.	Adequacy Evaluation and Comments
11.1 Does the final EIS discuss at appropriate points responsible opposing views not adequately addressed in the draft EIS and indicate the agency's responses to the issues raised? [40 CFR 1502.9(b)]					
11.2a Is the preferred alternative identified? [40 CFR 1502.14(e)]					
11.2b Does the document present the criteria and selection process for the preferred alternative?					
11.3 Does the final EIS demonstrate, through appropriate responses, that all substantive comments from other agencies, organizations, and the public were objectively considered, both individually and cumulatively (i.e., by modifying the alternatives, developing new alternatives, modifying and improving the analyses, making factual corrections, or explaining why the comments do not warrant agency response)? [40 CFR 1503.4]					
11.4 Are all substantive comments (or summaries thereof) and the agency's responses included with the final EIS? [40 CFR 1503.4(b)]					
11.5 Are any changes to the draft EIS clearly marked or otherwise identified in the final EIS?					
11.6 Is the final EIS suitable for filing with EPA, i.e. does it: – Have a new cover sheet?					
– Include comments and responses?					
– Include any revisions or supplements to the draft? [40 CFR 1503.4 and 1506.9]					

Source: Eccleston C.H., The EIS Book, CRC Press (2013).

Table C.12 Water Resources and Water Quality

	Yes	No	N/A	EIS Page Num.	Adequacy Evaluation and Comments
12.1 Does the EIS discuss potential effects of the proposed action/alternatives –On surface water quantity under normal operations?					
–Under accident conditions?					
–On surface water quality under normal operations?					
–Under accident conditions?					
12.2 Does the EIS assess the effect of the proposed action/alternatives on the quantity, quality, location, and timing of stormwater runoff? (e.g., will new impervious surfaces create a need for stormwater management or pollution controls)?					
12.3 Would the proposed action or alternatives require a stormwater discharge permit?					
12.4 Does the EIS evaluate whether the proposed action or alternatives would be subject to –Water quality or effluent standards?					
–National Primary Drinking Water Regulations?					
–National Secondary Drinking Water Regulations?					
12.5 Does the EIS state whether the proposed action/alternatives would involve –Work in, under, over, or having an effect on navigable waters of the United States?					
–Discharge of dredged or fill material into waters of the United States?					
–Deposit of fill material or an excavation that alters or modifies the course, location, condition, or capacity of any navigable waters of the United states?					
–Obtaining a Rivers and Harbors Act (Section 10) permit or a Clean Water Act (Section 402 or Section 404) permit?					
–Obtaining a determination under the Coastal Zone Management Act? If so, is such a determination included in the draft EIS?					
12.6 Does the EIS discuss potential effects of the proposed action and alternatives –On groundwater quantity under normal operations?					
–Under accident conditions?					
–On groundwater quality under normal operations?					
–Under accident conditions?					

(continued)

Table C.12 (Continued) Water Resources and Water Quality

	Yes	No	N/A	EIS Page Num.	Adequacy Evaluation and Comments
12.7 Does the EIS consider whether the proposed action or alternatives may affect any municipal or private drinking water supplies?					
12.8 Does the EIS evaluate the incremental effect of effluents associated with the proposed action and alternatives in terms of cumulative water quality conditions?					
12.9 If the proposed action may involve a floodplain, does the document discuss alternative actions to avoid or minimize impacts and preserve floodplain values?					

Source: Eccleston C.H., The EIS Book, CRC Press (2013).

Table C.13 Geology and Soils

	Yes	No	N/A	EIS Page Num.	Adequacy Evaluation and Comments
13.1 Does the EIS describe and quantify the land area proposed to be altered, excavated, or otherwise disturbed?					
13.2 Is the description of the disturbed area consistent with other sections (e.g., land use, habitat area)?					
13.3 Are issues related to seismicity sufficiently characterized, quantified, and analyzed?					
13.4 If the action involves disturbance of surface soils, are appropriate best management practices (e.g., erosion control measures) discussed?					
13.5 Have soil stability and suitability been adequately discussed?					
13.6 Does the EIS consider whether the proposed action may disturb or cause releases of any preexisting contaminants or hazardous substances in the soil?					

Source: Eccleston C.H., The EIS Book, CRC Press (2013).

Table C.14 Air Quality

	Yes	No	N/A	EIS Page Num.	Adequacy Evaluation and Comments
14.1 Does the EIS discuss potential effects of the proposed action on ambient air quality – Under normal operations?					
– Under accident conditions?					
14.2 Are any potential emissions quantified to the extent practicable (amount and rate of release)?					
14.3 Does the EIS evaluate potential effects to human health and the environment from exposure to any radioactive emissions?					
14.4 Does the EIS evaluate potential effects to human health and the environment from exposure to any hazardous chemical emissions?					
14.5 When applicable, does the EIS evaluate whether the proposed action and alternatives would – Be in compliance with the National Ambient Air Quality Standards?					
– Conform to the State Implementation Plan?					
– Potentially affect any area designated as Class I under the Clean Air Act?					
– Be subject to National Emissions Standards for Hazardous Air Pollutants?					
– Be subject to emissions limitations in an Air Quality Control Region?					
14.6 Does the EIS evaluate the incremental effect of emissions associated with the proposed action/alternatives in terms of cumulative air quality?					

Source: Eccleston C.H., The EIS Book, CRC Press (2013).

Table C.15 Wildlife and Habitat

	Yes	No	N/A	EIS Page Num.	Adequacy Evaluation and Comments
15.1 If the EIS identifies potential effects of the proposed action/alternatives on threatened or endangered species and/or critical habitat, has consultation with the USFWS or other applicable agencies been concluded?					
15.2 Does the EIS discuss candidate species?					
15.3 Are state-listed species identified, and if so, are results of state consultation documented?					
15.4 Are potential effects (including cumulative effects) analyzed for species other than threatened/endangered species and for habitats other than critical habitat (e.g., fish and wildlife)?					
15.5 Does the EIS analyze impacts on the biodiversity of the affected ecosystem, including genetic diversity and species diversity?					
15.6 Are habitat types identified and estimates provided by type for the amount of habitat lost or adversely affected?					
15.7 Does the EIS consider measures to protect, restore, and enhance wildlife and habitat?					

Source: Eccleston C.H., The EIS Book, CRC Press (2013).

Table C.16 Human Health Effects

	Yes	No	N/A	EIS Page Num.	Adequacy Evaluation and Comments
16.1 Have the following potentially affected populations been identified:					
– Involved workers?					
– Non-involved workers?					
– The public?					
– Minority and low-income communities (as appropriate)? [OE 12898]					
16.2 Does the EIS establish the period of exposure (e.g., 30 years or 70 years) for exposed workers and the public?					
16.3 Does the EIS identify all potential routes of exposure?					
16.4 When providing quantitative estimates of impacts, does the EIS use current dose-to-risk conversion factors that have been adopted by cognizant health and environmental agencies?					
16.5 When providing quantitative estimates of health effects due to radiation exposure, are collective effects expressed in estimated numbers of fatal cancers or cancer incidences?					
16.6 Are maximum individual effects expressed as the estimated maximum probability of a fatality or cancer incidence for an individual?					
16.7 Does the EIS describe assumptions used in the health effects calculations?					
16.8 As appropriate, does the EIS analyze radiological impacts under normal operation conditions for – Involved workers: - Population dose and corresponding latent cancer fatalities?					
- Maximum individual dose and corresponding cancer risk?					
– Non-involved workers: - Population dose and corresponding latent cancer fatalities?					
- Maximum individual dose and corresponding cancer risk?					
16.9 Does the EIS identify a reasonable spectrum of potential accident scenarios that could occur over the life of the action, including the maximum reasonably foreseeable accident?					
16.10 Does the EIS identify failure scenarios from both natural events (e.g., tornadoes, earthquakes) and from human error (e.g., forklift accident)?					

(continued)

Table C.16 (Continued) Human Health Effects

	Yes	No	N/A	EIS Page Num.	Adequacy Evaluation and Comments
16.11 As appropriate, does the EIS analyze radiological impacts under accident conditions for Involved workers: – Population dose and corresponding latent cancer fatalities?					
– Maximum individual dose and corresponding cancer risk?					
Non-involved workers: – Population dose and corresponding latent cancer fatalities?					
– Maximum individual dose and corresponding cancer risk?					
Public: – Population dose and corresponding latent cancer fatalities?					
– Maximum individual dose and corresponding cancer risk?					
16.12 Does the EIS discuss toxic and carcinogenic health effects from exposure to hazardous chemicals – For involved workers?					
– For non-involved workers?					
– For the public?					
– Under routine operations?					
– Under accident conditions?					
16.13 Does the EIS adequately consider physical safety issues for involved and non-involved workers?					

Source: Eccleston C.H., The EIS Book, CRC Press (2013).

Table C.17 Transportation

	Yes	No	N/A	EIS Page Num.	Adequacy Evaluation and Comments
17.1 If transportation of hazardous or radioactive waste/materials would be involved or if transportation is a major factor, are the potential effects analyzed (to a site, on site, and from a site)?					
17.2 Does the EIS analyze all reasonably foreseeable transportation links (e.g., overland transport, port transfer, marine transport, global commons)? [E.O. 12114]					
17.3 Does the EIS avoid relying exclusively on statements that transportation will be in accordance with all applicable state and federal regulations and requirements?					
17.4 Does the EIS discuss routine and reasonably foreseeable transportation accidents?					
17.5 Are the estimation methods used for assessing impacts of transportation among those generally accepted/recognized within the scientific community?					
17.6 Does the EIS discuss the annual, total, and cumulative impacts of all transportation actions, to the extent that such transportation can be estimated, on specific routes?					
17.7 Have transportation analyses adequately considered potential disproportionately high and adverse impacts to minority and low-income populations? [E.O. 12898]					

Source: Eccleston C.H., The EIS Book, CRC Press (2013).

Table C.18 Waste Management and Waste Minimization

	Yes	No	N/A	EIS Page Num.	Adequacy Evaluation and Comments
18.1 Are pollution prevention and waste minimization practices applied in the proposed action and alternatives (e.g., Is pollution prevented or reduced at the source when feasible? Would waste products be recycled when feasible? Are by-products that cannot be prevented or recycled treated in an environmentally safe manner when feasible? Is disposal only used as a last resort?)					
18.2 If waste would be generated, does the EIS examine the human health effects and environmental impacts of managing that waste, including waste generated during facility decontamination or decommissioning?					
18.3 Are waste materials characterized by type and estimated quantity, where possible?					
18.4 Does the EIS identify RCRA/CERCLA issues related to the proposed action and alternatives?					
18.5 Does the EIS establish whether the proposal would be in compliance with federal or state laws and guidelines affecting the generation, transportation, treatment, storage, or disposal of hazardous and other waste?					

Source: Eccleston C.H., The EIS Book, CRC Press (2013).

Table C.19 Socioeconomic Considerations

	Yes	No	N/A	EIS Page Num.	Adequacy Evaluation and Comments
19.1 Does the EIS consider potential direct, indirect, and cumulative effects on –Land use patterns?					
–Consistency with applicable land use plans, including site comprehensive plans; and any special designation lands (e.g., farmlands, parks, wildlife, conservation areas)?					
–Compatibility of nearby uses?					
19.2 Does the EIS consider possible changes in the local population due to the proposed action?					
19.3 Does the EIS consider potential economic impacts, such as effects on jobs and housing?					
19.4 Does the EIS consider potential effects on public water and wastewater services, stormwater management, community services, and utilities?					
19.5 Does the EIS evaluate potential noise effects of the proposed action and the application of community noise level standards?					
19.6 Does the EIS state whether the proposal could result in a disproportionately large adverse impact to minority or low-income populations? [EO 12898]					

Source: Eccleston C.H., The EIS Book, CRC Press (2013).

Table C.20 Cultural Resources

	Yes	No	N/A	EIS Page Num.	Adequacy Evaluation and Comments
20.1 Was the State Historic Preservation Officer consulted?					
20.2 Was a cultural resources survey conducted for both archaeological and historical resources (while maintaining confidentiality by not disclosing locations for sensitive sites)?					
20.3 Does the EIS discuss potential access conflicts and other adverse impacts to Native American sacred sites (while maintaining confidentiality by not disclosing locations)? [EO 13007]					
20.4 Does the EIS include a provision for mitigation in the event unanticipated archaeological materials (e.g., sites or artifacts) are encountered?					
20.5 Does the EIS address consistency of the proposal with any applicable or proposed cultural resources management plan?					

Source: Eccleston C.H., The EIS Book, CRC Press (2013).

Note

1. Modified from Environmental Impact Statement Checklist, US Department of Energy, 1997.

Table C.20 Cultural Resources [illegible]

[illegible]	Yes	[illegible]	[illegible]	EIS Req. [illegible]	Mitigation/ Explanation/ Comments
[illegible] (2007) [illegible] [illegible]					
[illegible] Would [illegible] buildings, [illegible] structures [illegible] [illegible] existing site?					
20. Does the [illegible] disturb potential [illegible] and other [illegible] impacts to [illegible] [illegible] sites [illegible] maintaining [illegible] [illegible]					
[illegible]					
[illegible]					

[illegible] (2013).

[illegible]

Index

Page numbers followed by f and t indicate figures and tables, respectively.

D

J

K

L

M

N

V

W

Y

* 9 7 8 1 4 6 6 5 8 3 6 3 4 *
12A02AF - #0004 - 120617 - C0 - 234/156/27 [29] - CB - 9781466583634